Marine Nuclear Power Technology

Junchong Yu

Marine Nuclear Power Technology

上海交通大学出版社
SHANGHAI JIAO TONG UNIVERSITY PRESS

Springer

Junchong Yu
Nuclear Power Institute of China
Chengdu, Sichuan, China

ISBN 978-981-15-2893-4 ISBN 978-981-15-2894-1 (eBook)
https://doi.org/10.1007/978-981-15-2894-1

Jointly published with Shanghai Jiao Tong University Press
The print edition is not for sale in China. Customers from China please order the print book from: Shanghai Jiao Tong University Press.

This Springer imprint is published by the registered company Springer Nature Singapore Pte Ltd.
The registered company address is: 152 Beach Road, #21-01/04 Gateway East, Singapore 189721, Singapore

Preface

Entrusted by the Editorial Board of the series of books for "Nuclear Energy and Nuclear Technology Publication Project", the editors began the preparatory work of the fascicle *Marine Nuclear Power Technology* at the beginning of 2014. As we know, with the rapid development of marine nuclear power technology and industry in recent years, lots of textbooks, monographs and popular science articles on marine nuclear power technology have been published all over the world. Making *Marine Nuclear Power Technology* a featured reference book becomes both an expectation of and a challenge for the editors.

Based on the analysis of the characteristics of existing related publications and the purpose of the "Nuclear Energy and Nuclear Technology Publication Project", the orientation of this book is a professional book on the fundamentals of the R&D process and the whole lifetime management of a marine nuclear power plant. This book consists of 18 chapters: Chapter 1 introduces basic types, design characteristics and development trends of marine nuclear power plants, providing the readers an overview of marine nuclear power plants; Chapter 2 elaborates comprehensively the design principles, methods and means for the reactor, which is the most critical part of the marine nuclear power plants, from such aspects as core physics, thermo-hydraulics, fuel design and reactor structure; Chapters 3–11 describe the design features and technological evolution of marine nuclear power plants, including the aspects of systems, equipment, instrumentation and control, radiation protection and shielding, vibration and noise reduction, mechanical analysis and evaluation, and reliability and maintainability design, etc.; Chapters 12–14 introduce the safety analysis, operation analysis and accident management related to marine nuclear power plants, and describe the safety design concept of marine nuclear power plants and the countermeasures and response for anticipated operational incidents and accidents; Chapter 15 focuses on the strategies, methods and application examples for aging management of marine nuclear power plants; Chapter 16 introduces the content, methods and procedures of relevant tests throughout the design of marine nuclear power plants; Chapter 17 is about the fuel loading and unloading process unique to marine reactors; and Chapter 18 goes to the schemes and methods for marine reactor decommissioning.

With the support by the National Key Laboratory on Reactor System Design Technology and Nuclear Power Institute of China (NPIC), a group of experienced experts and talented staff (please refer to the Editorial Board of this book for the detailed list) were organized to write the above chapters. This book was proofread by Director Yu Hongxing, and I made the final check and revision.

We would like to acknowledge the valuable inputs contributed by the experts from Nuclear Power Institute of China during the writing of this book, as well as the considerable supports and assistance provided by the related administrative departments, and to express our gratitude to them all.

While we have tried our best to bring the publishing process to a conclusion as satisfactory as possible, we regret any errors you may discover and appreciate any suggestions or comments.

Chengdu, China Junchong Yu

Acknowledgements

In the process of writing this book, the author does all chapter's designs and the content framework of the whole book. I would like to thank the following experts (in alphabetical order by surname) for contribution to the details of the book.

Xiaoming Chai	Zhi Chen	Xingdou Gao	Xiaoqiang He
Xindong Huang	Guangming Jiang	Changxiang Li	Yuanming Li
Longtao Liao	Wenjin Liu	Chuan Lu	Zongjian Lu
Ying Luo	Biao Quan	Danrong Song	Bin Tang
Hong Yang	Dong Yao	Hongxing Yu	Wei Zeng
Lin Zhang			

The translation of this book from its Chinese version is done by the following experts (in alphabetical order by surname), and I would like to extend thanks to them.

English translated by: Jiaquan Hu, Ming Lei, Yang Lan, Lihua Qin, Xue Xiong, Yue Zhang.

English proofread by: Xuedong Huang, Qiong Zhang.

Gratitude to Dr. Yingchun Yang from Shanghai jiaotong University Press for her effort to organize.

Junchong Yu

Contents

Chapter 1
Overview

1.1 Introduction

In December 1942, a group of American scientists led by Enrico Fermi used graphite and metallic uranium to set up a device that was later called "reactor" at the Tennis Hall of the University of Chicago, and they used this device to realize the controllable uranium nuclear fission reaction for the first time. Later, the United States, the Soviet Union and some other countries quickened their race to carry out the study of nuclear energy utilization, especially in the military field. The United States firstly proposed the concept of "pressurized water reactor" (herein after referred to as PWR) using pressurized water as the moderator and coolant, built the world's first land-based submarine prototype reactor, i.e. S1W, in 1948, and built the world's first nuclear-powered submarine "Nautilus" in 1954, opening up a new era in the development of marine nuclear power technology. The Soviet Union applied the nuclear power technology in nuclear power plants (herein after referred to as NPPs), built the world's first atomic power plant in 1954, vigorously carried out the study of the application of nuclear energy in the ships, launched the first nuclear-powered submarine of the Soviet Union in 1958, i.e. "Lenin Communist Youth League", and successively built over 200 nuclear-powered submarines. The United Kingdom started its nuclear-powered submarine development plan in 1954, and put its first multi-purpose nuclear-powered submarine "HMS Dreadnought" into service in 1963. France put its first nuclear-powered submarine "Le Redoutable" into service in 1971. China successfully built its first nuclear-powered submarine in the 1970s, becoming the world's fifth country with the nuclear power technology for submarines. In addition, some countries have been applying the nuclear power in the ships such as aircraft carriers, cruisers, destroyers, deep sea submersible vehicles, icebreakers and merchant ships, making full use of the advantages of nuclear power.

The marine nuclear power plant is a device converting the thermal energy generated by nuclear fission reaction into mechanical energy and electric energy to power the ships. It mainly consists of two parts, i.e. nuclear energy generation system and energy conversion system (see Fig. 1.1). Generally, the nuclear energy generation

J. Yu, *Marine Nuclear Power Technology*,
https://doi.org/10.1007/978-981-15-2894-1_1

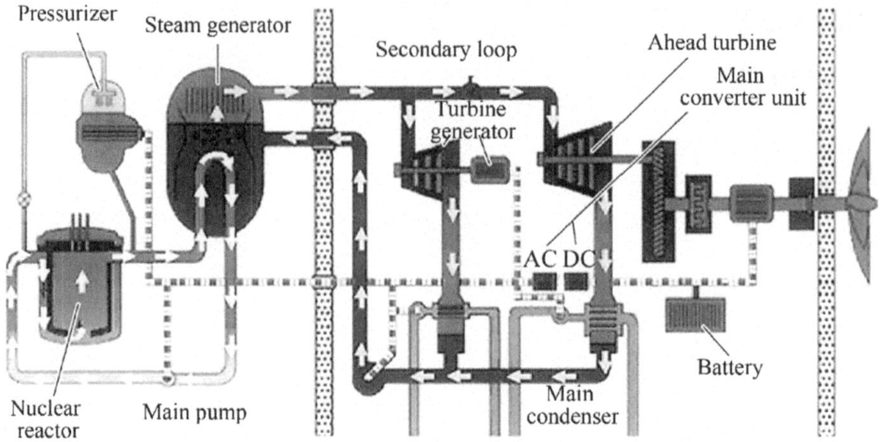

Fig. 1.1 Principle of marine nuclear power plant

system consists of reactor, primary loop, relevant shielding and nuclear instrument and control (hereinafter referred to as I&C) system, etc., and the energy conversion system consists of secondary loop, shaft system, electric system and relevant I&C system. Reactor is the device generating and maintaining the nuclear fission energy. Primary loop consists of reactor coolant system, relevant primary auxiliary systems and engineered safety systems and is the generic term of systems guaranteeing the safe operation and safety of nuclear reactor. Nuclear I&C system is the generic term of measuring equipment and control systems for temperature, pressure, flow, water-level and nuclear parameters of reactor and primary loop. Generally, such nuclear related systems and equipment are collectively called "nuclear power system".

Historically, the types of nuclear power plants that have been successfully applied in the ships include PWR type nuclear power plants and liquid-metal reactor (sodium and lead-bismuth) type nuclear power plants. Liquid-metal reactors, with the remarkable advantage of high heat transfer efficiency of liquid metal, can operate at the low-pressure environment and significantly reduce the volume and weight of the nuclear power plants. However, liquid-metal reactors pose high requirements for corrosion resistance of materials, and special heating devices are needed to maintain the liquid state of metals because of the rather high melting point of metals, bringing many new problems to the operation and maintenance of the reactors. For certain reasons, marine nuclear power plants with the liquid-metal reactors had been decommissioned in the 20th century. Currently, all marine nuclear power plants in service and under construction all over the world are PWR nuclear power plants. The working principle of such plants is as follows: the primary loop water removes the heat generated by nuclear fission in the reactor, and the steam generator transfers the heat to the water in the secondary loop and converts the water into steam to drive the turbine for power generation or directly drive the propeller (see Fig. 1.1).

Compared with conventional power plants, the nuclear power plant is characterized by extremely high nuclear fuel energy density, continuous high-power energy output and operation independent of oxygen and ship fuel, and has more obvious advantages in comprehensive combat capability of the ships such as self-supportability, endurance, maneuverability and concealing. However, the marine nuclear power plant also poses high requirements such as complex structure, intensive high and new technology and high safety and reliability requirements, and involves nuclear and radiation safety, so the lead time is long and capital investment is large.

1.2 Basic Types of Nuclear Power Ships

Since the building of the first nuclear-powered submarine in 1954, the technology for nuclear power plants has enjoyed the rapid development. In the military field, besides the nuclear-powered submarines, the ships using nuclear power plants include cruisers, destroyers, aircraft carriers, and deep-sea submersible vehicles. In the civilian field, the ships using nuclear power plants mainly include icebreakers and nuclear-powered merchant ships.

1.2.1 Nuclear Submarines

Nuclear submarine is the short name of nuclear-powered submarine, which is powered by the nuclear energy generated from nuclear fission. The first nuclear-powered submarine in the world, i.e. "Nautilus" of the United States, started its sea trial on January 24, 1954, showing the birth of nuclear-powered submarines. Currently, there are totally six countries all over the world avowing the possession of nuclear-powered submarines, i.e. the United States, Russia, the United Kingdom, France, China and India. The United States and Russia are the countries possessing the most nuclear-powered submarines. The application of nuclear strategic missiles brings the development of nuclear-powered submarines to a new stage. Nuclear-powered submarines carrying nuclear strategic missiles are a nuclear force with underwater deterrence.

According to the tasks and weapons, nuclear-powered submarines can be divided into the following types: a. nuclear-powered attack submarines that use torpedoes as the main weapons and are used for attacking surface ships and underwater submarines of the opponents; b. nuclear-powered ballistic missile submarines that use ballistic missiles as the main weapons, are also equipped with the torpedoes for self-defense and are used to attack the strategic targets; c. nuclear-powered cruise missile submarines that use cruise missiles as the main weapons and are used for strategic and tactical attacks.

1.2.2 Nuclear-Powered Aircraft Carriers

Nuclear-powered aircraft carriers are called nuclear aircraft carriers for short, using the nuclear energy as the power. They are large surface ships using the carrier-based aircrafts as the main combat weapons. A country can use the nuclear-powered aircraft carriers to put the military pressure on and fight against another country at a place far away from its territory, without relying on local airports.

After building nuclear submarines, the United States decided to use the nuclear power technology for submarines to develop that for aircraft carriers. In November 1961, the United States completed and put into service the nuclear-powered aircraft carrier "Enterprise", which is the first of this kind in the world. "Enterprise" was equipped with eight A2W nuclear reactors (with a single reactor power of 150 MW), providing it with larger maneuverability and amazing endurance. In addition, it could sail to any sea area in the world at a very high speed. Nuclear-powered aircraft carrier "Enterprise" brought the development of aircraft carriers into a new era.

Based on the nuclear-powered aircraft carrier "Enterprise", the United States developed the Nimitz-class nuclear-powered aircraft carriers, the second generation of nuclear-powered aircraft carriers of the United States. The lead ship "Nimitz" was completed and put into service in May 1975. "Nimitz" has a standard displacement of 72,916 tons, a length of 332.9 m and a beam of 40.8 m. It is equipped with two nuclear reactors (A4W/A1G) and four steam turbines. It has a speed of more than 30 knots. One time of fueling can maintain a continuous operation of 13 years and ensure a voyage of 1 million sea miles. It can carry 90 fighters of various types, with the maximum number of carrier-based aircrafts over one hundred. Nimitz-class nuclear-powered aircraft carriers are ten large nuclear-powered aircraft carriers of the United States Navy, with the side numbers from CVN-68 to CVN-77. With the design and building starting from the 1960s till the beginning of the 21st century, Nimitz-class nuclear-powered aircraft carriers were always the backbone of sea power and global strategy of the United States.

Currently, the United States has developed the third generation of Gerald R. Ford-class nuclear-powered aircraft carriers, with the lead ship CVN-78 using two brand-new A1B reactors. This type of nuclear power plant can further enhance the safety, increase the power density of the reactor compartment, extend the reactor core lifetime, enhance the power supply capacity, improve the equipment performance, simplify the systems and equipment, reduce the maintenance and personnel requirements, and increase the cost-effectiveness ratio.

Besides the United States, only France has a nuclear-powered aircraft carrier, i.e. "Charles de Gaulle" with a twin-reactor layout.

1.2.3 Nuclear-Powered Cruisers

The world's first nuclear-powered cruiser, i.e. the "USS Long Beach" nuclear-powered missile cruiser, was laid down by Bethlehem Steel Corporation in the United States in December 1957 and launched in 1960. The most important characteristic of the nuclear-powered cruiser is the strong cruising capability. The nuclear-powered cruiser can continuously cruise for several years and travel around the Earth for several circles without refueling. "USS Long Beach" has a full displacement of 1.8×10^4 t, and is equipped with cruise missiles, anti-ship missiles, anti-aircraft missiles and anti-submarine missiles. Later, the United States built other eight nuclear-powered cruisers of four types, all of which have been decommissioned.

Four Kirov-class nuclear-powered missile cruisers of the former Soviet Union are the only nuclear-powered cruisers other than those in the United States, and they are currently the largest cruisers in the world. Each of them has a full displacement of 2.8×10^4 t and is equipped with up to 250 missile-firing installations of various types and up to 296 missiles.

1.2.4 Nuclear-Powered Deep-Sea Facilities

Deep-sea facility is a kind of tool that can be directly manipulated at the deep sea for a long period and under all-weather conditions, without being influenced by the surface wind and waves. It can be used for underwater engineering operation, resource exploration, marine scientific research and maritime military confrontation. It can generally dive to the deep sea with a depth of 1,500–4,500 m, move independently at a maximum navigational speed of more than 10 knots, and work at the deep sea for more than 60 days, depending on the displacement. Currently, there are no underwater weapons that can effectively attack the deep-sea facility.

Nuclear-powered deep-sea facility can make its endurance and deep-sea mobility far much higher than those of the conventional ones, making it an underwater military platform with unique advantages. The United States has built one nuclear-powered deep-sea submersible vehicle (NR-1), and Russia has built at least seven nuclear-powered deep-sea submersible vehicles (such as 10831, 1910 and 1851) and one hybrid-power test submersible vehicle. Such vehicles are used for intelligence reconnaissance, weapon system test, marine research, search and salvage, etc.

1.2.5 Nuclear-Powered Icebreakers

Nuclear-powered icebreakers are the icebreakers using the nuclear power as the main energy source.

From the late 1950s, the Soviet Union began to use the nuclear-propelled ships for civilian development. Civilian nuclear-powered ships, including icebreakers and ice-breaking cargo ships, are used for assisting in the transportation at the Arctic coasts in the north of the Soviet Union. Since the world's first nuclear-powered icebreaker "Lenin" was put into service in 1959, the Soviet Union/Russia has successively built totally nine nuclear-powered icebreakers of three generations, and is currently developing the fourth generation of nuclear-powered icebreakers.

1.2.6 Nuclear-Powered Merchant Ships

Nuclear-powered merchant ships are the large merchant ships propelled by the nuclear power plants. The world's first nuclear-powered merchant ship is "Savannah" of the United States that was completed and put into operation in May 1962. In December 1968, Germany finished the building of the world's first nuclear-powered ore carrier "Otto Hahn", and Japan completed the building of its first nuclear-powered merchant ship "Mutsu" cargo ship.

Nuclear-powered merchant ships have the following advantages: a. large endurance: one time of fueling can maintain the continuous navigation of several years; b. high navigational speed: large power of nuclear power plants can realize the rather high navigational speed; and c. high economic efficiency: although the initial investment is high, the cost of fuel can be saved and the economic efficiency is generally high. However, nuclear-powered merchant ships require special wharfs for maintenance, service and loading and unloading, bringing extremely large limitations for the nuclear-powered merchant ships. Therefore, all above nuclear-powered merchant ships were decommissioned, and there is no nuclear-powered merchant ship in service at present.

1.3 Design Characteristics and Development Trends of Marine Nuclear Power Plants

1.3.1 Design Characteristics

The marine nuclear power plant, on the one hand, has its unique characteristics because of its construction environment and operating conditions that are quite different from those of land-based NPPs. On the other hand, each type of marine nuclear power plant also is with its own obviously unique characteristics because of its mission and operating environment.

Marine nuclear power plants are installed on the ships. Because the ships often have the weight and size limitations, the building space reserved for the nuclear power plant is usually very limited. Therefore, an important characteristic of the marine nuclear power plant is its small size and light weight. Nuclear-powered ships often travel on the sea and are subject to the negative effects of long-term exposure to marine conditions on the safety of the core. In addition, it is also necessary to consider the effects of possible strong impact caused by hull collisions and torpedo explosions on marine nuclear power plants. Due to their different missions, nuclear-powered ships have different operating conditions. For example, the nuclear power icebreakers require the high power at a short-time step, so there are special requirements for the power output of nuclear power plants.

1.3.2 Development Trends

1.3.2.1 Development History

From the development history of marine nuclear power technology, it is not difficult to find that the nuclear power technology for submarines is always the key parts in the development of the marine nuclear power technology, with several generations of upgrade. In terms of the evolution of reactor technology, the United States has developed several generations of nuclear power technology for submarines, which have gone through start-up stage, mass construction stage, performance improvement stage and performance optimization stage. The Soviet Union/Russia, the United Kingdom and France have also developed several generations of nuclear power technology for submarines.

On the basis of the development of nuclear power technology for submarines, various countries have continuously expanded the application of nuclear power technology for submarines to that for aircraft carriers, destroyers and frigates, deep-sea facilities, and icebreakers. The United States has successively developed three generations of nuclear power technology for aircraft carriers, and Russia has done the same for icebreakers.

In general, the development of marine nuclear power technology in various countries has experienced the process of increasing the power, reducing the quantity of reactors in individual ship, enhancing the natural circulation capacity of reactors, extending the reactor core lifetime and continuously improving the reactor safety and reliability.

1.3.2.2 Development Characteristics

(1) Continuous development is supported by national strategy.

Certain countries take the development of marine nuclear power technology as the national strategy. For example, the United States Navy Reactor Program implements a special study on marine reactors and gives priority to the development of reactor technology. They pay high attention to the development of marine nuclear power technology, ensure long-term adequate financial support, focus on the implementation of medium and long-term development plans, and establish and maintain the fundamental nuclear science and technology capabilities.

(2) PWR is the preferred reactor type for marine nuclear power plants.

Marine nuclear power plant is still one of the priorities of current development. According to different military requirements, various types of marine nuclear power plants with different power levels and technical characteristics have been developed for submarines, aircraft carriers and large surface ships, to meet different power and electricity requirements by various weapons. PWR is still the preferred reactor type for the new generation of marine nuclear-powered plants.

(3) Improvement and enhancement of the overall performance of marine nuclear power plants are persistent and ongoing efforts.

In the research and development of marine nuclear power plants, steady progress is essential while striving to break through single key technology, giving full play to the role of technology promotion, and gradually enhancing the overall performance of nuclear powered plants. Regarding the demerits of existing nuclear power plants, improvement plans are formulated to further improve the reliability and the safety, and to realize the sustainable development of nuclear power technology.

(4) Well-developed technology based on full verification is widely used.

The development of marine nuclear power plants follows a path from test reactor to test ship for improvement and optimization. Only well-developed technology based on full verification is widely used on operational ships. For example, for more than 50 years of the development of the submarine nuclear power plants, the United States has always put priority on the technical feasibility and the reliability of the plants, and has successively built several land-based prototype reactors.

1.3.2.3 Technological Development Trends

Based on the available current information on the development of marine nuclear power technology, it can be concluded that the technological development trends are mainly as follows: PWR dominates the technical route with continuous improvement of its performance; and new technologies are explored actively.

(1) PWR technology

PWR technology is still the marine nuclear power technology that will be unswervingly developed by various countries. Its development trends are as follows: safety and reliability are continuously improved, reactor core lifetime prolonged, natural circulation capacity of reactors raised, and vibration and noise reduction studied intensively [1].

(2) New technologies

While keeping the development of PWR technology, new technologies are constantly explored to further significantly improve the overall performance of marine nuclear power plants.

Reference

1. Minjun, Peng. 2008. *Marine Nuclear Power Unit*. Beijing: Atomic Energy Press.

Chapter 2
Nuclear Reactors

2.1 Overview

Pressurized water reactor (PWR) generally refers to a reactor and related systems that use pressurized light water (H_2O) as the moderator and coolant. PWR was originally designed by Westinghouse Electric Corporation of the United States for nuclear submarines, and subsequently developed in the civilian nuclear power field. After more than 60 years of improvement, PWR has now become the reactors with most mature technology for worldwide military and civil nuclear power plants, and also the dominating reactors in the civil nuclear power industry in China. The schematic diagram of a PWR is shown in Fig. 2.1.

Enriched uranium is used as the nuclear fuel in PWRs. The enrichment varies from low value (1–5%) to medium or high value (20–97%), depending on the specific designs. Light water is used as the moderator to moderate fast neutrons into thermal neutrons, and as the coolant to remove the heat in the core, for its favorable price and excellent capability for heat transfer. In order to ensure a high thermal energy conversion efficiency of the thermodynamic system, the nuclear reactor core should have a rather high outlet temperature, based on the thermodynamic principle. However, light water has a low boiling point at atmospheric pressure. For this reason, the system pressure of the coolant has to be increased. Therefore, PWR is a light-water cooled and moderated reactor in a high-pressure state (about 15 MPa), and the reactor coolant outlet temperature is around 300 °C.

Based on the PWR principle, the reactor design research mainly involves nuclear reactor physics, reactor thermal-hydraulics, fuel assemblies and related components, reactor pressure vessels, control rod drive mechanisms, reactor internals, reactor roof structure, reactor support and shielding.

J. Yu, *Marine Nuclear Power Technology*,
https://doi.org/10.1007/978-981-15-2894-1_2

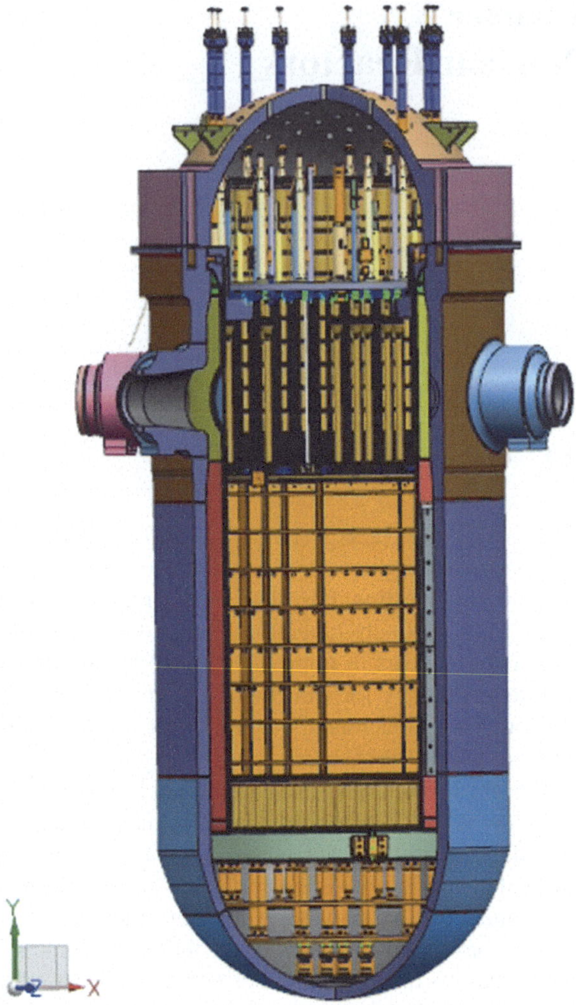

Fig. 2.1 Pressurized water reactor

2.2 Nuclear Reactor Physics

A nuclear reactor is a device in which neutrons are used to induce nuclear fission reactions in heavy nuclei and the released energy can be converted into mechanical energy or electric energy. Nuclear reactor physics is a theory describing the methods to maintain the nuclear fission chain reactions in a controllable, sustainable and steady state. Meanwhile, in order to utilize the fission energy effectively and safely, nuclear reactor physics involves the study of the generation and distribution of fission energy in the reactor core, and the types and distribution of the generated radioactive substances and radioactive particles.

According to the design procedure of a nuclear reactor, the design of the nuclear reactor core related to neutronics, fission reactions and fission energy is generally referred to as the reactor nuclear design. It mainly consists of nuclear reactor reactivity control (such as built-in reactivity control and reactivity regulation), power distribution, lifetime, resistance to the disturbance of nuclear reactor (negative feedback under disturbances to maintain the stability of fission chain reactions), startup and shutdown of the nuclear reactor, and the behavior of the nuclear reactor under transient conditions. In order to ensure the accuracy and rationality of the reactor nuclear design, numerical simulations and tests are conducted for verification.

This part introduces the theory, the design and the verification of nuclear reactor physics from four aspects, i.e. theory of nuclear reactor physics, reactor nuclear design, software for reactor nuclear design, and reactor test verification.

2.2.1 Theory of Nuclear Reactor Physics

2.2.1.1 Interaction Between Neutrons and Atomic Nuclei

Nuclear physics is an independent academic discipline to describe the interaction mechanism between neutrons and atomic nuclei. The theory of nuclear reactor physics is based on the research results of nuclear physics. This part briefly introduces the fundamentals of nuclear physics.

In a nuclear reactor core, a large number of neutrons interact with atomic nuclei during the flight, such as in the manner of collision. Interactions between neutrons and atomic nuclei can be divided into two major categories: scattering and absorption.

Scattering reaction is the reaction in which the incident particles are neutrons and the emitted particles resulted from the interaction with the atomic nuclei are still neutrons. According to the change of the internal energy of atomic nucleus before and after the interaction, the scattering is classified as inelastic scattering (after which the energy level of the atomic nuclei changes) and elastic scattering (after which the energy level of the atomic nuclei does not change). Inelastic scattering occurs only when the energy of the incident neutrons is higher than a certain threshold, so it has the characteristic of threshold energy. In a thermal neutron nuclear reactor, the fast neutrons (high-energy neutrons) are moderated into thermal neutrons (low-energy neutrons), during which elastic scattering plays a dominating role.

Absorption reaction is the reaction that the incident particles are neutrons, and after the interaction of incident neutrons with atomic nucleus, the incident neutrons disappear, and new particles (such as γ, ρ or α particles) or two or three neutrons are emitted, or the atomic nucleus undergoes fission and two or more nuclei of intermediate mass are formed. In a nuclear reactor, the most important absorption reactions are radiative capture reaction (during which the atomic nuclei absorb neutrons and then emit γ particles) and fission reaction (during which the atomic nucleus absorb neutrons and split into two or more atomic nuclei of intermediate mass). Some important radiative capture reactions are as follows:

$$^{238}_{92}\mathrm{U} \xrightarrow{(n,\gamma)} {}^{239}_{92}\mathrm{U} \underset{23\mathrm{min}}{\xrightarrow{\beta^-}} {}^{239}_{93}\mathrm{Np} \underset{56h}{\xrightarrow{\beta^-}} {}^{239}_{94}\mathrm{Pu}$$

$$^{232}_{92}\mathrm{Th} \xrightarrow{(n,\gamma)} {}^{233}_{90}\mathrm{Th} \underset{22\mathrm{min}}{\xrightarrow{\beta^-}} {}^{233}_{91}\mathrm{Pa} \underset{27d}{\xrightarrow{\beta^-}} {}^{233}_{92}\mathrm{U}$$

Through this reaction and subsequent decay chains, nuclide ^{238}U (or ^{232}Th) in the reactor may spontaneously undergos a transformation into a fissile nuclide ^{239}Pu (or ^{233}U).

In a nuclear reactor, only a few nuclides can undergo fission reactions. Nuclides which can be caused to undergo fission, regardless of the energy level of incident neutrons, are referred to as *fissile nuclides*, such as ^{233}U, ^{235}U, ^{239}Pu and ^{241}Pu. Nuclides which will undergo fission only when the neutron energy is higher than a certain threshold are referred to as *fissionable nuclides*, such as ^{232}Th and ^{238}U. In a nuclear reactor, the fission caused by fissile nuclides is the main part of the fission process for the release of nuclear energy.

2.2.1.2 Nuclear Reaction Cross Section and Nuclear Reaction Rate

Nuclear reaction cross section is a physical quantity describing the probability of interaction between neutrons and atomic nuclei, and is classified as microscopic cross section and macroscopic cross section.

Assume that a beam of parallel monoenergetic neutrons with an intensity of I incident normally upon and uniformly across the face of a target with a unit area, a thickness of Δx and an atomic nucleus number per unit volume of N, and a neutron detector is placed behind the target. The detector obtains a neutron beam intensity of I′ (see Fig. 2.2).

Based on the experimental results, the following mathematical relationship can be established:

$$\frac{I - I'}{I} = \sigma N \Delta x \tag{2.1}$$

where, σ is a proportional constant correlating the nature of the target nuclei and the neutron energy, called a *microscopic cross section*. From the equation, we can see

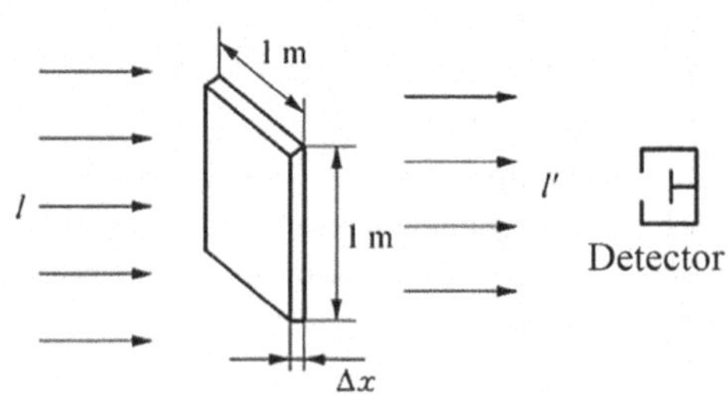

Fig. 2.2 Attenuation of parallel neutron beams penetrated a thin target

that microscopic cross section is a measure of the probability of interaction between an incident neutron with given energy and a target nucleus. Its unit is m^2. In the actual use, σ is usually on the order of 10^{-28} m^2, and this unit is referred to as a *barn* (b), where $1b = 10^{-28}$ m^2. Different nuclear reactions have different microscopic cross sections. Depending on various interactions between neutrons and atomic nuclei, the nuclear reaction cross section has the following relationships:

$$\sigma_t = \sigma_s + \sigma_a \tag{2.2}$$

$$\sigma_s = \sigma_e + \sigma_{in} \tag{2.3}$$

$$\sigma_a = \sigma_\gamma + \sigma_f + \sigma_{n,p} + \sigma_{n,\alpha} + \cdots\cdots \tag{2.4}$$

where, the subscripts t, s, a, e, in, γ, f, (n, p) and (n, α) respectively mean total cross section, scattering cross section, absorption cross section, elastic scattering cross section, inelastic scattering cross section, radiative capture cross section, fission cross section, (n, p) cross section and (n, α) cross section.

In the nuclear reactor engineering, when describing the reaction between neutrons and a large number of atoms, according to Eq. (2.1), σN is defined as a physical quantity Σ,

$$\Sigma = \sigma N \tag{2.5}$$

Thus, Eq. (2.1) can be rewritten as follows:

$$\frac{I - I'}{I} = \Sigma \Delta x \tag{2.6}$$

The physical quantity Σ is called a *macroscopic cross section*. It characterizes the probability of interaction of neutron with all nuclei in a unit volume and also can be interpreted as the probability per unit path length traveled that the neutron will undergo a reaction with a nucleus in the medium. Its physical unit is m^{-1}. Similar to the microscopic cross section, the macroscopic cross section also has the relationships similar to Eqs. (2.1), (2.2), (2.3) and (2.4).

According to the physical significance of macroscopic cross section, for a mixture of different nuclides, the calculation equation for macroscopic cross section is as follows:

$$\Sigma_x = \sum_i N_i \sigma_{x,i} \tag{2.7}$$

where, x means the type of nuclear reaction and i indicates a nuclide.

In the theory of nuclear reactor physics, an important physical quantity is neutron fluence rate, which is also generally called neutron flux density or neutron flux in the reactor engineering field. It means the total path length traveled by all neutrons in a unit volume within a unit time. Neutron flux is generally represented by the symbol ϕ, and its calculation equation is as follows:

$$\phi = vN \tag{2.8}$$

where, υ means the neutron velocity and N means the neutron density.

According to the physical significances represented by macroscopic cross section and neutron flux respectively, the nuclear reaction rate (generally indicated by the symbol R) for the interaction between neutrons and substances in a unit volume within a unit time is the macroscopic cross section multiplied by neutron flux, that is:

$$R = \Sigma\phi \tag{2.9}$$

Nuclear reaction rate is an important variable for a nuclear reactor. For different nuclear reactions, the nuclear reaction rates are different. For a homogeneous mixture of several nuclides, the nuclear reaction rate is the sum of reaction rates for the interactions between neutrons and individual nuclides:

$$R = \phi\Sigma_1 + \phi\Sigma_2 + \phi\Sigma_3 + \cdots = \phi\Sigma \tag{2.10}$$

where, Σ means the total macroscopic cross section of the homogeneous mixture.

Physical quantities such as microscopic cross section, macroscopic cross section, neutron flux and nuclear reaction rate are the basis of nuclear reactor physics. The physical laws governing the change of the cross sections with energy are given below.

The value of microscopic cross section depends on the incident neutron energy and the nature of the target nuclei. For a considerable number of heavy nuclides, three regions exist in the variation of the absorption cross section with the energy of the incident neutron on a curve of absorption cross section versus neutron energy. Low energy region (generally lower than 1 eV): in this energy region, absorption cross section σ_a gradually increases as the neutron energy decreases, and is roughly inversely proportional to the neutron velocity. Therefore, this region is also called $1/\upsilon$ region of absorption cross section. Intermediate energy region (1 eV < E < 10^4 eV): cross sections of many heavy nuclides show a large number of peaks in this energy region, and these peaks are called resonance peaks. High energy region (E > 10^4 eV): the neutron cross section is rather small, usually less than 10b, and the energy dependence of cross section also tends to be smooth. Figures 2.3, 2.4 and 2.5 respectively show the changes of corresponding microscopic cross sections of nuclides ^{235}U, ^{238}U and ^{1}H with the energy.

The above description of nuclear cross section law is a great approximation, and can only be used for qualitative understanding. With the increasing of the accuracy in nuclear cross section data measurement and the opening up of global nuclear data, the nuclear reactor researchers and designers can download the latest evaluation cross section data libraries directly from the Internet. The data libraries contain all

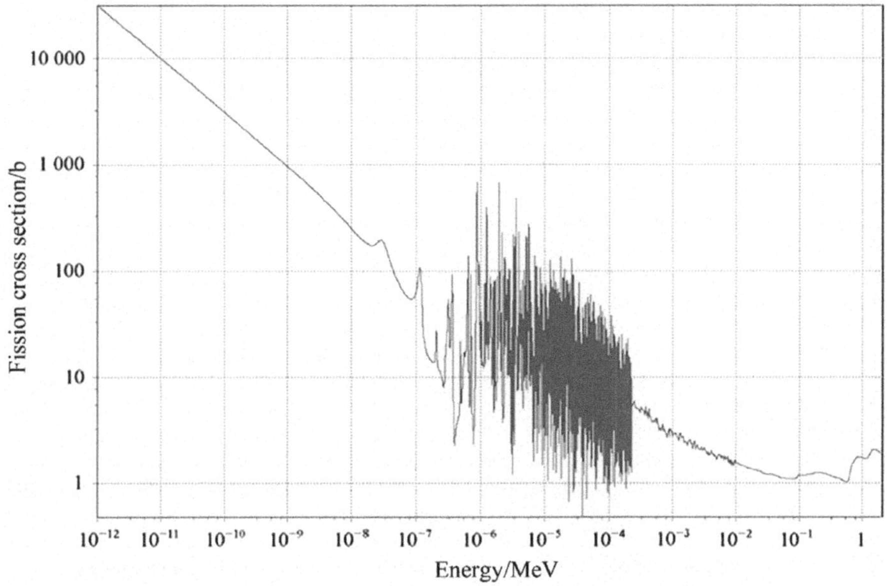

Fig. 2.3 Change of microscopic fission cross section of nuclide ^{235}U with energy

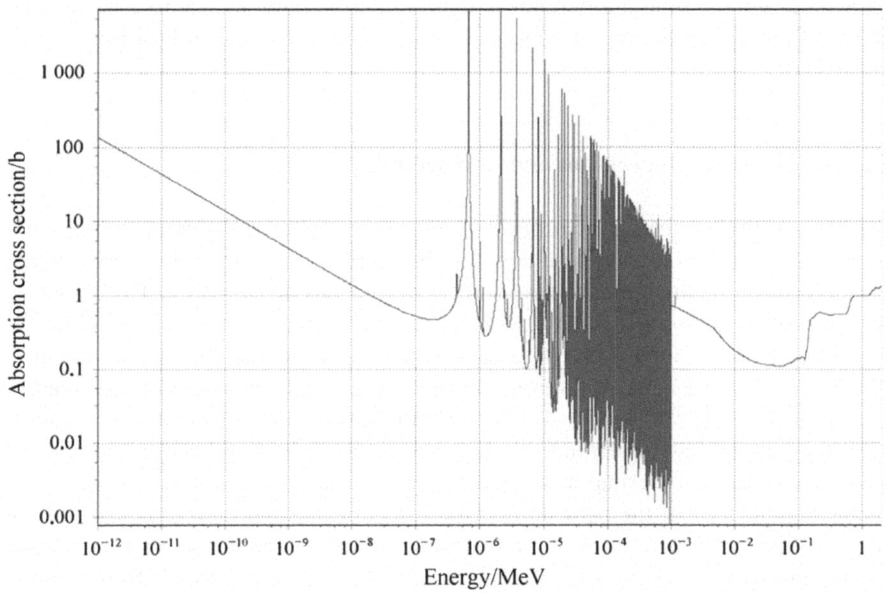

Fig. 2.4 Change of microscopic absorption cross section of nuclide ^{238}U with energy

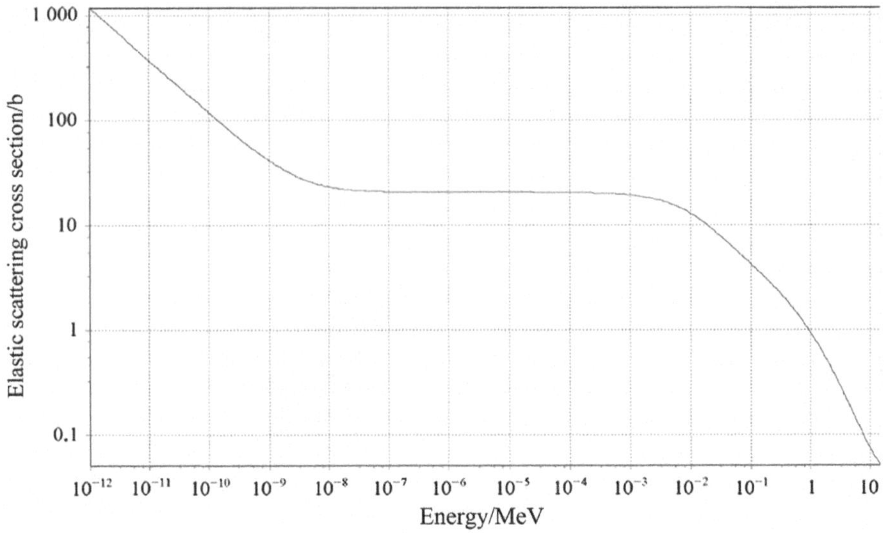

Fig. 2.5 Change of microscopic elastic scattering cross section of nuclide ^{1}H with energy

the important nuclear data for neutron reactions of nuclides necessary for reactor nuclear design. Internationally renowned nuclear cross section data libraries include ENDF of the United States, CENDL of China, JENDL of Japan, JEF of Europe and BROND of Russia.

2.2.1.3 Neutron Moderation and its Spectrum

In a reactor, the number of neutrons with different energy at different spatial locations varies. Understanding the change of the number of neutrons with energy and spatial locations is the important purpose of reactor physics research. The results of the variation of neutrons in a PWR with energy and spatial locations can be obtained by using advanced numerical computation codes to solve the Boltzmann equation which describes the neutron motion. However, the reactor physics is introduced in a traditional manner in this book, allowing the researchers to qualitatively understand the general law governing the distribution of neutrons related to energy and spatial locations in a thermal neutron reactor. To be specific, the law governing the energy dependence of the neutron distribution is introduced under the assumption of spatial independence (i.e. assume that the medium is infinite and homogeneous); and the law governing the spatial dependence of the neutron distribution is introduced under the assumption of energy independence (i.e. one-velocity assumption). Through the study of these laws, a deeper understanding of the reactor characteristics can be obtained, such as self-shielding effect of energy caused by resonance absorption, allowing the designers to have more perceptual knowledge during their reactor design. This part mainly introduces the neutron moderation process and the neutron moderation spectrum in an infinite homogeneous medium.

As we can see from Fig. 2.3, the fission cross sections of the fissile nuclide ^{235}U in the low energy region is much larger than those in the intermediate and high energy regions. For this reason, a PWR generally uses thermal neutrons to realize the chain fission reaction. The average energy of the neutrons generated by fission reaction is 2 MeV. In order to moderate such high-energy fast neutrons into thermal neutrons, a PWR uses water as the moderator, utilizing the rather large elastic scattering cross sections of nuclides ^{1}H and ^{16}O to moderate fast neutrons into thermal neutrons.

According to the laws of Newtonian mechanics, a neutron loses more energy per collision with a light nucleus than a heavy nucleus. After one collision with a hydrogen nucleus, a neutron may lose all its kinetic energy, but after one collision with ^{238}U, the maximum energy lost is about 2% of the original kinetic energy. Therefore, the moderator has to be the light nucleus. Water is often used as the moderator in a PWR, and the hydrogen nuclei plays a dominating role in the moderation. Other materials that can be used as moderators include heavy water, graphite and beryllium. In order to measure the moderating power of materials, "moderating power" and "moderating ratio" are commonly used for measurements.

Moderating power is the product of the moderator's macroscopic scattering cross section Σ_S and average lethargy gain ξ (i.e. the average logarithmic energy decrement per collision) after the scattering collision. The larger the Σ_S is, the more likely the neutrons and the moderator will scatter. The larger the ξ is, the more kinetic energy per collision will be lost on the average by the neutron. The product of them reflects the power of the moderator to moderate the neutrons. However, the interaction between nuclides and neutrons not only involve the scattering, but also the absorption. If there are too much absorption interaction, it will cause excessive loss of neutrons in the reactor and, as a result, the moderator will not be suitable any longer. Therefore, the moderating ratio can more fully reflect the effectiveness of the moderation. Among several common moderators, water has the strongest moderating power, so the volume of reactor core can be rather small when using water as the moderator. However, the moderating ratio of water is rather small because of its large absorption cross section. As a result, a PWR has to use the enriched uranium as the fuel. Heavy water and graphite have rather large moderating ratios and very small absorption cross sections, so they can use natural uranium as the nuclear fuel. However, the moderating power of heavy water and graphite is much weaker than that of water, so the core volume of heavy water reactors and graphite reactors is much larger than that of the light water reactors.

In an infinite homogeneous medium, according to the law governing the energy loss after the collision of neutrons with atomic nuclei and the neutron conservation relation in a certain small energy region, the *neutron moderation equation* in steady state can be obtained. Assuming that the absorption cross section in an infinite homogeneous medium is very small, the analytical solution to neutron moderation spectrum can be obtained. This neutron spectrum is in the form of $1/E$, which is also called *Fermi spectrum* distribution and often used as an approximation of the moderation spectrum in a reactor. The derivation of moderation spectrum requires a large number of mathematical equations, which are not intended to be detailed in this book. Readers can read the reference books [1, 2].

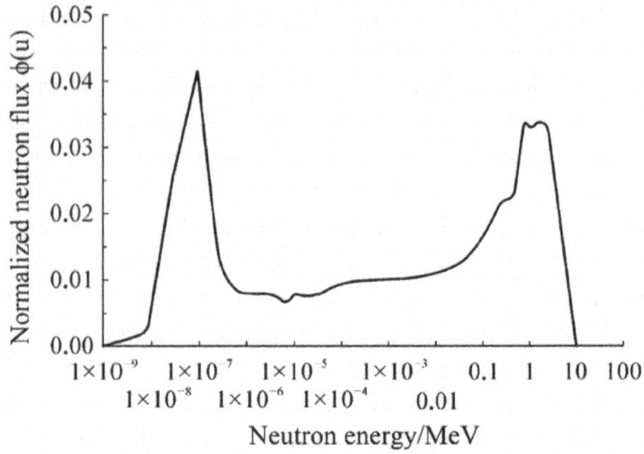

Fig. 2.6 Neutron spectrum in PWR

Neutron moderation spectrum mainly describes the variation of the flux with energy for intermediate-energy neutrons. For fast neutrons, the neutron spectrum can be approximated by the fission neutron spectrum. For thermal neutrons, the neutron spectrum can be modeled by the Maxwell spectrum. The shape of neutron spectrum for the entire energy range in a PWR is illustrated in Fig. 2.6. In this figure, the abscissa is a logarithmic coordinate, so the moderation spectrum (the intermediate-energy spectrum) becomes a constant. In the low energy range of the moderation spectrum, the curve is not smooth. Instead, there are many "sinking points." Such flux sink is caused by the resonance absorption peaks of the resonance nuclides in this energy range. Therefore, during the numerical simulation of reactor physics, special attention should be paid to the influence of these resonance peaks.

2.2.1.4 Critical Condition of Nuclear Reactor and Neutron Space Distribution

In a reactor, *neutron transport equation* can statistically describe the variation of neutrons with space, time and energy. Under certain assumptions, *neutron diffusion equation* can also be used to approximately describe the statistical distribution of neutrons [3]. For any given device, the variation of neutron number with space, energy and time can be obtained by solving the neutron transport equation. For most devices without external neutron source, the total number of neutrons will change with time, either decrease or increase.

For a nuclear reactor, controlled fission chain reactions should be maintained in a steady state, that is, the fission reaction rate should not change with time. Since the nuclides, including the fissile nuclides, will not have any great change in a relatively short time, the neutrons should not change with time in a short time. In order to achieve the condition of stable reactor operation, there are strict requirements for the mass of fissile nuclides in the reactor, the reactor volume, the neutron absorber nuclides and the layout of the reactor.

Generally, the physical quantity *effective multiplication factor*, k_{eff}, is used to describe the extent to which a nuclear reactor deviates from the steady-state operation. The physical significance of k_{eff} is that: for a given system without any external neutron source,[1] it is the ratio of the number of neutrons in two successive fission neutron generations:

$$k_{\text{eff}} = \frac{\text{Number of neutrons in one generation}}{\text{Number of neutrons in preceding generation}} \tag{2.11}$$

Another definition for k_{eff}: for a given system without any external neutron source, it is the ratio of the neutron production rate to the total neutron loss (absorption and leakage) rate:

$$k_{\text{eff}} = \frac{\text{Rate of neutron production in the system}}{\text{Rate of neutron loss (absorption + leakage) in the system}} \tag{2.12}$$

The above two definitions of the effective multiplication factor is equivalent. The physical significance of k_{eff} shows that, if $k_{\text{eff}} = 1$, the neutron production and loss rates will be the same, and hence the chain fission reaction will be time-independent, that is, the chain fission reaction process will be in a steady state, and such a system is referred to as being *critical*; if $k_{\text{eff}} > 1$, the neutron production rate will exceed the loss rate, leading to the growth of neutron population and power generation, and such a system is said to be *supercritical*; if $k_{\text{eff}} < 1$, the neutron production rate will be less than the loss rate, causing a decrease in neutron population and power generation, and such a system is said to be *subcritical*.

For the complexity of the real nuclear reactor structure, simple equations cannot clarify the important problems such as how to achieve the criticality condition and how to determine the flux distribution or power distribution when the reactor is critical. The way to judge whether the system is critical is to judge whether the real reactor k_{eff} is equal to 1, in which k_{eff} can be obtained by solving the k-eigenvalue steady-state neutron transport equation using the numerical simulation method. If the system is supercritical or subcritical, it can be restored to critical by changing the content of fissile nuclides or poisons. When the calculated $k_{\text{eff}} = 1$, the neutron flux and power distribution obtained by numerical simulation calculation are exactly the neutron flux and power distribution of the critical system.

In order to make a qualitative understanding of the critical conditions of nuclear reactor and the neutron flux distribution under criticality, we simplify the nuclear reactor into the problems of homogeneous materials and simple geometries (including spheres, cylinders and rectangular cubes), and assume that all neutrons have the same velocity (i.e. one-velocity assumption), so as to analytically obtain the critical conditions and the neutron flux distribution under criticality.

---

[1] If the neutron source is not started up by an external source placed inside a nuclear reactor, the startup of neutron source during the power operation of the reactor, in fact, has little effect on the steady-state operation of the entire reactor.

For the convenience of further introduction, let's first introduce several physical quantities.

① k_∞ is the infinite multiplication factor of the material, representing the neutron balance characteristic of the material itself:

$$k_\infty = \frac{\nu\Sigma_f}{\Sigma_a} \tag{2.13}$$

② M^2 is the migration area of the material, representing 1/6 of the mean square distance traveled by a neutron from its birth in fission to its eventual demise via thermal absorption.

③ B_m^2 is the critical buckling of the material, which is defined as follows:

$$B_m^2 = \frac{k_\infty - 1}{M^2} \tag{2.14}$$

Since both k_∞ and M^2 are the intrinsic properties of the material, the material buckling B_m^2 is also an inherent property of the material and is independent of the reactor geometry. The physical significance of the material buckling B_m^2 is that: if the relative neutron leakage of the reactor is B_m^2, the reactor will be critical.

From the analytical solution of the neutron diffusion equation [1, 2], the geometric buckling and neutron flux distribution for homogeneous bare reactors in the geometries of sphere, cylinder and rectangular cube can be obtained (see Table 2.1). Geometric bucklingB_g^2 represents the relative leakage of the reactor under given geometric conditions. This variable is just a function of geometry and is independent of reactor materials. The larger the B_g^2, the smaller the reactor geometry and the greater the relative leakage, and vice versa.

The critical condition of the reactor is that the geometric buckling B_g^2 of the reactor is equal to the material buckling B_m^2, that is, the relative neutron leakage caused by the reactor geometry is exactly equal to the leakage needed for the criticality of the material. In general, for a reactor with given dimensions, the geometric buckling B_g^2 is not necessarily equal to the material buckling B_m^2. When $B_g^2 > B_m^2$, it means that the relative neutron leakage caused by the reactor geometry is too large and that the reactor dimensions are rather small, so the reactor cannot reach criticality and is subcritical; when $B_g^2 < B_m^2$, it means that the relative neutron leakage caused by

Table 2.1 Geometric buckling and neutron flux distribution of homogeneous bare reactors in several geometries

Geometry	Geometric buckling	Neutron flux distribution
Sphere (radius R)	$B_g^2 = (\frac{\pi}{R})^2$	$\phi(r) = \frac{1}{r}\sin\frac{\pi}{R}r$
Right-angled cuboid	$B_g^2 = (\frac{\pi}{a})^2 + (\frac{\pi}{b})^2 + (\frac{\pi}{c})^2$	$\phi(x, y, z) =$ $\cos\frac{\pi}{a}x \cos\frac{\pi}{b}y \cos\frac{\pi}{b}z$
Cylinder (radius R, height H)	$B_g^2 = (\frac{2.405}{R})^2 + (\frac{\pi}{H})^2$	$\phi(r, z) = J_0(\frac{2.405}{R}r)\cos(\frac{\pi}{H}z)$

the reactor geometry is too small and that the reactor dimensions are rather large, so the reactor is supercritical. For a reactor with given dimensions, the effective multiplication factor can be defined as follows:

$$k_{eff} = \frac{k_\infty}{1 + M^2 B_g^2} \quad (2.15)$$

For a critical reactor ($B_g^2 = B_m^2$), the relative distribution of its neutron flux is shown in Table 2.1. It should be noted that the relative neutron flux distribution of the cylindrical device with homogeneous medium, which is nearest to the shape of the real reactor, is a cosine distribution in the axial direction, and a zero-order Bessel function distribution in the radial direction. In the analysis of the real reactor, the expressions in Table 2.1 are often utilized for the conditions of lower requirements for the accuracy of the neutron flux distribution. Table 2.1 only gives the relative neutron flux distribution. The absolute value of neutron flux can be obtained by the total reactor power (i.e. the total fission reaction rate).

For homogeneous bare reactors mentioned above, the relative neutron leakage of the reactor only depends on the geometry. In the real reactor, there is reflector surrounding the core to reduce the leakage by scattering back the leaked neutrons. Therefore, a good moderator is generally also a good reflector material. Common reflector materials are water, graphite and beryllium. With the reflector, the critical volume of the reactor is reduced. Such geometric reduction is called "reflector saving". In a real PWR, common reflector is baffle-water reflector or steel-water reflector.

2.2.1.5 Reactor Fuel Depletion

In a reactor, after the interactions between the nuclides (including fissile nuclides and nuclides in structural materials) and neutrons, the isotopic composition will change. For example, the fissile nuclides in the fuel continuously undergo the fission reactions and transmutations. At a rough estimate, a reactor with unit thermal power (1 MW) consumes 1.23 g of nuclide ^{235}U per day. Fertile nuclides (such as ^{238}U or ^{232}Th) are converted to fissile nuclides (such as ^{239}Pu or ^{233}U) by neutron capture and subsequent decay. Meanwhile, fission events produce more than 300 types of fission products (the fission fragments generated at the time of fission and the isotopes produced via subsequent decay are collectively called "fission products"). The fission products can undergo neutron transmutation, and unstable fission products may undergo radioactive decay to be transmuted into other nuclides. The nuclides in burnable poisons or structural materials can also be converted to other isotopes by neutron transmutation and decay. Generally, the transmutation-decay chain is used to describe the transmutation-decay process. The transmutation-decay chains are illustrated in Fig. 2.7.

The equation describing the change of the number of burnup nuclides is the burnup equation, as follows:

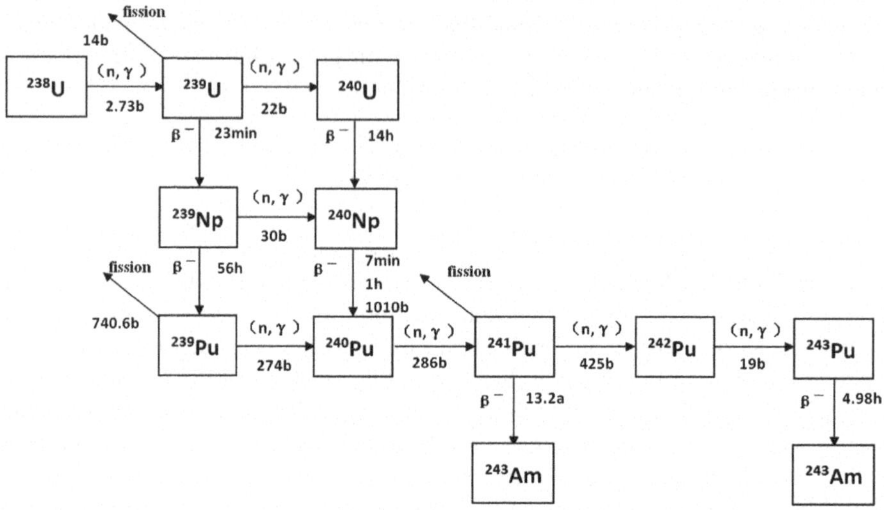

Fig. 2.7 Transmutation-decay chains

Concentrations of various fuel isotopes in a reactor are described by a coupled set of production-destruction equations, called *depletion equations*:

$$\frac{dN_i(t)}{dt} = -\alpha_i(t)N_i(t) + \sum_j \beta_{j\to i}(t)N_j(t) + \sum_k f_{k\to i}(t)N_k(t) \tag{2.16}$$

where, j means the non-fission nuclides; k means the fission nuclides; $N_i(t)$ means the nucleus density of the ith nuclide; $\alpha_i = \sum_{g=1}^{G} \sigma_{a,g,i}(t)\phi_g(t) + \lambda_i$ means the disappearance rate of nuclide i, including disappearance rates related to absorption reaction and decay; $\beta_{j\to i}(t)$ means the production rate of nuclide i by nuclide j, including neutron transmutation and decay; and $f_{k\to i}(t)$ means the production rate of nuclide i by fission nuclides k via fission reaction.

Combining the transmutation-decay chains for the fuel isotopes, the approximate solution of Eq. (2.16) can reveal the variation regularity of the concentrations of burnup nuclides with time. In order to provide an intuitive concept and a better understanding of the burnup characteristics of some important nuclides in a PWR, the following part describes the variation regularity of important fission product nuclides with time, and the concepts of reactor lifetime and burnup.

(1) Poisoning of fission products ^{135}Xe and ^{149}Sm

For a thermal neutron reactor, among the fission products, nuclides ^{135}Xe and ^{149}Sm are particularly important. On the one hand, they and their precursors have relatively large fission yields, so their concentrations increase rapidly with the start-up of the reactor. On the other hand, they have enormous thermal neutron absorption cross sections (see Fig. 2.8), so their buildup and loss can appreciably affect the reactor reactivity.

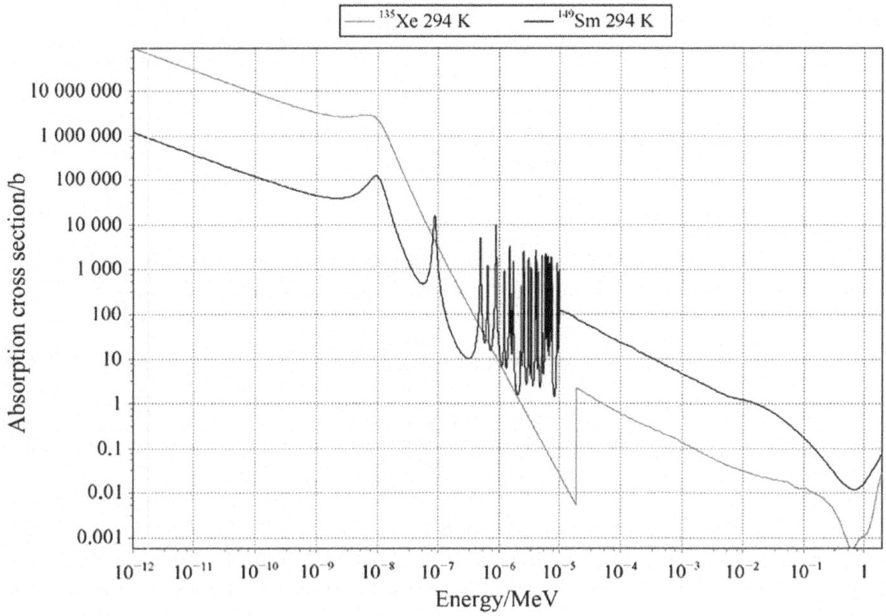

Fig. 2.8 Absorption cross sections of nuclides ^{135}Xe and ^{149}Sm

Fig. 2.9 Simplified transmutation-decay scheme for I–Xe

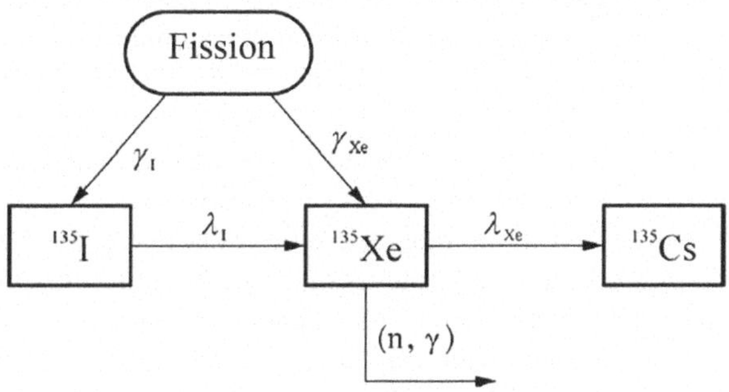

To make an intuitionistic result, the burnup chain is simplified (see Fig. 2.9). Since the half-lives of ^{135}Sb and ^{135}Te are very short, their effects on the intermediate process can be ignored. The sum of the fission yields of ^{135}Sb and ^{135} and the direct fission yield of ^{135}I is taken as the fission yield of ^{135}I, i.e. $\gamma_I = \gamma_{sb} + \gamma_{Te} + \gamma_I'$ (where γ_I' means the direct fission yield of ^{135}I). The thermal neutron absorption cross section of ^{135}I is only 8b, and its half-life is only 6.7 h. When the thermal neutron flux density is 10^{14} cm^{-2} s^{-1}, the loss of ^{135}I caused by neutron absorption is much less than that caused by decay. Therefore, the thermal neutron absorption of ^{135}I can be ignored, and it is deemed that ^{135}I completely decays to ^{135}Xe. Based on the above simplification, the values of fission yields and decay constants for ^{135}I and ^{135}Xe are tabulated in Table 2.2. As ^{135}Xe has a large absorption cross section and a short half-life, its

Table 2.2 Fission yields and decay constants for ^{135}I and ^{135}Xe

Fission products	Fission yield γ/%				Decay constant λ/s^{-1}
	^{233}U	^{235}U	^{239}Pu	^{241}Pu	
^{136}I	4.884	6.386	6.100	7.694	2.87×10^{-5}
^{135}Xe	1.363	0.228	1.087	0.255	2.09×10^{-5}

concentration rapidly increases and approaches to saturation following startup. The concentrations of ^{135}I and ^{135}Xe eventually saturate at equilibrium values:

$$N_{^{135}I}(\infty) = \frac{\gamma_{^{135}I}}{\lambda_{^{135}I}}\Sigma_f\phi \tag{2.17}$$

$$N_{^{135}Xe}(\infty) = \frac{\gamma_{^{135}I} + \gamma_{^{135}Xe}}{\lambda_{^{135}Xe} + \sigma_{a^{135}e}\phi}\Sigma_f\phi \tag{2.18}$$

After 40 h of the steady-state operation of the reactor, ^{135}Xe reaches saturation. It can be seen from the Eq. 2.18 that both the concentration of saturated ^{135}Xe and the negative reactivity resulted from ^{135}Xe are related to the thermal neutron flux of the reactor. When the value of thermal neutron flux density in the reactor is small, the xenon reactivity is also small. When the thermal neutron flux is high, the xenon reactivity will approach a maximum and become independent of the thermal neutron flux. The thermal neutron flux density of power reactor at the rated power is generally high, and the xenon reactivity is independent of the thermal neutron flux. However, if the reactor operates at a low power, the xenon reactivity will depend on the power level.

When the reactor is shut down, the neutron flux can be deemed as approximately equal to zero, and the direct production rate of ^{135}Xe by fission reaction is zero. However, ^{135}I in the reactor continues to decay to ^{135}Xe, and the removal of ^{135}Xe via β^-decay also continues. Since the half-life of ^{135}Xe is longer than that of ^{135}I, following a shutdown, ^{135}Xe concentration in the reactor may initially build up, reach a maximum value and then gradually decay out. For a reactor operating at high flux prior to shutdown, ^{135}Xe concentration will peak approximately 11 h after the shutdown.

Figure 2.10 illustrates the curves for the variations of ^{135}Xe concentration and excess reactivity (i.e. the reactivity of the core without any method to control the reactivity) with time following a reactor start-up, steady-state operation and then sudden shutdown. As we can see from the figure, since the thermal neutron absorption cross section of ^{135}Xe is enormous, the behavior of excess reactivity of the entire reactor is opposite to the behavior of ^{135}Xe concentration. When ^{135}Xe concentration after shutdown reaches the maximum value, the excess reactivity of the reactor reaches the minimum value. The magnitude by which the excess reactivity decreases to the minimum value after shutdown is referred to as *iodine well depth*. The iodine well depth is closely related to the thermal neutron flux prior to shutdown. The larger

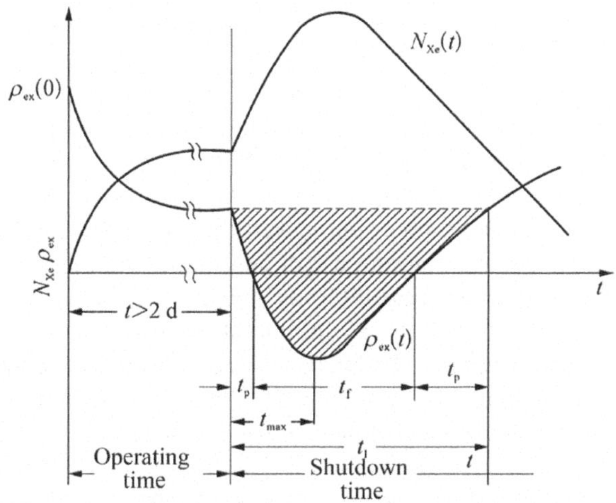

Fig. 2.10 Changing of ^{135}Xe concentration and excess reactivity with time before and after shutdown

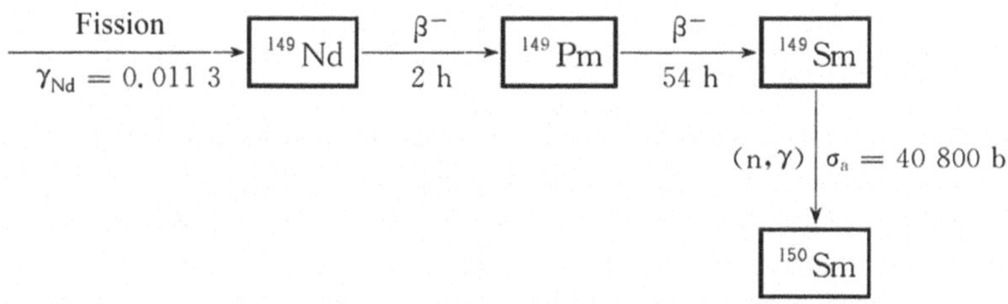

Fig. 2.11 Relevant simplified transmutation-decay chain for ^{149}Sm

the thermal neutron flux is, the greater the iodine well depth will be. It is worth mentioning that for marine nuclear power plants, in order to ensure the maneuverability, the excess reactivity of the marine reactor should always be greater than zero (if the excess reactivity is less than zero, the reactor cannot be restarted timely and thus fail to provide energy for nuclear-powered ships). Therefore, the design of marine nuclear power reactors requires that the excess reactivity at the maximum iodine well depth should always be greater than zero.

Among all fission products, the effect of nuclide ^{149}Sm on thermal neutron reactor is second only to that of ^{135}Xe. The transmutation-decay chain for nuclide ^{149}Sm is illustrated in Fig. 2.11. Following a reactor startup, ^{149}Sm will be produced by the β^-decay of ^{149}Pm and removed from the core by neutron absorption (n, γ). About 5 days of operation after the start-up of the reactor, ^{149}Sm concentration will build up to an equilibrium value:

$$N_{Pm-149}(\infty) = \frac{\gamma_{Pm-149}}{\lambda_{Pm-149}} \Sigma_f \phi \tag{2.19}$$

$$N_{Sm}(\infty) = \frac{\gamma_{Pm-149}\Sigma_f}{\sigma_a^{Sm-149}} \tag{2.20}$$

It can be seen from the Eq. 2.20 that the equilibrium concentration of ^{149}Sm is independent of the neutron flux while that of ^{149}Pm (i.e., the precursor of ^{149}Sm) depends on the neutron flux.

When a reactor is shut down after running sufficiently long to build up equilibrium concentrations, the ^{149}Sm concentration will continuously increase as the accumulated ^{149}Pm decays into ^{149}Sm and the only means of removal for ^{149}Sm ceases. After more than 200 h of decay, the ^{149}Sm nuclide density approximates the maximum value. This maximum value is related to the equilibrium concentration of ^{149}Sm and the neutron flux level before shutdown. The higher the neutron flux level is, the greater the maximum value will be. If the neutron flux before the shutdown is greater than 0.87×10^{14} cm^{-2} s^{-1}, the maximum concentration of ^{149}Sm will be twice the equilibrium concentration. If the neutron flux level before shutdown is rather low, there will be little change of the ^{149}Sm concentration after the shutdown. As there is an inverse relationship between the ^{149}Sm concentration and the excess reactivity, the behavior of the ^{149}Sm during post-shutdown and restart periods requires special considerations in the design of marine reactors.

(2) Reactor lifetime and burnup

As PWR core burnup progresses, fissile nuclides are continuously consumed, and the excess reactivity of the reactor gradually decreases since some fission products have rather large absorption cross sections (such as ^{135}Xe). If the excess reactivity of the reactor decreases to 0, the reactor cannot reach the criticality. Generally, the time from the beginning of the reactor operation to the point at which the excess reactivity becomes zero (or the effective multiplication factor decreases to one) is called *reactor core lifetime*. For marine reactors, the maximum xenon concentration should be considered. On this basis, the time from the beginning of the reactor operation to the point at which the excess reactivity becomes zero (considering the maximum xenon concentration) is called *marine reactor core lifetime*.

For PWR cores, in order to obtain a longer reactor core lifetime, it is necessary to increase the inventory of fissile nuclides in the core. However, too much inventory of fissile nuclides will lead to a rather large initial excess reactivity. Therefore, supplemental burnable poisons have to be used to compensate for the excess reactivity.

Generally, fuel burnup is used to express the retention time and service life of fuel in the reactor, and it is a measure of the total energy produced per unit mass of fuel in the initial loading, and also a measure of the degree of fuel depletion. The most common representations for the burnup are as follows.

① The energy generated per unit mass of fuel, with unit as MW · d/tU:

$$BU = \frac{\int_0^T P(t)dt}{W_U} \tag{2.21}$$

The numerator in this measuring unit is the generated energy, and P(t) is the power. The denominator W_U is the mass of heavy metals (uranium, plutonium and thorium) in the fuel.

② The ratio of the mass of burnt fissile nuclides (W_B) to the mass of loaded fissile nuclides (W_F):

$$\alpha_F = \frac{W_B}{W_F} \times 100\% \tag{2.22}$$

③ The ratio of the mass of burnt fissile nuclides (W_B) to the mass of loaded fuel (W_U):

$$\alpha_U = \frac{W_B}{W_U}\ \text{kg/t} \tag{2.23}$$

④ The total number of fissions in a unit volume of fuel. This measure is called *fission density*, and the unit is m^{-3}:

$$Fd = \frac{\int_0^T \Sigma_f(t)\phi(t)dt}{V_F} \tag{2.24}$$

Above-mentioned measuring units for fuel burnup can be mutually converted.

2.2.1.6 Boltzmann equation for neutron transport and its computational methods

The equation describing the law governing the statistical distribution of neutrons in a nuclear reactor is the Boltzmann equation for neutron transport. Its general form is as follows:

$$\frac{1}{v}\frac{\partial \varphi}{\partial t} = \int_0^{+\infty}\int_{\Omega'} \Sigma_s(r, E')f(r, E' \to E, \Omega' \to \Omega)\varphi(r, E', \Omega', t)dE'd\Omega' \\ +S(r, E, \Omega, t) - \Omega \cdot \nabla\varphi - \Sigma_t(r, E)\varphi \tag{2.25}$$

where, υ means the neutron velocity; Ω means the direction of neutron motion; r means the position; E means the neutron energy; φ, which is short for φ (r, E, Ω, t), means the neutron angular flux and is related to the direction Ω. $S(r, E, \Omega, t)$ means the external neutron source and the fission neutron source.

Equation (2.25) is the conservation equation for the neutron number at a certain position and certain time. The left side of the equation means the rate of the change of neutron number in this phase space (the phase space is composed of position, energy and direction) with time. The first item of the right side means the number of neutrons scattered from other phase spaces into this phase space. The second item of the right side means the number of neutrons in this phase space produced by the external source or the fission source. The third item of the right side means the loss rate of neutrons via neutron motion. The fourth item of the right side means the loss rate of neutrons via interactions with atomic nuclei.

By strictly solving the Eq. (2.25), we can obtain the number of neutrons in the reactor at any time and any spatial location, and in any energy range and any direction. If this number is known, we can calculate the rates of all possible nuclear reactions throughout the region and understand the physical properties of the reactor. However, Eq. (2.25) is so complicated that the analytical solution is only available for very simple problems and cannot be obtained for real reactors.

After decades of development, various numerical methods have been proposed to simplify the solution of neutron transport equations. A brief and comprehensive introduction to some methods is given below. Readers who are interested in the details of these methods can refer to relevant reference books [2–4].

The computational methods for neutron transport are discussed below from five aspects, i.e. time variable, energy variable, direction angle variables, spatial variables and steady-state eigenvalue k_{eff}.

(1) Time variable

The reactor generally operates in a steady state or quasi-steady state. Therefore, during the design of a reactor, it is generally assumed that the neutron flux in the reactor does not change rapidly with time (slow change with burnup is a quasi-steady process and can be solved by the steady-state solving methods). The effective multiplication factor k_{eff} is introduced in the equation to consider the deviation of the reactor from criticality (actually, k_{eff} not equal to one indicates a transient process). The steady-state neutron transport equation with the introduction of k_{eff} is as follows:

$$\begin{aligned}\Omega \cdot \nabla\varphi + \Sigma_t(r, E)\varphi = &\int_0^{+\infty}\int_{\Omega'} \Sigma_s(r, E')f(r, E' \to E, \Omega' \to \Omega)\varphi(r, E', \Omega')dE'd\Omega' \\ &+ \frac{1}{k_{eff}}\chi(E)\int_0^{+\infty} \nu\Sigma_f(r, E')dE' \int_{\Omega'} \varphi(r, E', \Omega')d\Omega' \end{aligned} \quad (2.26)$$

where, $\chi(E)$ means the fission neutron spectrum.

In the actual operation of the reactor, if the effective multiplication factor k_{eff} is not equal to one, the control reactivity can be introduced in the core to make k_{eff} equal to one. For a real transient process in the reactor, such as the rod ejection accident, it is necessary to discretize time derivatives in the transport equation, usually using implicit difference scheme.

(2) Energy variable

The neutron energy in the reactor ranges from 10^{-5} eV to 20 MeV, spanning 12 orders of magnitude. Meanwhile, the cross sections of various nuclides change drastically with energy, especially in the resonance energy region. Therefore, it is very difficult to process the energy variable. At present, two methods are adopted. One method is to select a large number of energy points (generally tens of thousands or even hundreds of thousands) in the entire neutron energy range. The energy variable is discretized on these energy points assuming that the neutron cross section at each energy point represents the cross section of its neighboring region. With dense energy points, this method is endowed with rather high accuracy, and is adopted in Monte Carlo neutronics solution method. The other method is to disperse the entire energy range into tens or even hundreds of regions. In the field of reactor physics, each energy region is generally called a group. That is to say, the entire energy range is divided into tens or even hundreds of groups. Since the cross section changes with neutron energy, it is necessary to use a typical neutron spectrum in advance to obtain the weighted average of the cross section for each energy group. For the PWRs, the commonly used weighting spectrum is "Maxwell thermal neutron spectrum—1/E moderation spectrum—fission neutron spectrum" previously introduced in this section. For the resonance energy regions, since the cross sections change drastically and affect the neutron spectrum, it is impossible to obtain the average cross sections of the resonance energy groups by prior calculation. Generally, the resonance integrals for the typical environment and temperature are pre-tabulated. Then the resonance processing will be based on actual problems to solve the average cross section in the resonance energy group. The resonance calculation method is still one of the hot topics in the numerical investigation of reactor physics. The use of multi-group method for processing the energy variable greatly reduces the computational cost. Therefore, this method is the main stream in the numerical calculation of reactor physics, and widely used in the numerical programs.

(3) Direction angle variables

In the neutron transport equation, the neutron angular flux is coupled by scattering and fission reactions. Currently, there are three types of methods for processing the direction angle variables. The first type of method, as used in the Monte Carlo method, uses the random numbers to select the neutron angles during the simulation of neutron movement. This type of method requires a very large number of samples, but it has a high accuracy and can simulate the real neutron angle distribution. The second type of method is to disperse the angle directions, and only solve the neutron transport equation in the discrete angle directions. The scattering source term and the fission source term are used to couple different discrete directions to obtain the neutron angular flux in the discrete directions. Based on the relationship between the neutron scalar flux and the neutron angular flux, the neutron scalar flux is solved. This type of method includes discrete ordinate method (Sn) and method of characteristics (MOC). The third type of method is to expand the neutron angular flux by the polynomials, generally by spherical harmonics (Pn). With this method, the neutron transport equation is converted into an equation set for the neutron angular flux with different expansion orders. Solving this equation set can obtain the neutron angular flux component in different order. Based on the spherical harmonics expansion relationship, the real neutron angular flux can be obtained. It is worth mentioning that the neutron diffusion equation can be obtained by reserving the first-order expansion relationship:

$$-\nabla D(r,E)\nabla\phi(r,E) + \Sigma_R(r,E)\phi(r,E) = \frac{1}{k_{eff}}\chi(E)\int_0^{+\infty} \nu\Sigma_f(r,dE')\phi(r,dE')dE' \quad (2.27)$$

Since there is no angle variable in the diffusion equation and the solution result is the neutron scalar flux ϕ required by the nuclear reactor, this type of method is widely used in the numerical simulation calculation for reactor cores. However, it is assumed that, in the neutron diffusion equation, the neutron flux is uniformly distributed at all angles. Of course, the transport approximation can be used to consider part of anisotropy [1]. However, when the neutron flux anisotropy is strong, the calculation results obtained by the diffusion equation have rather large errors, such as in the vicinity of strong absorbers and strong neutron sources, or at the boundary of the reactor core.

(4) Spatial variables

Similar to the processing methods for spatial variables in other partial differential equations, there are many processing methods for spatial variables. Several common methods are described below. The first method is the finite difference method, which uses the difference to replace the derivative of the neutron angular flux to the spatial variable in the transport equation (or neutron diffusion equation). It is mainly used in the structured grids. Various kinds of fine-grid finite difference neutron programs use this method. The second method is the finite element method. It is based on the variation principle and the weighted residual method, and can be used in the unstructured grids. The third method is the coarse-grid nodal method. In this method, the integral in a coordinate direction of the neutron diffusion equation (or the neutron transport equation) is performed to obtain the one-dimensional equation in this direction. The equations in different coordinate directions are coupled by the leakage terms of other directions. To obtain the final solution, the polynomial expansion is performed for the neutron flux, and continuity of various orders is used to obtain the neutron flux. This method is mainly used in the structured coarse grids, and widely used for the diffusion solution of PWR cores in current NPPs. The fourth method is the spectral method, which uses the analytic basis function satisfying the neutron diffusion equation to expand the neutron flux, and then uses the boundary conditions between the nodes to solve the expansion coefficient. The fifth method is the parallel line sweeping method, which uses the intersections between a group of parallel rays and the grids to describe the grid geometry information. Therefore, this method has no special requirements for the grids, and is mainly adopted in the collision probability method and method of characteristics. This method gains wider and wider application in the reactor core calculation. All the above methods have their own advantages and disadvantages. The parallel line sweeping method has the strongest geometric adaptability, but the highest computation cost; the coarse-grid nodal method has the fastest calculation speed, but the worst geometric adaptability; the fine-grid finite difference method is intermediate in both computation load and geometric adaptability; and the finite element method has better geometric adaptability than the fine-grid finite difference method.

(5) Steady-state eigenvalue k_{eff} variable

In the steady-state neutron transport equation, it is required to solve the eigenvalue k_{eff}. Since the general calculation for reactor physics only requires the maximum eigenvalue, the power iteration method is generally used. This method can only obtain the maximum eigenvalue. In recent years, some scholars have used the Krylov subspace method to solve the eigenvalue k_{eff}, which can obtain multiple eigenvalues.

2.2.2 Reactor Nuclear Design

A nuclear reactor is a very complicated device. The nuclear reactor design involves many professional fields with quite different professional knowledge and design methods, and is generally divided into several design packages for synergetic work by making full use of various professionals. Each design package involves a relatively single discipline and the results of different design packages are interrelated through the interfaces, so as to achieve the goals of "breaking up the whole into parts" and improving the design efficiency. It is worth mentioning that during the reactor design, iterations shall be performed continuously with the inputs from multiple disciplines. For example, the repeated iteration shall be performed for reactor core nuclear design, thermo-hydraulic design and fuel assembly design to obtain an optimized scheme for a reactor core. Finally, the repeated iteration shall be performed for this scheme and other designs such as system design, reactor structure design and mechanical design, to obtain an optimized overall scheme for a reactor (including structural systems and parameters).

The tasks in marine PWR reactor nuclear design mainly include: the determination of the overall parameters (such as power level, lifetime, volume, coolant temperature and safety) for the nuclear reactor design and the basic structural forms of reactor fuel assemblies and reactor core based on the overall ship requirements with the cooperation with other professional fields at the scheme design stage; the design of fuel loading method, burnable poisons, control rods, reflector and neutron sources for the reactor, and the design and optimization of the control method for the change of burnup and power at the detailed design stage, based on the determined overall parameters and basic structural form of the reactor.

Main design criteria are as follows:

① the reactor core reactivity control shall be reasonable, meeting the reactor core lifetime requirements (considering the impact of iodine well), compensating for the reactivity fluctuation caused by temperature change of the coolant in the reactor core, and meeting the requirements for rapid change in reactor power and for the safe shutdown of the reactor;

② there shall be a self-stabilization function for the reactor core reactivity fluctuation, that is, the moderator temperature coefficient and the fuel temperature coefficient shall be negative;

③ the reactor power oscillation shall have the inherent stability, or the power oscillation can be effectively controlled by reasonable means;

④ the power distribution of the reactor in high power operation shall be as flat as possible to provide favorable conditions for the coolant to remove the heat, and the power distribution distortion during the regulation of reactor core power change shall be as small as possible. The common measures used in reactor core nuclear design are briefly introduced for each design criterion.

In a PWR, the commonly used reactivity control methods are generally as follows:

① Fixed burnable poisons are arranged in the reactor core to "override" the excess reactivity in the initial reactor. The fixed burnable poisons are continuously consumed with the burnup of the fuel in the reactor to control the reactor excess reactivity.
② A certain number of control rods are arranged in the reactor, and the reactivity is controlled by inserting or withdrawing the control rods.
③ Soluble burnable poisons, such as soluble boron, are added in the reactor coolant during the operation of the reactor, and the reactivity is controlled by diluting or decreasing the concentration of soluble burnable poisons.

Among the above control methods, if the soluble burnable poisons are used for the dynamic regulation of the reactivity, additional corollary equipment shall be provided in the primary loop. For marine reactors, such additional equipment will accordingly increase the volume and weight, and thus this method is avoided in the marine reactors.

Generally, the reactivity coefficient can be regulated by adjusting the ratio of fuel to moderator in the reactor or changing the reactor leakage.

During the dynamic regulation of the reactivity by using the control rods, the movement of the control rods will cause the reactor power distortion. Therefore, during the marine reactor design, the flattening of the power distribution of rod-controlled reactor core is rather difficult. Generally, the following methods can be used for regulation: ① fuel zoning by enrichment; ② reasonable arrangement of burnable poisons; and ③ reasonable control rod regulation.

2.2.3 Software for Reactor Nuclear Design

During the nuclear design of nuclear reactor core, the nuclear performance of the designed reactor core can be roughly estimated by the theory of reactor physics. However, as the reactor core state is described by complex partial differential equations, an accurate prediction requires the discrete of the partial differential equations with the assistance of computer numerical simulation software, to provide guidance to the detailed core design.

Nuclear design software for reactor cores of marine NPPs and land NPPs are similar in calculation models and algorithm. However, the characteristics of fuel assemblies for marine reactors shall be considered: rather small reactor core, strong neutron leakage and frequent movement of control rods during the operation. The commonly used nuclear design software can be divided into the following two types.

(1) Monte Carlo neutronics analysis software
The Monte Carlo neutronics analysis software uses computer to simulate the motion of fundamental particles such as neutrons in the reactor, so as to simulate the reactor core performance. This type of calculation software has strong geometric adaptability, and meets the requirements for the design of reactor cores with different spectra. The main disadvantage of this type of software is that its calculation speed is rather slow (especially for calculating the distribution in nuclear reactor core, such as power distribution and neutron flux distribution). Although the calculation speed can be significantly increased by parallel computing, the speed is still slower than that of the deterministic software. At present, the Monte Carlo neutronics analysis software commonly used in the world includes MCNP [5], KENO [6], RMC [7], MC.21 [8], JMCT [9], OPENMC [10] and MOI [11].
(2) Deterministic neutronics software
In the deterministic neutronics software, the direct transport calculation of 3D reactor core is very costly, so the "two-step method" was often used for calculation. The first step is to perform the 2D lattice transportation calculation to obtain rather accurate results for the interior of the assemblies and then homogenize the assembly. During the 3D calculation for the reactor core, it is assumed that the interior of the assembly is homogeneous, regardless of the internal structure of the assembly. After the core calculation is completed, the refined results for the interior of assembly are reconstructed. The two-step design software systems commonly used in the world include the APA system [12] developed by Westinghouse of the United States and the SCIENCE system [13] developed by AREVA of France. At present, with the development of computers, the calculation methods based on 3D neutron transport have been gradually used in the reactor design. The one-step software packages used in the world include DeCART [14], nTRACER [15], MPACT [16] and TIGER-3D [17].

2.2.4 Design Verification

The reactor nuclear design is based on the theory of nuclear reactor physics. Although the numerical simulation means have been used to perform simulations, there are a large number of approximations and uncertainties in current nuclear data, numerical simulation methods themselves and simulation process, leading to rather large uncertainty in the numerical simulation results. The reliability of the numerical simulation results shall be fully validated when there is insufficient test verification.

The main reasons for rather large numerical simulation errors or rather large uncertainty in reactor core nuclear design are as follows: ① uncertainty of nuclear cross section data; ② the errors in the calculation methods themselves resulting from the simplification of neutron transport theory and burnup equation (such as the use of neutron diffusion models and the simplification of lumped fission products) to improve the computational efficiency of the computer programs; ③ errors resulting from the simplification or equivalent treatment in the neutronics simulation for the actual reactor structure and the surrounding environment (for example, the structure of spacer grids commonly used for PWR fuel assemblies), which are too complicated to be described precisely by the neutronics calculation software; and ④ deviations resulting from the design software, such as computational grid division and calculation model selection.

In order to save the cost and time for tests, the nuclear design scheme for a real reactor core will not be directly verified. However, before the engineering production, assembling or comprehensive experiment of the real reactor core, a series of typical small-scale zero-power test schemes are designed and the tested to directly verify the accuracy of numerical simulation methods, and thus to improve the accuracy of the reactor core nuclear design scheme. The numerical simulation methods can be verified with smaller-scale tests and fewer tests. The test conditions are not necessarily identical to the actual operating conditions of the reactor core; meanwhile, the verification of the accuracy of numerical simulation methods can indirectly verify the accuracy of the actual reactor core schemes. If the numerical simulation results of a test are quite different from the test results, the problems causing the calculation deviations of numerical simulation software can be detected and corrected, and thus the numerical simulation methods can achieve a higher calculation accuracy. If the numerical simulation results of a series of related tests are considered to have rather high calculation accuracy, we can believe that the performance of the reactor core designed by using this numerical simulation program has rather small deviation from the simulation results and that the engineering requirements can be met. The zero-power physical tests of the reactor core generally include criticality tests for various loading conditions and various preset control rod positions, control rod value tests, temperature coefficient tests and power distribution tests. The design of various zero-power physical test schemes is a comprehensive work. On the one hand, the actual loading conditions of real reactors shall be considered. On the other hand, the test schemes shall cover various functions and calculation of the numerical simulation software. Finally, the test schemes shall be continuously adjusted according to the numerical simulation results of the tests to meet the requirements for zero-power physical verification of the reactor core.

After the real reactor core is manufactured, assembled and commissioned, the comprehensive tests can be performed for the reactor core to fully and actually verify the reactor core nuclear design. The tests involving reactor physics mainly include various physical tests at different stages such as reactor core loading (refueling), first criticality test, zero-power physical test, physical tests with power increasing at various power levels and power distribution tracking and monitoring during the power operation. The main objectives of the comprehensive physical tests of the reactor core are to: ① make the reactor safely reach the criticality for the first time; ② in the zero-power state and at various power levels during the level-by-level power increasing, verify that the deviation of the designed, calculated and estimated values of core physical parameters from the measured values meets the test acceptance criteria, including design criteria and safety criteria; ③ measure, inspect or scale the performance of in-core or ex-core nuclear instrument system to provide correct and effective monitoring of the operating state of reactors, and ensure that the protection system provides the correct monitoring signals; and ④ by tracking and monitoring the reactor core rod positions and power distribution, ensure the reactor core safety during the power operation of reactors, and verify the accuracy of reactor core nuclear design calculation.

2.3 Reactor Thermo-Hydraulics

2.3.1 Overview

The reactor thermo-hydraulics mainly studies the problems in temperature field, flow field and pressure field in the reactor, and thus ensures that the heat generated by the fission reaction in the reactor core is efficiently and safely removed. The temperature field problems mainly involve the temperature field problems in the reactor core fuel, the reactor core coolant, the coolant in the reactor systems, the critical components, and the problems of the heat dissipation and insulation of systems and components. The flow field problems mainly involve the coolant flow distribution at the reactor core inlet, reactor core bypass flow, flow stability, coolant flow and mixing, and reactor system coolant flow. The pressure field problems mainly involve the pressure drop in the reactor system and the pressure distribution of the coolant in the reactor system. All these thermo-hydraulic problems are further divided into two categories, i.e. steady-state thermo-hydraulic problems and transient thermo-hydraulic problems. Steady-state thermo-hydraulics refers to a state under which none of the macroscopic operating parameters change with time, such as conditions at rated power level and other power levels, steady-state natural circulation conditions and single-loop operating conditions of the reactor. Transient thermo-hydraulics refers to specific conditions or abnormal conditions under which the operating parameters change with time, such as power increase/reduction conditions, conditions for inter-conversion between forced circulation and natural circulation and accident conditions of the reactor. Specifically, the reactor thermo-hydraulics mainly focuses on the temperature distribution in fuel elements, flow and heat transfer characteristics of the coolant, and predicts the thermal parameters of the reactor under various operating conditions and the change of pressure, temperature, flow and other thermal parameters with time under various transient and accident conditions.

The main tasks in the reactor thermo-hydraulic design are to carry out the argumentation and optimization of design parameters (including main parameters, structure parameters and parameters of main and auxiliary systems) and the research and design (including reactor core design, protection setpoint design, device maneuverability design, reactor design transient analysis, device operation analysis and typical accident analysis) at different design stages. The purpose of reactor thermo-hydraulic design is to ensure that the reactor can, under normal operating conditions, safely and efficiently remove the heat from the reactor core and generate the steam meeting the plant requirements, and, under accident conditions, ensure the thermal safety of the reactor core and meet the safety limit criteria.

2.3.2 *Reactor Heat Transfer Theory*

There may be various heat transfer phenomena during the reactor operation, including heat conduction, heat convection and radiation heat transfer. The heat generated by the nuclear fission reaction of the fuel is transferred to the outer surface of the fuel cladding by heat conduction. The outer surface of the fuel cladding transfers the heat to the coolant by heat convection. Under certain abnormal conditions such as severe accident conditions, the radiation heat transfer shall be considered. In steam generators, electric heating elements in pressurizers and main pumps, the heat transfer between primary side and secondary side of the reactor and that between the heating equipment and the coolant are also realized by heat conduction and heat convection.

2.3.2.1 Heat Conduction

Heat conduction of fuel elements refers to a process in which the heat generated by nuclear fission in the fuel elements is transferred, by the means of conduction of heat, from the inside of fuel pellets with a higher temperature to the outer surface of their cladding with a lower temperature. For marine reactors, the commonly used fuel elements in the reactor core are in the form of rod, plate or sphere. The heat conduction usually includes the heat conduction in the fuel pellet, the air gap (if any) between the fuel pellet and the cladding, and the cladding [18–20].

As the carriers for heat transfer between primary side and secondary side of the reactor, steam generators utilize its tubes to transfer the heat of the coolant at the primary side to the feedwater at the secondary side. Heat conduction also occurs in the tubes. Similar phenomenon also occurs inside the electric heating elements in the pressurizers and the main pumps. The heat conduction equation used in the thermal calculation is as follows:

$$\rho c_p \frac{\partial T(r,t)}{\partial t} = \frac{1}{r}\frac{\partial}{\partial r}\left(\lambda r \cdot \frac{\partial T(r,t)}{\partial r}\right) + q_v(r,t) \tag{2.28}$$

where, ρ means the density of material (kg/m^3); c_p means the specific heat capacity at constant pressure of material [J/(kg °C)]; λ means the heat conductivity coefficient of material [W/(m °C)]; q_v means the heat release rate per unit volume (W/m^3); and r means the radial dimension (m).

In the analysis of specific heat conduction problems of fuels, the above heat conduction equation can be rewritten to reflect the specific forms and boundary

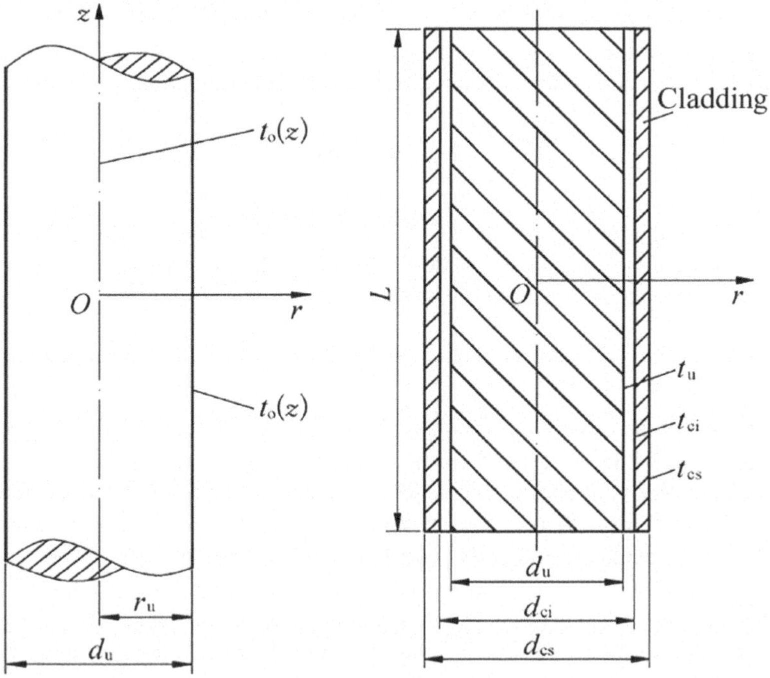

Fig. 2.12 Heat conduction model for rod fuel elements

conditions of the fuel elements, so as to obtain the equations for the heat conduction in cylindrical, plate and sphere fuel elements and for heat conduction in the fuel pellet, the air gap and the cladding. For fuel elements with air gap (such as rod fuel elements containing UO_2 pellets), the fuel pellets may crack under irradiation and heating conditions, and the gap distance between the pellet centerline and the inner wall of the fuel cladding becomes irregular. During the fission reaction, the content and distribution of fission gas released by the pellets will not be constant. Therefore, it is difficult to determine the heat conductivity coefficient of the air gap. Some countries have performed relatively sufficient research on this area for NPPs and obtained relevant empirical and experimental values. However, there is still a lack of research for marine nuclear power plants.

The heat conduction model for rod fuel elements is illustrated in Fig. 2.12.

The heat conduction equation of the fuel pellet is as follows:

$$t_0 - t_u = q\frac{r_u}{2k_u}, \tag{2.29}$$

where, t_0 means the centerline temperature of fuel pellet (°C); t_u means the surface temperature of fuel pellet (°C); r_u means the radius of fuel pellet (m); q means the surface heat flux of fuel pellet (W/m^2); and k_u means the thermal conductivity [W/(m °C)].

The fuel pellets of rod fuel elements generally use UO_2 ceramic fuel. The thermal conductivity k_u of the cold press sintered UO_2 ceramic fuel with a density of 95% (theoretical value) is usually calculated by the following equation:

$$k_u = \frac{38.24}{t + 402.55} + 4.788 \times 10^{-13}(t + 273.15)^3, \tag{2.30}$$

where, t means the fuel temperature (°C). The applicable range of this equation is 0 to 2,450 °C (temperature) and 0–40,000 MW · d/tU (fuel burnup).

The heat conduction equation of the fuel cladding is as follows:

$$t_{ci} - t_{cs} = \frac{q_l}{2\pi k_c} \ln \frac{d_{cs}}{d_{ci}}, \tag{2.31}$$

where, t_{ci} means the temperature of the inner wall of the cladding (°C); t_{cs} means the temperature of the outer wall of the cladding (°C); q_l means the linear power density of the outer surface of the cladding (W/m); k_c means the thermal conductivity of the cladding [W/(m °C)] (If zirconium alloy is used, the thermal conductivity value increases gradually with temperature (38–482 °C) in the range of 11.8–13.5 W/(m °C); if the stainless steel is used, the thermal conductivity value increases with temperature (38–538 °C) in the range of 14.8–21.5 W/(m °C)); d_{ci} means the inner diameter of the cladding (m); and d_{cs} means the outer diameter of the cladding (m).

The gap heat conduction equation for the helium-filled rod fuel element is as follows:

$$q = \eta(t_u - t_{ci}), \tag{2.32}$$

where, q means the heat flux of outer surface of pellet (W/m^2); t_u means the temperature of outer surface of pellet (°C); t_{ci} means the temperature of inner surface of cladding (°C); and η means the equivalent heat transfer coefficient of gap [W/(m^2 °C)].

The equivalent heat transfer coefficient of the gap η shall not be less than 1 W/(cm^2 °C) under normal operating conditions. When the burnup is around 10,000 MW·d/tU, the equivalent heat transfer coefficient of the gap may be minimum but will not be less than 0.5 W/(cm^2 °C). Generally, the value of 0.7 W/(cm^2 °C) is used. Under the high burnup where the pellet and the cladding contact each other, the value of 2 W/(cm^2 °C) can be used as the equivalent heat transfer coefficient of the gap. At present, in the designs of PWR NPPs, the value of 5,678 W/(m^2 °C) is generally used as the equivalent heat transfer coefficient of the gap for calculation. Of course, for specific cases, the equivalent heat transfer coefficient of the gap can be determined through relevant experimental study.

The heat conduction model for plate fuel elements is illustrated in Fig. 2.13.

The heat conduction equation of fuel pellet is as follows:

$$t_0 - t_u = q_v \frac{\delta_u}{2k_u}, \tag{2.33}$$

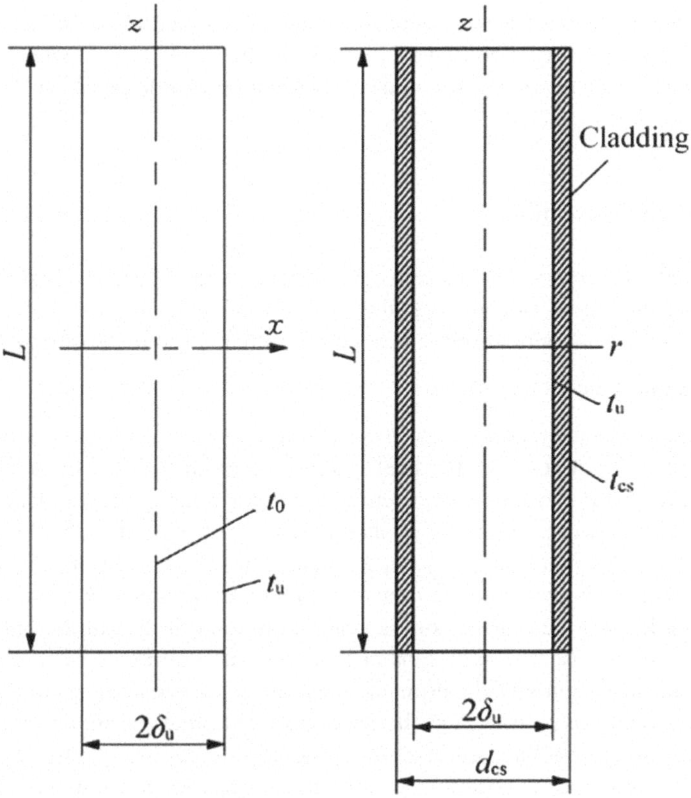

Fig. 2.13 Heat conduction model for plate fuel element

where, t_0 means the centerline temperature of pellet (°C); q_υ means the average volumetric power density of fuel pellet (W/m^3); δ_u means the half thickness of pellet (m); and k_u means the thermal conductivity of pellet [W/(m °C)].

For dispersion fuel commonly used in plate fuel elements, the thermal conductivity ku can be calculated by using the following equation recommended by Jokob:

$$k_u = k_s \frac{1 - (1 - ak_{dp}/k_s)b}{1 + (a - 1)b}, \tag{2.34}$$

where, ks means the thermal conductivity of dispersion fuel matrix [W/(m °C)]; kdp means the thermal conductivity of dispersion particles [W/(m °C)]; a = $3k_s/(2k_s + k_{dp})$; b = Vdp/(Vs + Vdp); Vs means the total volume of matrix (m^3); and Vdp means the total volume of dispersion particles (m^3).

The heat conduction equation of the cladding is as follows:

$$t_{ci} - t_{cs} = \frac{q}{k_c}\delta_c \tag{2.35}$$

where, t_{ci} means the temperature of inner wall of cladding (°C); t_{cs} means the temperature of outer wall of cladding (°C); q means the heat flux of outer surface of cladding (W/m^2); k_c means the thermal conductivity of cladding [W/(m °C)]; and δ_c means the thickness of cladding (m).

2.3.2.2 Heat Convection

For marine PWR reactors, the heat convection occurs during the heat transfer between the solid and the liquid. According to the state of the liquid, the heat convection can be further divided into single-phase heat convection and boiling heat transfer.

(1) Single-phase heat convection

Under normal operating conditions, most of the marine reactor core coolant is in a single-phase liquid state, so the heat transfer between the fuel element and the coolant is the single-phase heat convection. For some reactor core coolant channels with rather high power fraction, the coolant may be in a local subcooled boiling state or saturated boiling state. In this case, the heat transfer between the fuel elements and the coolant is the boiling heat transfer. Under certain abnormal conditions and accident conditions, the reactor core is very likely to be in overall two-phase flow conditions. In this case, the heat transfer is mainly the boiling heat transfer. Under normal operating conditions, the primary side of the steam generator is usually under single-phase heat convection, and the secondary side may be under single-phase heat convection and boiling heat transfer at the same time. The pressurizer may be under single-phase heat convection or boiling heat transfer, depending on the actual heating process. The heat transfer from the main pump to the coolant is usually the single-phase heat convection.

According to the flow conditions of the coolant, heat convection can be further divided into forced heat convection and natural heat convection. Marine reactors may be in forced circulation condition or natural circulation condition, so the above two heat convection conditions occur frequently. According to the types of steam generators (saturated natural-circulation steam generator or once-through steam generator), the secondary side feedwater may be under the above two heat convection conditions.

According to the types of coolants, heat convection can be further divided into heat convection of water and heat convection of steam. For example, the superheated steam section of once-through steam generator is under the heat convection of steam.

Generally, Newton's law of cooling is used for the calculation of heat convection, i.e., $Q = hA\Delta T$ where, Q means the heat transfer power (W); h means the heat convection coefficient [W/(m^2 °C)]; A means the heat transfer area (m^2); and ΔT means the heat transfer temperature difference (°C). Among the heat transfer calculations for the reactor core, the heat convection calculation is the most complicated, because it is difficult to determine the heat convection coefficient, and different relational expressions are used for different flow conditions.

For single-phase forced convection, such as the calculation of the forced convection of fluid in a circular channel, the Dittus-Boelter relational expression can be used:

$$Nu = 0.023\mathrm{Re}^{0.8}\mathrm{Pr}^{n} \tag{2.36}$$

where, Nu means the Nusselt number $Nu = \frac{hl}{\lambda}$, where, h means the heat convection coefficient [W/(m^2 °C)], l means the characteristic length of flow channel, and λ means the heat conductivity coefficient of fluid [W/(cm °C)]; during the determination of Re and Pr, the average fluid temperature is used as the qualitative temperature, and the equivalent diameter of the flow channel is used as the characteristic length; when the fluid is heated, $n = 0.4$; when the fluid is cooled, $n = 0.3$; the scope of application of the above equation is $10^4 < Re < 1.2 \times 10^5$, $0.6 \le Pr \le 120$ and L/d (pipe length/inner diameter) $\ge$50; there is a membrane temperature difference below average (for water, the difference is between 20–30 °C) between the fluid and the wall surface.

During the calculations for other types of flow channels (such as rectangular flow channel, rod bundle flow channel and pebble bed flow channel), suitable forced heat convection relational expressions can be selected for different types of flow channels.

For single-phase natural convection, the heat transfer calculation is extremely complicated, with rather large influence of the geometries of channels. For certain types of flow channels, the empirical relational expressions for specific conditions can be obtained only through the experiments. For example, for a vertical-wall flow channel, Hoffmann recommended that the following equations be used for calculations when the heat flux q of the wall is a constant:

$$\text{When } 10^5 < Gr_x^* < 10^{11} \text{ (laminar flow)},\ \mathrm{Nu}_{\mathrm{x,m}} = 0.6\left(Gr_x^* \cdot \mathrm{Pr}\right)_m^{1/5}; \tag{2.37}$$

$$\text{When } 2 \times 10^{13} < Gr_x^* < 10^{16} \text{ (turbulent flow)},\ \mathrm{Nu}_{\mathrm{x,m}} = 0.17\left(Gr_x^* \cdot \mathrm{Pr}\right)_m^{1/4} \tag{2.38}$$

where, Gr_x^* means the modified Grashof number, the expression of which is as follows:

$$Gr_x^* = g \cdot \beta \cdot q \cdot x^4/(k\nu^2) \tag{2.39}$$

where, g means the gravitational acceleration (m/s^2); β means the coefficient of cubical expansion of water (1/°C); q means the surface heat flux (W/m^2); χ is the characteristic length, i.e. the vertical distance calculated from the heat transfer starting point (m); k means the thermal conductivity [W/(m °C)]; and υ means the kinematic viscosity (m^2/s).

(2) Boiling heat transfer

The flow of matters of multiple phases in the same system is called multiphase flow. For marine reactors, there may be flow conditions with single-phase water and water/vapor co-existence sometimes. This is a typical multiphase flow condition: two-phase flow. Marine reactors using PWRs may have two-phase flow phenomenon in which subcooled boiling occurs on the surface of fuel elements and saturated boiling occurs in the channels. The existence of the two-phase flow significantly changes the heat transfer performance and the flow characteristics of the coolant and has important influence on the safe operation of the reactor [21].

In the heated channels, the vapor phase matters and the liquid phase matters in the steam-water mixture are flowing at the same time, forming a variety of flow conditions that are called "flow patterns". There are generally four (4) main flow patterns: bubble flow, slug flow, annular flow and droplet flow. The heat transfer under two-phase flow condition is called "boiling heat transfer". According to different conditions, the boiling heat transfer can be further divided into subcooled boiling, saturated nucleate boiling, transition boiling and film boiling. The development of two-phase flow pattern and the development of boiling state are illustrated in Fig. 2.14.

When the fluid flows in the heated channel, with the increasing of fluid temperature, the bubbles are locally generated on the surface of the heating element. At this time, the main flow region of fluid is still in a subcooled state, so this boiling flow condition is called "subcooled boiling". $x = 0$ is defined as the turning point between single-phase liquid state to subcooled boiling state, where, x means the true mass quality, i.e. the ratio of the cross section area of gaseous phase to the entire cross section area of the entire flow channel. As the temperature of the fluid continues to increase, more and more bubbles are generated on the surface of the fuel element, gradually depart from the surface of the element and enter the main flow region. In the subcooled state, some bubbles entering the main flow region condense again and disappear. As the temperature of the fluid continues to increase, when the fluid in the main flow region reaches saturation, the subcooled boiling state ends and the saturated boiling state begins. $x_e = 0$ is defined as the turning point between subcooled boiling and saturated boiling, where, x_e means the equilibrium quality, $x_e = \frac{(h - h_{fs})}{h_{fg}}$, where, h means the specific enthalpy of gas-liquid two-phase mixture, h_{fs} means the specific enthalpy of saturated liquid, and h_{fg} means the latent heat of vaporization. The units of both the specific enthalpy and the latent heat of vaporization are J/kg. With the further evolution of the fluid state, the transition boiling and film boiling may also occur. For different boiling types, different heat transfer relational expressions are used for calculation.

When the fluid is in the transition from a single-phase flow condition to a state in which bubbles are generated, the flow condition will change. Accordingly, the relational expressions for the calculation of flow and heat transfer shall be adjusted. The determination of bubble generation point (bubbling starting point) is especially important. There are many ways to determine the bubbling starting point. The commonly used methods are those suggested by Zuber and Saha. The relational expressions derived by them are as follows:

$$\text{When } Pe = \frac{G D_e c_{p,f}}{k_f} \leq 70000,\ h_{fs} - h_d = 0.0022 \frac{q D_e c_{p,f}}{k_f}; \tag{2.40}$$

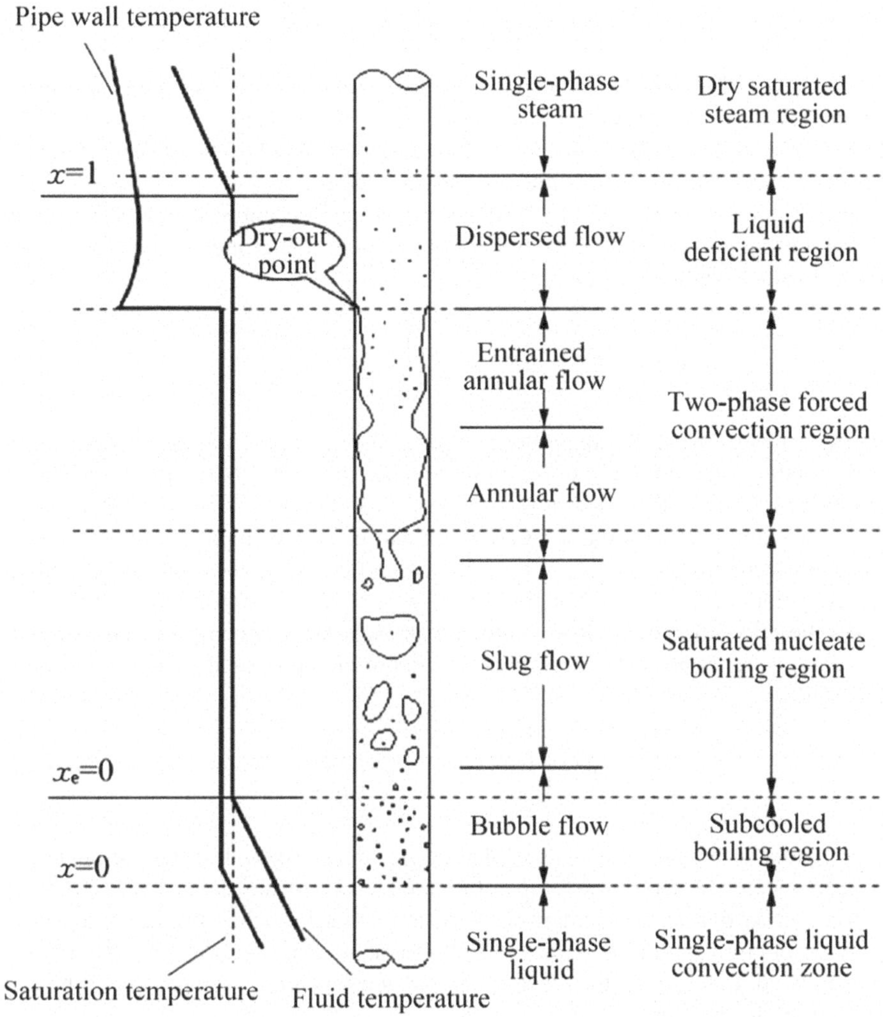

Fig. 2.14 Development of two-phase flow pattern and development of boiling state

$$\text{When } Pe \geq 70000,\ h_{fs} - h_d = 154q/G \tag{2.41}$$

where, *Pe* means the Peclet number, which is dimensionless; *G* means the mass flowrate [kg/(m^2 s)]; *D*e means the equivalent diameter (m); $c_{p,\mathrm{f}}$ means the specific heat capacity at constant pressure of liquid [J/(kg °C)]; k_f means the thermal conductivity of liquid [W/(m °C)]; *q* means the heat flux at the bubbling starting point (W/m^2); h_fs means the specific enthalpy of liquid (J/kg); and h_d means the specific enthalpy of the fluid (liquid) at the bubbling starting point (J/kg).

The heat balance equation, based on the fact that the heat obtained by the fluid along the hot channel is equal to the enthalpy increase of the fluid, is given by

$$qP_h(z_d - z_{in}) = GA(h_d - h_{in}) \tag{2.42}$$

When h_d is known, the axial position z_d of bubbling starting point D can be obtained. Where, P_h means the heating perimeter of the channel (m); z_{in} means the channel inlet distance (m); A means the flow area of channel (m^2); and h_{in} means the specific enthalpy of the fluid at the channel inlet (J/kg).

In addition, for water at 0.1–13.6 MPa, Bergles and Rohsenow obtained the criteria for the starting point of subcooled boiling based on the experimental data for several types of smooth industrial tubes:

$$q_{ONB} = 1.798 \times 10^{-3} p^{1.156} \left[\frac{9}{5}(t_w - t_{sat}) \right]^{2.828/p^{0.0234}} \tag{2.43}$$

where, t_w means the wall temperature (°C); t_{sat} means the saturation temperature at corresponding pressure (°C); q_{ONB} (w/m^2) means the heat flux required for boiling initiation with the wall superheat ($t_w - t_{sat}$) (°C) and system pressure p (Pa).

By the simultaneous solution of above equation and the single-phase forced heat convection equation $q = h[t_w - t_f(z)]$, we can obtain the location of the boiling starting point at certain fluid temperature $t_f(z)$, and the corresponding heat flux q.

For the calculation of nucleate boiling, many scholars have proposed corresponding heat transfer relational expressions. The expression proposed by Jens and Lottes is as follows:

$$q = \left(\frac{t_w - t_{sat}}{25} \right)^4 \exp\left(\frac{4p}{6.2} \right) \tag{2.44}$$

where, q means the surface heat flux (MW/m^2); t_w means the wall surface temperature (°C); t_{sat} means the saturation temperature (°C); and p means the pressure (MPa).

After a loss of coolant accident (LOCA) in the reactor, when the safety injection system is put into operation to inject water into the reactor core, the reactor core may be re-flooded. In this case, the reactor core surface depends on transition boiling and film boiling for heat transfer and cooling.

For the transition boiling, the relational expression proposed by Tong is as follows:

$$h_{TB} = 5054 + 95736 \exp(-0.018 \Delta t_{sat}) \tag{2.45}$$

where, h_{TB} means the heat transfer coefficient for transition boiling (W/m^2); Δt_{sat} means the local superheat (°C); the experimental range is $p = 1.379 \times 10^7$ Pa, $G =$ 1,193.5–3,390.8 kg/(m^2 s), and $t_w = 347.8$–593.8 °C.

For the film boiling, the calculation formula proposed by Groeneveld is as follows:

$$Nu = 3.27 \times 10^{-3}\mathrm{Pr}^{1.32}\left\{\mathrm{Re}\left[x + \frac{\rho_g}{\rho_f}(1-x)\right]\right\}^{0.901} Y^{-1.5} \tag{2.46}$$

where, $Y = 1 - 0.1(1-x)^{0.4}(\frac{\rho_f}{\rho_g} - 1)$; *Nu*, *Re* and *Pr* are dimensionless numbers calculated according to the characteristics of gaseous phase; and x means the mass quality, the calculation formula of which is as follows:

$$x = \frac{\rho_g V_g A_g}{\rho_f V_f A_f + \rho_g V_g A_g} \tag{2.47}$$

where, ρ_g and ρ_f are the densities of the steam and the liquid in the cross section of the channel (kg/m^3), respectively; V_g and V_f are the average velocities of the steam and the liquid (m/s) in the cross section, respectively; A_g and A_f are the average cross section areas of the steam and the liquid (m^2), respectively.

The film boiling calculation equation proposed by Groeneveld can be used for the vertically or horizontally placed circular or annular pipes with water as the medium and with a mass flowrate of $G \geq 800\ kg/(m^2\ s)$.

2.3.2.3 Radiation Heat Transfer

Radiation is a phenomenon of which the electromagnetic waves transfer the energy. The electromagnetic wave radiation generated by heat is called "thermal radiation". The electromagnetic waves of thermal radiation are excited during the change of thermal motion state of microscopic particles in the object. As long as the temperature of the object is higher than the "absolute zero", the object constantly changes thermal energy to radiant energy, causing the thermal radiation [22].

For marine PWRs, in most conditions, the radiation heat transfer plays a very minor role. Only when the temperature difference between the heat source and the coolant is large enough will the radiation heat transfer produce rather large amount of heat transfer. For example, when the reactor is re-flooded in a LOCA due to a large break, the reactor core is in a state of dryout because of the lack of coolant, and the temperature of fuel elements can be more than one thousand degrees. In this case, the radiation heat transfer will account for a large portion in the heat transfer between the fuel elements and the steam.

For the calculation of radiation heat transfer, the Boltzmann law is usually used:

$$E = \varepsilon C_0 \left(\frac{T}{100}\right)^4 \tag{2.48}$$

where, E means the radiation power (W/m^2); C_0 means the black-body radiation coefficient, the value of which is 5.67 W/(m^2 K^4); ε means the spectral emissivity, which depends on the properties of the specific heat transfer material and the shapes of the heat transfer structure; and T means the material temperature (K).

2.3.3 Reactor Hydraulics

For marine PWRs, a series of hydraulic phenomena will occur because of the flow of coolant, including coolant pressure drop, coolant flow distribution, flow stability, reactor core bypass flow and hydraulic loads. Among these phenomena, some are closely related to the safety of reactors, some are important references for the structural design of reactors, and some are the inputs for the design of equipment such as main pumps.

2.3.3.1 Pressure Drop

During the flow of coolant in the reactor, frictional pressure drop occurs because of the viscous effect of coolant and the friction with the wall of the flow channel; partial form resistance pressure drop occurs because of the flow through irregular flow channel; acceleration pressure drop occurs because of the coolant density change which is caused by the heating and cooling of the reactor core and steam generators; and gravitational pressure drop occurs because of the change in geopotential difference. In addition, two-phase pressure drop occurs when there is phase change in coolant [18, 20].

The calculation of reactor pressure drop mainly provides an important reference for the design of main pump lift.

(1) Elevation pressure drop

The elevation pressure drop of the coolant mainly occurs due to the difference in vertical height between two cross sections concerned in the system. The relational expression for calculation is as follows:

$$\Delta p_{el} = \int_{z_1}^{z_2} \rho g \sin\theta dL \tag{2.49}$$

where, Δp_{el} means the elevation pressure drop; ρ means the fluid density; g means the acceleration of gravity; θ means the included angle between the channel axis and the horizontal plane; z_1 means the axial coordination of cross Sect. 2.1; z_2 are the axial coordination of cross Sect. 2.2; and L means the flow channel length.

The flow of coolant in the marine reactor under natural circulation conditions is driven by elevation pressure drop of the system.

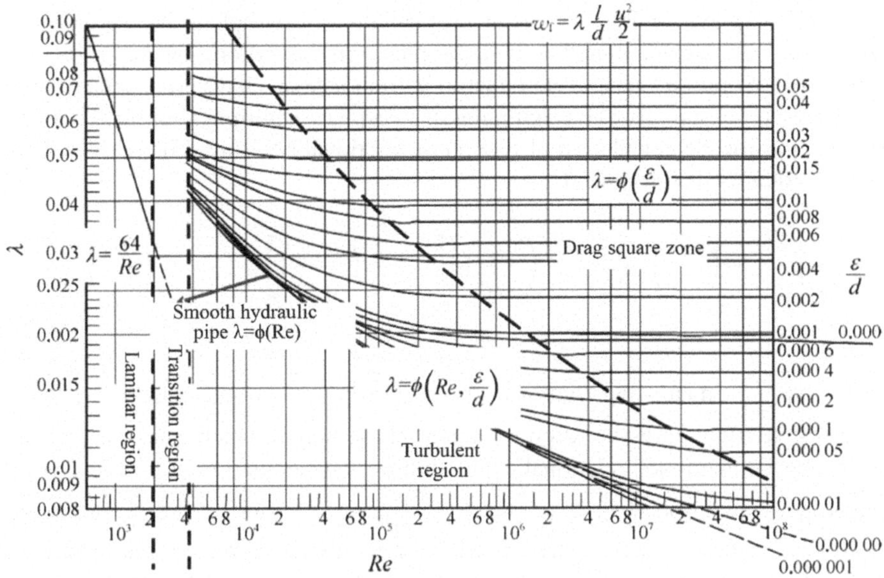

Fig. 2.15 Moody curve for friction coefficient

(2) Frictional pressure drop

Since the coolant is viscous, it will form a velocity gradient near the boundary layer of the flow channel. As a result, frictional resistance exists during the flow of coolant, resulting in the frictional pressure drop. The Darcy equation is commonly used to calculate the frictional pressure drop:

$$\Delta p_f = f \frac{L}{D_e} \frac{\rho V^2}{2} \tag{2.50}$$

where, Δp_{f} means the frictional pressure drop; f means the frictional resistance coefficient; L means the channel length; D_{e} means the equivalent diameter of channel; ρ means the fluid density; and V means fluid flowrate.

f is related to many factors such as fluid flow pattern, flow condition, heating condition, channel geometry and surface roughness. It can be calculated by different suitable empirical relational expressions. In engineering application, it can be determined by the Moody diagram (see Fig. 2.15).

When the coolant has a phase change, a new pressure drop occurs because of the interaction between the liquid coolant and the gaseous coolant. This pressure drop is called "two-phase pressure drop". During the normal operation of a marine reactor, except some possible local reactor core under subcooled boiling and saturated boiling in the reactor core, the entire primary circuit is almost in the single-phase flow condition. In this case, the two-phase pressure drop plays a very minor role. However, when the reactor is in abnormal or accidental conditions, the reactor core is likely to be under overall saturated boiling condition. In this case, the two-phase pressure drop plays an obvious role. The calculation of two-phase pressure drop is much more complicated than that of single-phase pressure drop. The key is the calculation of two-phase multiplication factor. Various scholars have proposed different relational expressions for the calculation of two-phase multiplication factor. Here, the equation for the calculation of two-phase frictional pressure drop derived from Fanning equation is introduced:

$$-\left(\frac{dp}{dz}\right)_f = -\left(\frac{dp}{dz}\right)_{f0}\left[1+x_e\left(\frac{v_{fg}}{v_{fs}}\right)\right]\left[1+x_e\left(\frac{\mu_{fg}}{\mu_{gs}}\right)\right]^{-0.25} = -\left(\frac{dp}{dz}\right)_{f0}\varphi_{f0}^2, \tag{2.51}$$

where, $-\left(\frac{dp}{dz}\right)_f$ means the frictional pressure drop for two-phase flow; $-\left(\frac{dp}{dz}\right)_{f0}$ means the frictional pressure drop for single-phase liquid, which can be calculated by the above–mentioned relational expression for the calculation of frictional pressure drop for single-phase flow; $\left[1+x_e\left(\frac{v_{fg}}{v_{fs}}\right)\right]\left[1+x_e\left(\frac{\mu_{fg}}{\mu_{gs}}\right)\right]^{-0.25} = \varphi_{f0}^2$ means the multiplication factor for two-phase frictional pressure drop. Where, μ_{fg} means the difference between the viscosity coefficients for liquid phase and gaseous phase, i.e. $\mu_{fg} = \mu_{fs} - \mu_{gs}$; υ_{fg} means the specific volume difference between saturated steam and saturated liquid, i.e. $v_{fg} = v_{gs} - v_{fs}$; and x_{e} means the equilibrium mass quality, $x_e = \frac{(h-h_{fs})}{h_{fg}}$ (where, h means the specific enthalpy of gas-liquid two-phase mixture, h_{fs} means the specific enthalpy of saturated liquid, and h_{fg} means the latent heat of vaporization).

(3) Acceleration pressure drop

The pressure drop due to the density change caused by coolant heating and cooling is called "acceleration pressure drop", the relational expression of which is as follows:

$$\Delta p_a = \int_{V_1}^{V_2} \rho V dV \tag{2.52}$$

where, V_1 means the fluid velocity at cross Sect. 2.1; V_2 means the fluid velocity at cross Sect. 2.2; and ρ means the fluid density.

For marine PWRs, the acceleration pressure drop plays a rather minor role because the coolant density usually does not change very much.

(4) Partial pressure drop

The coolant pressure drop caused by the change of flow channel shape during the flow of coolant is called "partial pressure drop", including the changes occurred in suddenly expanded flow, suddenly contracted flow, orifice flow and bend flow. Marine reactors are compact and complex in structure. The coolant flow in primary circuit includes various types of flow under the partial pressure drop. For example, the flow from the lower plenum to the reactor core zone includes the orifice flow in the lower plate of the reactor core and the suddenly contracted flow into the reactor core; the flow from the reactor core to the ascent section is the typical suddenly expanded flow; the flow in the inverted U-shaped tube of the steam generator is the typical bend flow; the flow in the bottom head of the reactor is the 180° reversed flow. Partial pressure drop occurs in the valves when the fluid flows through the gate valves and check valves on the main pipe. The equation for the calculation of partial pressure drop is as follows:

$$\Delta p_e = \varepsilon \frac{\rho V^2}{2} \tag{2.53}$$

where, ρ means the fluid density; V means the fluid flowrate; ε means the local form resistance coefficient which is related to the types of local form resistance. There are various empirical relational expressions for the calculation of ε, which can be obtained from the fluid resistance manual.

For the partial pressure drop of the flow channel structures with rather complex shapes, it may be impossible to use an empirical relational expression for the calculation, or no engineering manual can be used as references. In this case, the experiments can be performed for measurement, or the numerical methods (such as methods of computational fluid dynamics) can be used for the determination.

2.3.3.2 Flow Distribution

In the flow field of reactor coolant, the top concern is the flow distribution at the reactor core inlet, because the distribution results have direct effects on the thermal safety of the reactor core. Generally, theoretical calculation and experimental investigation are combined to obtain the characteristics of the flow distribution at the reactor core inlet [18, 20].

During the actual thermo-hydraulic design of the reactor core, the minimum flow distribution factor is defined according to the design requirements, and the structure design of fuel assemblies and experimental verification are performed to ensure that the flow distribution factor at reactor core inlet meets the design requirements.

2.3.3.3 Flow Stability Analysis

The coolant flow instability may occur during the operation of marine reactors. This flow instability may result in the severe deterioration of the core heat exchange conditions that threaten the reactor safety, or lead to the hydraulic impact that threaten the stability of the reactor structure. During the design of PWRs, attention shall be paid to two flow instability phenomena of the coolant: the flow instability in the system and the flow instability in parallel channels of the reactor core.

(1) Flow instability in the system

The flow instability in the system of a PWR, also called the Ledinegg flow instability, is a typical flow drift type instability in which the flow of the system suddenly changes from one steady state to another steady state. This flow instability may occurs only when the slope $\left(\frac{\partial \Delta P}{\partial G}\right)_{internal}$ of the pressure drop-flow curve (commonly called the internal flow characteristics) of the reactor coolant system is less than or equal to the slope $\left(\frac{\partial \Delta P}{\partial G}\right)_{external}$ of the driving head flow curve (commonly called the external flow characteristics, such as the lift of the coolant pump and the elevation pressure drop of the system). Therefore, the criterion for determining that there will be no Ledinegg instability is as follows:

$$\left(\frac{\partial \Delta P}{\partial G}\right)_{internal} > \left(\frac{\partial \Delta P}{\partial G}\right)_{external} \tag{2.54}$$

(2) Flow instability in parallel channels of the reactor core

The mechanism of this instability is as follows: in parallel channels of the reactor core, because of the disturbance of external factors, the disturbance of flow or heating power in some channels may occur, resulting in the change in coolant state (e.g. flowrate, temperature, density and flow form pattern) in these channels; in turn, this results in the pressure drop or the heating power change in these channels, and thus changes the flow of the coolant in these channels. Finally, the repeated changes in the flow of coolant may occur in these channels and expand to a number of channels in the reactor core. Generally, this flow instability does not affect the flow stability of the entire system. In the case of the flow instability in parallel channels of the reactor core, the total flow of the system and the pressure drop of upper plenum and lower plenum will not be significantly changed. Therefore, this phenomenon is hidden and not easy to be monitored. This flow instability can easily result in the burning of the reactor core channels, so it is a key point for the thermo-hydraulic design of the reactor core. The main factors influencing the flow instability in parallel channels are as follows: ① pressure: the higher the system pressure is, the smaller the density difference between steam and water will be, the less likely the partial pressure rise will take place, and the less likely the flow pulsation will occur; ② mass quality at the outlet: the less the mass quality at the outlet is, the smaller the volume change of the steam-water mixture will be, and the steadier the flow will be; ③ heat flux: the smaller the heat flux is, the smaller the volume change of the steam-water mixture caused by the fluctuation of the heat flux will be, and the smaller the possibility of pulsation will be; and ④ flowrate: the greater the flowrate at the inlet is, the less likely the increasing of the steam volume blocking the fluid flow will be, thus mitigating or preventing the pulsation among tubes. For the elimination of the flow instability in parallel channels of the reactor core, the most effective method, in addition to the regulation of the parameters related to above factors, is to install a throttling element at the inlet of the heating section to increase the inlet resistance.

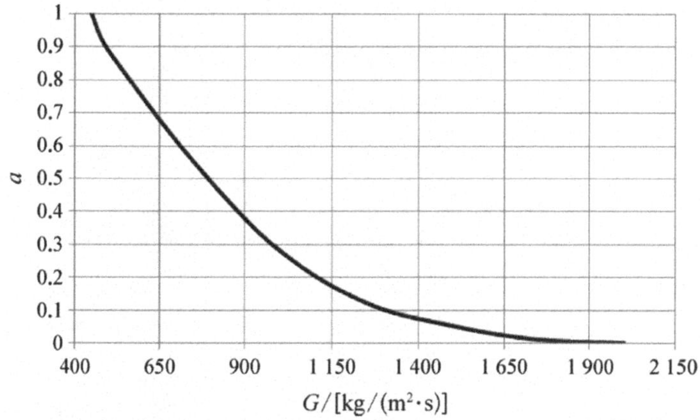

Fig. 2.16 Relationship between *a* and mass flowrate

Experiments indicate that the following criteria shall be met to prevent the flow instability in parallel channels of the reactor core:

$$\frac{\Delta p_{no} - \Delta p_j}{\Delta p_B} \geq a \tag{2.55}$$

where, Δp_{no} means the pressure drop of heating section (Pa); Δp_{j} means the pressure drop caused by the throttling element (Pa); Δp_{B} means the pressure drop of boiling section (Pa), including the pressure drop of the superheating section when there is superheated steam at the outlet; a is a constant which depends on the operating pressure P of the system and the mass flowrate G of fluid in the channel. Figure 2.16 illustrates the relationship between a and the mass flowrate.

2.3.3.4 Reactor Core Bypass Flow

Reactor core bypass flow refers to the portion of the reactor system coolant flow that does not contribute to the cooling of the fuel elements. Some of the reactor core bypass flow is specifically considered during the design, such as the flow for cooling of control rod assemblies, cooling of peripheral components of the reactor core, cooling of the reactor top head and cooling of reactor core baffles. Other part of reactor core bypass flows is undesirable during the design. However, it occurs due to the inherent characteristics of the structure, such as the bypass flow of nozzles at the reactor inlet and outlet. This part of bypass flow is also called "leakage flow" or "invalid bypass flow".

Figure 2.17 illustrates the typical schematic diagram of the reactor core bypass flow in various parts of a PWR. The bypass flow for guide tube (Q5) refers to the flow of coolant flowing through the core guide tubes and partly for cooling the control

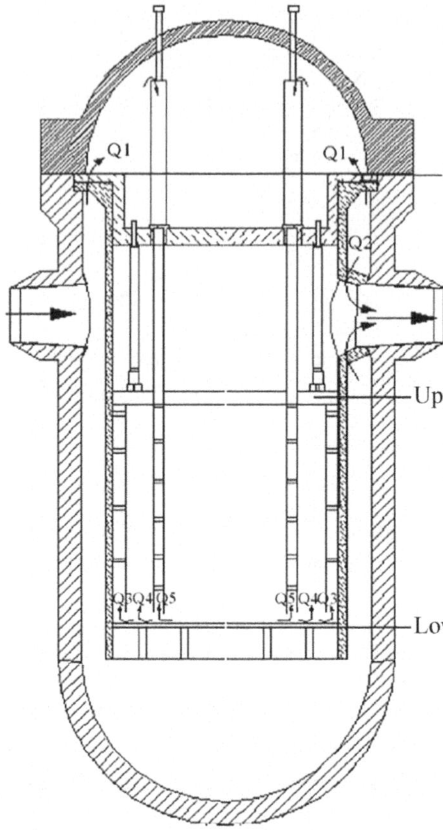

Fig. 2.17 Reactor bypass flow

rods, which is usually determined based on the heat share of the control rods. The bypass flow for peripheral components of reactor core (Q4) refers to the flow of coolant in the flow channels between the outermost peripheral components and the baffles and mainly for cooling the baffles. The bypass flow for reactor core baffles (Q3) refers to the flow of coolant flowing through the flow channels between reactor core baffles and the barrels and for cooling the barrels and baffles. The bypass flow for upper plenum (Q1) refers to the flow of coolant entering the pressure vessel annulus from the reactor inlet nozzle, directly flowing upwards into the reactor top head, getting mixed with reactor core outlet coolant and flowing out of the reactor pressure vessel from the outlet nozzle. This part of coolant is mainly for cooling the reactor top head and controlling the temperature of the upper plenum coolant. The bypass flow for reactor nozzles (Q2) is the flow of coolant entering the reactor from the reactor inlet nozzle, directly flowing through the gap between the reactor outlet nozzle and the barrel and clamp sleeve and flowing out of the pressure vessel from the reactor outlet nozzle. This part of bypass flow is undesirable during the reactor design but cannot be eliminated because of the inherent characteristics of the structure. It can only be minimized by various means.

For different design objects, corresponding bypass flow share limits shall be determined according to engineering practices and general requirements. In the hydraulic design, the structural design shall be optimized through theoretical calculation according to the limits for each part of the bypass flows, so as to preliminarily meet the requirements. The hydraulic tests shall be conducted for verification and validation. The commonly used theoretical analysis methods include special bypass leakage calculation programs and CFD calculation programs, and the commonly used tests include partial hydraulic test and overall hydraulic simulation test.

2.3.3.5 Hydraulic Loads

The hydraulic loads of the reactor mainly include the lifting force of reactor core fuel assemblies and the hydraulic loads of structural components of the reactor under accident conditions [18, 19].

(1) Lifting force of reactor core fuel assemblies
The lifting force of reactor core fuel assemblies is resulted from the flushing of fuel assemblies by the upward flow of the coolant in the reactor core zone. It is mainly composed of the frictional force between the coolant flow and the fuel assemblies, and the local resistance in local fuel assemblies (including the inlet and outlet of fuel assembly and the spacer grids) generated by the coolant flow. It is typically characterized by the product of the pressure drop in fuel assembly area, the equivalent cross section area of fuel assembly and the corresponding empirical coefficient.
The resultant force of the lifting force, the buoyancy force and the gravity of fuel assembly shall be balanced with the force of the holddown spring of fuel assemblies. Therefore, the calculation value of the lifting force of reactor core fuel assemblies is an important input for the design of the holddown spring of fuel assemblies. The pressing force of the holddown spring shall ensure that the fuel assemblies are stable under the worst conditions.
(2) Hydraulic loads of structural components under accident conditions
Under certain accident conditions, the reactor internals may have abnormal hydraulic loads because of a large change in coolant state. These hydraulic loads may pose a great threat to the structural safety of the reactor. Typical accident conditions include LOCA because of a large break in the reactor. Under the accident conditions, the coolant state changes drastically in a short time, resulting in the dramatic change of the coolant pressure distribution and flow conditions, further generating the hydraulic loads of structural components unfavorable to the safety. In the accident analysis of the reactor, the hydraulic loads of structural components under accident conditions require thorough analysis and evaluation.

2.3.4 Reactor Thermo-Hydraulic Design

2.3.4.1 Tasks of Thermo-Hydraulic Design

For a specific reactor system and its core structure, the reactor thermo-hydraulic design mainly involves ① using relevant calculation models (usually the calculation programs developed for special purposes) based on heat transfer theory and hydraulics to conduct thermo-hydraulic calculations of the reactor system and the core so as to obtain thermo-hydraulic parameters of the temperature, pressure and flow fields; ② in addition to those models, using thermo-hydraulic design criteria to evaluate the thermal safety properties of the reactor system and its reactor core; and ③ providing necessary thermo-hydraulic parameters for other relevant disciplines. For example, at the schematic design stage, the reactor thermo-hydraulic design mainly involves the verification and validation of main reactor parameters, the verification and optimization of reactor structure parameters, and the verification and related designs of the main system and relevant safety systems specifically engineered into it. At the detailed design stage, it mainly deals with the calculation of thermal parameters under different operating conditions, analysis of natural-circulation capability, design of protection setting values, evaluation of safety analysis, maneuverability analysis, design transient analysis and provision of inputs to mechanical design.

2.3.4.2 Thermo-Hydraulic Design Criteria

As for PWRs, the thermo-hydraulic design criteria mainly involve departure from nucleate boiling ratio, highest fuel central temperature and core flow instability [18, 23].

(1) Departure from nucleate boiling ratio (DNBR)

When the flow and heat-transfer characteristics of the fuel element surface reach a certain condition, some part of the fuel element surface will be deficient of the cooling by the coolant due to the accumulation of bubbles, making the heat transfer switch from nucleate boiling to partly dryout and thus causing burnout. Therefore, the heat flux at the time when the departure from nucleate boiling starts is defined as the critical heat flux. The critical heat flux of fuel elements is related with the structure size of elements, cell dimensions and fluid parameters, and it is generally obtained through tests and can be characterized by a relational expression. For example, the W-3 equation for axial uniform heating is as follows:

$$\begin{aligned} q_{DNB,eu} = {} & 3.154 \times 10^6\{(2.022 - 6.238 \times 10^{-8}p) + (0.1722 - 1.43 \times 10^{-8}p) \\ & \times \exp[(18.177 - 5.987 \times 10^{-7}p)x_e]\}[(0.1484 - 1.596x_e + 0.1729x_e|x_e|) \\ & \times \frac{0.2048G}{10^6} + 1.037](1.157 - 0.869x_e)[0.2664 + 0.8357\exp(-124D_e)], \\ & \times [0.8258 + 0.341 \times 10^{-6}(h_{fs} - h_{f,in})] \end{aligned} \tag{2.56}$$

where, $q_{DNB,eu}$ is the critical heat flux of axial uniform heating (W/m^2); p is the coolant working pressure (Pa); G is coolant mass flowrate [kg/(m^2 h)]; De is the equivalent diameter of coolant channel (m); h_{fs} is the saturated enthalpy of coolant (J/kg); $h_{f,in}$ is the coolant enthalpy at the core inlet (J/kg); x_e is the equilibrium vapor content at the calculated point z and $|x_e|$ is the absolute value.

The calculation expression for the equilibrium vapor content x_e is:

$$x_e = \frac{h_f(z) - h_{fs}}{h_{fg}} \tag{2.57}$$

where, h_{fs} is the latent heat of vaporization (J/kg).

The range in which W-3 equation is applicable is: $p = (6.895–15.86) \times 10^6$ Pa, $x_e = -0.15 \sim +0.15$, $G = (4.9–24.5) \times 10^6$ kg/(m^2 h), $De = 0.005\,08–0.017\,8$ m, heating length $L = 0.254–3.668$ m, $h_{f,in} \geq 930$ kJ/kg, heated perimeter/wetting perimeter = 0.88–1.00, the geometry of the channel can be circular, rectangular or like rod cluster.

In the thermo-hydraulic design of reactor core the ratio of the critical heat flux calculated by a relational expression under certain conditions to the actual heat flux on the surface of fuel elements is deemed as an important parameter to determine whether the reactor core is safe or not, which is called the departure from nucleate boiling ratio (DNBR). The calculation of DNBR also needs to consider the influences brought about by maritime condition, engineering manufacture deviation, actual operation deviation and uncertainty of design models.

(2) Highest fuel central temperature

The temperature level of fuel pellets is also an important parameter to determine the thermal safety of the reactor core. At the thermo-hydraulic design, the maximum reactor core temperature obtained through the calculation of fuel temperature field is compared with the fuel temperature limit to determine the thermal safety of the reactor core. While in the practical calculation, the maximum fuel central temperature is usually calculated by the equation for fuel temperature field in Sect. 2.3.2 with the coolant temperature and the heat flux of fuel elements as the inputs.

(3) Core flow instability

As for traditional marine reactors, if the core adopts the open flow path structure (e.g., rod bundle flow path), the core flow instability is normally not considered under the forced coolant circulation condition; in the case of the closed parallel flow path structure, such instability shall be considered, whether under forced coolant circulation condition or natural circulation condition. For the evaluation and determination of the core flow instability, refer to the methods for analyzing the flow instability in Sect. 2.3.3. The core flow instability boundary can be obtained firstly through theoretical analysis and experimental study and then applied in the engineering design. Such instability is normally validated based on the outlet temperature of the core heat hot channel or vapor content.

2.3.4.3 Steady State Thermo-Hydraulic Analysis Model

In the thermo-hydraulic design of reactors, the steady state thermo-hydraulic analysis of the reactor core shall be conducted to gain information on temperature field and flow field of various core channels at steady state so as to perform the thermal safety evaluation of the reactor core, optimize the core design and provide thermo-hydraulic parameters necessary for other disciplines.

The steady state thermo-hydraulic analysis model of the reactor core falls into two categories: single-channel model and sub-channel model, and the basic calculation models of both models follow three conservation equations for fluids, i.e., mass conservation equation, momentum conservation equation and energy conservation equation.

The mass conservation equation is:

$$\frac{dW}{dZ} = \frac{d(\rho VA)}{dZ} = 0 \tag{2.58}$$

where, W is the fluid mass flow (kg/s); Z is the reference coordinate direction; ρ is the fluid density (kg/m^3); V is the fluid flowrate (m/s); A is the flow area of flow path (m^2).

The momentum conservation equation is:

$$-\frac{dP}{dZ} = \rho g\cos\theta + \frac{1}{A}\frac{d}{dZ}\left(\frac{W^2}{\rho A}\right) + \frac{W^2 f U_e}{2\rho A^3} = \frac{dP_g}{dZ} + \frac{dP_a}{dZ} + \frac{dP_f}{dZ} + \frac{dP_c}{dZ} \tag{2.59}$$

where, Z is the reference coordinate direction, ρ is the fluid density (kg/m^3); g is the gravity acceleration (m/s^2); θ is the included angle between the direction of fluid flow and the direction of reference coordinate Z; A is the flow area of flow path (m^2); W is the fluid mass flow (kg/s); f is the friction coefficient of fluid flow; U_e is the wetted perimeter (m); $\frac{dP_g}{dZ}$ is the elevation pressure drop gradient; $\frac{dP_a}{dZ}$ is the acceleration pressure drop gradient; $\frac{dP_f}{dZ}$ is the friction pressure drop gradient; $\frac{dP_c}{dZ}$ is the partial pressure drop gradient.

The energy conservation equation is:

$$W\frac{dH}{dZ} = qU_h \tag{2.60}$$

where, W is the fluid mass flow (kg/s); H is the fluid specific enthalpy (J/kg); Z is the reference coordinate direction; q is the heat flux (W/m^2); U_h is the heated perimeter (m).

(1) Single-channel model

The prominent feature of the single-channel model is that the heat exchange transfer and mass transfer that might exist between channels are not considered. The main principle of such model is that the whole entire reactor core is simulated with the average channel to simulate the mean parameters of the whole core, which include average power distribution, mean mass flow rate, equivalent diameter of the average channel of the core; the hot channel is used to simulate the worst condition of the core, i.e., the worst thermal conditions are taken into consideration, such as the worst flow distribution factor at reactor core inlet, the worst core power distribution (including the maximum channel factor for specific enthalpy rise and the maximum hot spot factor of the core), the adverse engineering factor (engineering manufacture deviation and engineering installation deviation are considered in a conservative or relatively conservative way). In addition, all channels are considered as the closed flow paths, without consideration of the coolant mixing effect between channels (including the mixing of energy, mass and momentum). The aforesaid computing method, called "one-step method", has evolved into "two-step method" to reduce the conservatism of single channel model. Firstly, the most unfavorable assembly inside the core is deemed as the "hot channel" in the model and the "one-step method" mentioned above is used to calculate its mean parameters, and then such parameters are taken as the mean channel parameters with the most unfavorable and minimum calculation unit as the "hot channel", which represents the worst condition of the core. In addition, in order to further reduce the conservatism of the single-channel model, the enthalpy field for the hot channel at each step is multiplied by a mixing factor, the empirical value deduced from tests or the calculation of enthalpy field for hot channel.

For the single-channel model, the biggest difficulty is to accurately calculate the average flow of coolants for "hot channel". Because subcooled boiling or saturated boiling might occur in the hot channel when heated, which will affect the flow resistance inside the "hot channel". On the other hand, it is generally assumed that the pressure of each channel (hot channel and average channel) at the core outlet is identical, but the channel pressure of "hot channel" at the inlet is relatively lower than that of average channel due to the influence of flow field distribution of the lower plenum, which makes the driving pressure head of "hot channel" lower than that of the average channel. All the above factors will have an impact on the calculation of the flow of "hot channel".

According to the single-channel theory, the equation for calculating the driving pressure head of the hot channel is:

$$\Delta P_{h,e} = K_{f,h}\Delta p_{f,m} + K_{a,h}(\Delta p_{in,m} + \Delta p_{a,m} + \Delta p_{gd,m} + \Delta p_{ex,m}) + \Delta p_{el,m} \tag{2.61}$$

where, $K_{\mathrm{f,h}} = (1 - \delta)^{2-b}$ is the modified factor for frictional pressure drop of hot channel due to the lower plenum (δ is the uneven flow distribution coefficient for hot channel due to the lower plenum, b is the empirical coefficient); $K_{\mathrm{a,h}} = (1 - \delta)^2$ is the modified lower plenum factor for form resistance pressure drop and acceleration pressure drop; $\Delta P_{\mathrm{h,e}}$ is the effective driving pressure head at both ends of hot channel; $\Delta p_{\mathrm{f,m}}$ is the friction pressure drop of average channel; $\Delta p_{\mathrm{in,m}}$ and $\Delta p_{\mathrm{ex,m}}$ are the form resistance pressure drop for average channel inlet and outlet respectively; $\Delta p_{\mathrm{a,m}}$ is the acceleration pressure drop of average channel; $\Delta p_{\mathrm{gd,m}}$ is the form resistance pressure drop of average channel caused by the spacer grid; $\Delta p_{\mathrm{ex,m}}$ is the elevation pressure drop of average channel.

The calculation equation for pressure drop of hot channel is:

$$\Delta p_h = \Delta p_{f,h} + \Delta p_{in,h} + \Delta p_{a,h} + \Delta p_{gd,h} + \Delta p_{ex,h} + \Delta p_{el,h} \tag{2.62}$$

where, Δp_{h} is the pressure drop of hot channel; what on the right side are friction pressure drop, form resistance pressure drop of inlet, acceleration pressure drop, grid pressure drop, form resistance pressure drop of outlet and elevation pressure drop for hot channel, respectively. For the computing method for pressure drop of different channels, refer to Sect. 2.3.3.

During the calculation of hot channel flow, firstly, the driving pressure head $\Delta P_{\mathrm{h,e}}$ is calculated with the parameters of average channel as the inputs, and then a flow rate is assumed for the hot channel and the pressure drop Δp_{h} is calculated from the above calculation equation for pressure drop of hot channel. Next, the convergence criteria, $\left|\frac{\Delta p_h - \Delta p_{h,e}}{\Delta p_{h,e}}\right| \leq \varepsilon$ is used to judge whether it meets requirements, where ε is the iterative convergence value. If the above convergence criterion is met, the flow rate of hot channel can be obtained; if not, it is necessary to perform the calculation again by assuming a new flow rate for the hot channel. Figure 2.18 shows the thermo-hydraulic calculation process based on the single-channel theory.

(2) Sub-channel model

With the development of thermo-hydraulic design theory, the traditional single-channel model is too conservative in core simulation and causes a large deviation from core physical phenomenon of "open" channel (core coolant can experience traverse mixing between channels), so sub-channel model was developed. The sub-channel model is different from the single-channel model mostly because of its consideration of traverse mixing (exchange of mass, monument and energy) between coolant channels during the simulation of the reactor core, and each core channel can be stimulated more realistically with reference to the actual core power distribution.

For large PWRs, given the same thermal parameters, it is found that the minimum DNBR value of the fuel element surface increases by about 5–10% after comparing the results calculated from the sub-channel model with those from the single-channel model. Therefore, calculation with the sub-channel model can improve not only the accuracy of thermal design, but also the reactor economy. However, unlike the single-channel model, where only a few channels are stimulated, calculation with sub-channel model needs to analyze a large number of channels, which leads to heavy computation. Figure 2.19 provides an example to divide sub-channels in a typical lattice. In fact, the division of sub-channels is artificially determined and one sub-channel can be composed of one or several fuel assemblies, or can be just the flow path formed by adjacent fuel elements inside a fuel assembly.

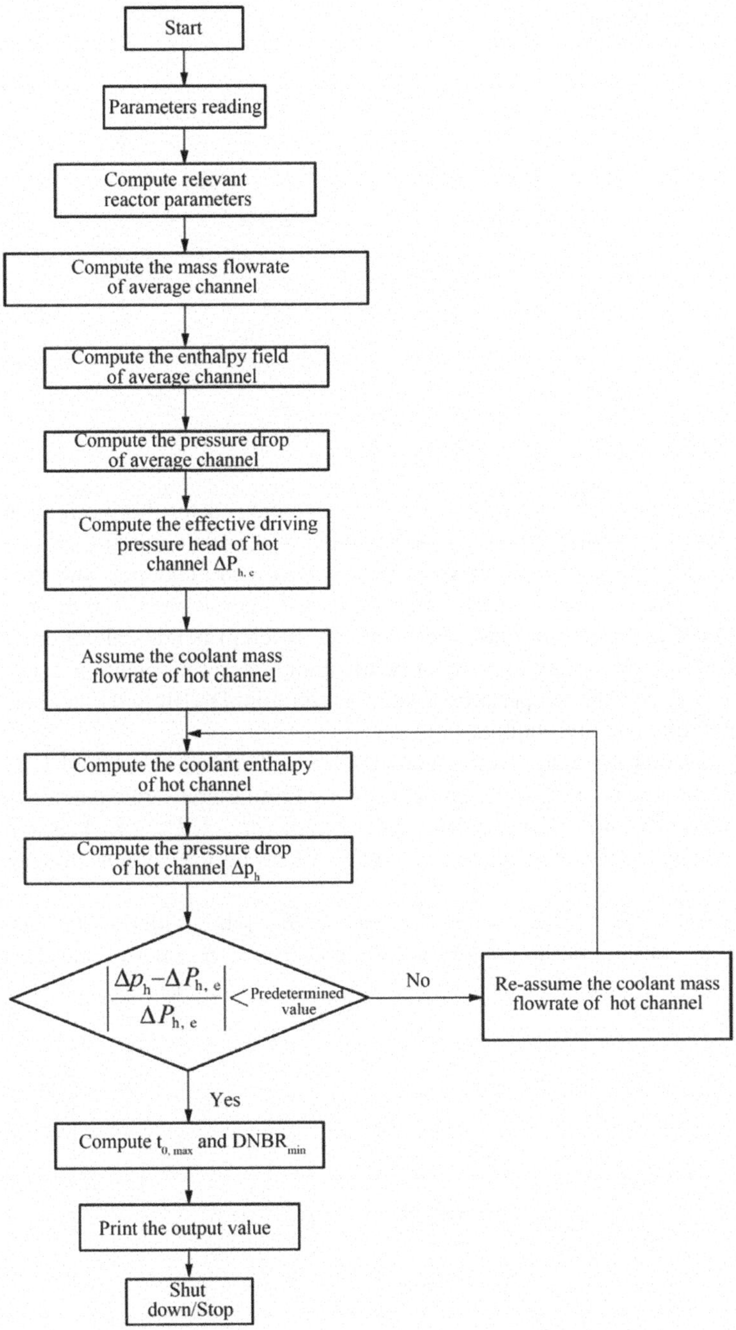

Fig. 2.18 Thermo-hydraulic calculation process based on single-channel theory

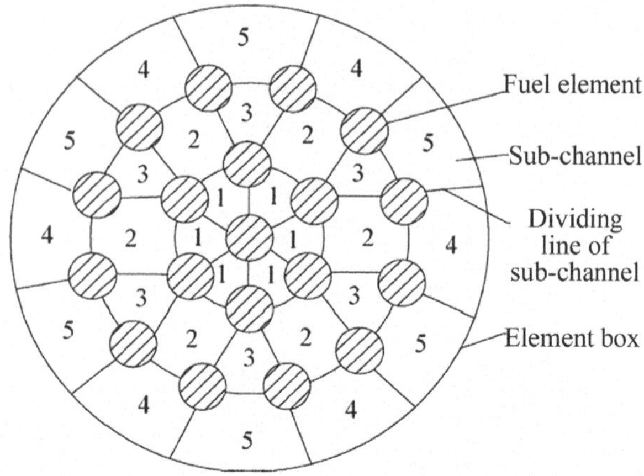

Fig. 2.19 Channel division of sub-channel model

There are many different sub-channel models and reactor thermo-hydraulic calculation codes for applying these models in the world, and their major difference lies in the processing model for transverse flow between sub-channels and the empirical relational expression used. But all they have in common is that the flow field, enthalpy field and pressure field of coolants at different heights along the axis can be obtained by solving the mass, energy, axial and lateral momentum conservation equations, and then the temperature field, the minimum DNBR and other parameters of each fuel element are obtained.

Assuming that there are n sub-channels to be calculated, numbered 1, 2, 3, …i, j,… to n respectively; each sub-channel is divided into L nodes of equal length along the axial height, consecutively numbered 0, 1, 2, …, $l-1$, l, …, L. The step length l for sub-channel i is taken as an example to illustrate the coolant flow shown in Fig. 2.20.

In this figure, $W_{i,l}$ is the coolant flow rate (kg/s) at the outlet of step length l of sub-channel i; $W_{i,l-1}$ is the coolant flow rate (kg/s) at the inlet of step length l of

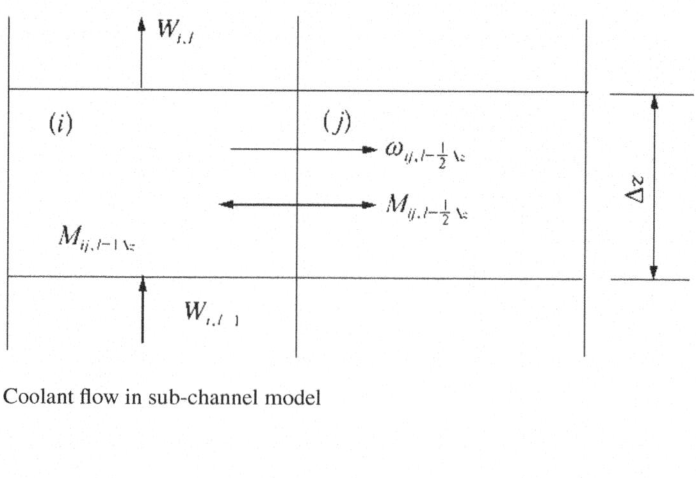

Fig. 2.20 Coolant flow in sub-channel model

sub-channel i; Δz is the length (m) for one step length along the channel longitudinal axis, $\omega_{ij,l-\frac{1}{2}}$ is the transverse flowrate [kg/(m s)] of coolants from sub-channel i to the adjacent sub-channel j within the step length l; $M_{ij,l-\frac{1}{2}}$ is the mixing flowrate [kg/(m s)] of transverse turbulent flow.

The mass conservation equation for the lth step length of sub-channel i is:

$$W_{i,l} - W_{i,l-1} = -\sum_{j=1}^{n_j} \Delta\omega_{ij,l-\frac{1}{2}} \times \Delta z \tag{2.63}$$

where, n_j is the number of sub-channels adjacent to sub-channel i; $\Delta\omega_{ij,l-\frac{1}{2}}$ is the net transverse velocity [kg/(m s)] from sub-channel i to sub-channel j.

The energy conservation equation for the lth step length of sub-channel i is:

$$W_{i,l}h_{i,l} - W_{i,l-1}h_{i,l-1} = \overline{q}_{i,l-\frac{1}{2}}\Delta z - \sum_{j=1}^{n_j} Q_{ij,l-\frac{1}{2}}\Delta z - \sum_{j=1}^{n_j} M_{ij,l-\frac{1}{2}}(h_{i,l-\frac{1}{2}} - h_{j,l-\frac{1}{2}})\Delta z, \tag{2.64}$$

where, $h_{i,j}$ is the outlet fluid specific enthalpy (J/kg) of lth step length of Sub-channel i; $h_{i,l-1}$ is the inlet fluid specific enthalpy (J/kg) of lth step length of Sub-channel i; $\overline{q}_{i,l-\frac{1}{2}}$ is the average linear power density [J/(m s)] of fuel elements for lth step length of Sub-channel i; $Q_{ij,l-\frac{1}{2}}$ is the heat per unit time (J/s) taken away by transverse flow between sub-channel i and adjacent channel j for lth step length; where,

$$Q_{ij,l-\frac{1}{2}} = \frac{1}{2}\left[(h_{i,l-\frac{1}{2}} - h_{j,l-\frac{1}{2}})\left|\omega_{ij,l-\frac{1}{2}}\right| + (h_{i,l-\frac{1}{2}} + h_{j,l-\frac{1}{2}})\omega_{ij,l-\frac{1}{2}}\right] \tag{2.65}$$

The axial momentum conservation equation for the lth step length of Sub-channel i is:

$$p_{i,l-1} - p_{i,l} = \Delta p_{i,l-\frac{1}{2},el} + \Delta p_{i,l-\frac{1}{2},a} + \Delta p_{i,l-\frac{1}{2},f} + \Delta p_{i,l-\frac{1}{2},c} + \Delta p_{ij,l-\frac{1}{2}} \tag{2.66}$$

where, $p_{i,l-1}$ and $p_{i,l}$ are the inlet and outlet pressure (Pa) for the lth step length of sub-channel i respectively; the first item to fourth item on the right side of the equation are the elevation pressure drop, acceleration pressure drop, frictional pressure drop and form resistance pressure drop for the lth step length of sub-channel i respectively; $\Delta p_{ij,l-\frac{1}{2}}$ is the flow resistance pressure drop (Pa) of axial momentum caused by the net transverse flow of coolants between sub-channels i and j for the lth step.

The transverse pressure gradient of adjacent channels is the driving force for transverse flow, so the transverse momentum equation of the lth step length of sub-channel i can be given as follows:

$$p_{i,l} - p_{j,l} = c_{ij}\omega_{ij,l-\frac{1}{2}}\left|\omega_{ij,l-\frac{1}{2}}\right| \tag{2.67}$$

where, $p_{i,l}$ and $p_{j,t}$ are separately the pressure (Pa) at the outlet of the lth step length of sub-channels i and j, $\omega_{ij,l-\frac{1}{2}}$ is the crossflow flowrate [kg/(m s)] of coolant from sub-channels i to j within the entire step length l; c_{ij} is the transverse flow resistance coefficient between the sub-channels i and j within the step length l, which is determined by tests and relative to the geometric condition of channels.

2.3.4.4 Transient Thermo-Hydraulic Analysis Model

Compared with the steady state thermo-hydraulic analysis model, the transient thermo-hydraulic analysis model has incorporated the changes of physical parameters over time into the mass, momentum and energy conservation equations and thus has the capability to conduct core thermo-hydraulic calculations and analysis under different transient conditions. This model can be applied in the analysis of various transient conditions [18, 23].

The equations for such model are presented in this section by taking homogeneous flow model as an example, where it is assumed that the velocities of vapor phase and liquid phase are equal, both phases stay in the thermodynamics balance state, and mean parameters of the two-phase flow are used in this the equations. Conservation equation system for the homogeneous flow model is identical to that for single-phase flow. The one-dimensional equations are normally used in the transient analysis of the reactor system, specifically as follows.

The mass conservation equation is:

$$\frac{\partial(\rho A)}{\partial \tau} + \frac{\partial(GA)}{\partial z} = 0 \tag{2.68}$$

where, A is the flow area of the channel (m^2); G is the mass flowrate [kg/(m^2 s)]; ρ is the fluid density (kg/m^3); τ is the time (s); z is the axial height (m).

The momentum conservation equation is:

$$\frac{\partial(\rho A)}{\partial \tau} + \frac{\partial(vG^2 A)}{\partial z} = -A\frac{\partial p}{\partial z} - \frac{fvAG|G|}{2D_e} - \rho g A \sin\varphi \tag{2.69}$$

where, p is the pressure (Pa); $v = \frac{1}{\rho}$; f is the frictional resistance coefficient; De is the equivalent diameter (m); g is the acceleration of gravity (m/s^2); φ is the angle between the direction of flowrate and horizontal direction.

The energy conservation equation is:

$$\frac{\partial(\rho e A)}{\partial \tau} + \frac{\partial}{\partial z}\left[GA\left(h + \frac{V^2}{2} + \Phi\right)\right] = (q + q_e)U_h \tag{2.70}$$

where, Uh is the heated perimeter (m); q is the linear heating power density of fuel (W/m). For two-phase flow, the definitions of above equations are as follows:

average density of gas-liquid mixture $\rho = \rho_f(1-\alpha) + \rho_g\alpha$, where, a is the void fraction; mass flowrate $G = \rho_f V_f(1-\alpha) + \rho_g V_g\alpha$; average specific enthalpy of gas-liquid mixture $h = h_f(1-x) + h_g x$, where, x is the mass quality; total specific energy of gas-liquid mixture $e = u + \frac{V^2}{2} + \varphi$, where, u is the average specific thermodynamic energy u = uf (1 − x) + ugx; φ is the gravity potential function $g = \frac{\partial\varphi}{\partial z}$; and surface heat flux equivalent to volume heat-release rate $q_e = \frac{1}{U_h}\int_A q_v dA$.

For the transient temperature field of fuel elements, the equation describing the relation of the temperature field versus time is:

$$\rho c_p \frac{\partial t}{\partial \tau} = \nabla \cdot k\nabla t + q_v \tag{2.71}$$

For rod fuel elements, it is normally assumed that heat conduction is axially symmetric. If it is assumed that the rod material is uniform, the coefficient of heat transfer is equal around the rod and the cylindrical coordinate is adopted, the differential equation for rod fuel elements is as follows:

$$\rho_u c_u \frac{\partial t}{\partial \tau} = \frac{1}{r}\frac{\partial}{\partial r}\left(k_u r \frac{\partial t}{\partial r}\right) + \frac{\partial}{\partial z}\left(k_u \frac{\partial t}{\partial z}\right) + q_v \tag{2.72}$$

The cladding temperature changes within a small range, and its thermal conductivity can be regarded as a constant with the internal heat source being ignored. The differential equation of heat conduction for the cladding is:

$$\rho_c c_c \frac{\partial t}{\partial \tau} = k_c \left(\frac{\partial^2 t}{\partial r^2} + \frac{1}{r}\frac{\partial t}{\partial r}\right) \tag{2.73}$$

where, subscripts u and c represent the fuel pellet and the cladding respectively; ρ, c, k are the density (kg/m^3), specific heat capacity (J/kg) and thermal conductivity [W/(m °C)] respectively; t is the temperature (°C); τ is the time (s); r is the radial length (m); z is the axial length (m); q_v is the volume heat-release rate (W/m^3).

2.3.4.5 Design of Reactor Main Parameters and Relevant Parameters

For the design of marine reactors, at the conceptual design stage and schematic design stage, the verification and validation of main reactor parameters are performed based on the overall demands and general technical specification, after the reactor type is determined. For the PWRs, main parameters, including reactor power, system pressure, coolant flow, average temperature of coolant, major parameters of the core (such as core height, flow area and heat transfer area), steam pressure, steam temperature, quantity and type of main equipment (components) and reactor layout, shall be verified and validated with the help from the professionals responsible for overall design and those from other relevant disciplines at such stages to optimize those parameters. The subsystem parameters of the reactor system and main parameters of equipment are usually determined according to the functions and objectives to be achieved by each subsystem and equipment.

2.3.4.6 Design of Protection Setting Values

The safety of marine reactors is guaranteed by relevant safety systems via protection signals, so one of the important tasks is the design of alarm and actuation setting values of each protection signal.

The primary protection signals of marine PWR vary from one operating condition to another, such as the main pump speed regulating operation condition and natural circulation condition of the main pump, and those signals include the shutdown protection signal for high outlet temperature of reactor coolant, shutdown protection signal for high pressurizer pressure, shutdown protection signal for low pressurizer pressure, reverse insertion signal for high nuclear power, shutdown signal for high neutron flux change rate, short-cycle shutdown signal, shutdown protection signal for low main pump speed, protection signal for high steam radiation dose, and protection signal for high radiation dose of reactor compartment. The setting values of above protection signals are determined through a large amount of transient analysis and design verification.

As the nuclear reactor provides power for ships, the basic requirements for the design of protection settings are to guarantee the reactor safety and reduce the scram as less as possible, i.e., "shutdown" signals are set as less as possible.

2.3.4.7 Design of Natural Circulation Capability

For marine reactors, due to the strict restrictions on the dimensions and weight of reactor compartments and the influence of marine conditions such as ship tilting, swaying and undulation on the natural circulation of reactors, one research task to carry out the reactor natural circulation capability design under severe external conditions [18, 23].

For the reactor natural circulation, the main driving force comes from the gravity head caused by the difference in the coolant density of the hot and cold legs in the reactor system loop. When the natural circulation flow reaches steady state, this gravity head will be equal to the total resistance pressure drop resulting from the flow of coolants in the loop, i.e.:

$$\Delta p_d = \Sigma \Delta p \tag{2.74}$$

$$\Sigma \Delta p = \Delta p_{core} + \Delta p_{sg} + \Delta p_{pump} + \Delta p_{valve} + \Delta p_{other} \tag{2.75}$$

where, $\Delta P_{\rm d}$ is the driving force of natural circulation (Pa); $\Sigma \Delta p$ is the total resistance (Pa) of the reactor coolant system; $\Delta p_{\rm core}$ is the total flow resistance (Pa) for the reactor core zone; $\Delta p_{\rm sg}$ is the total flow resistance for the steam generator zone; $\Delta p_{\rm calve}$ is the total flow resistance (Pa) for the main pump zone; $\Delta p_{\rm calve}$ is the total flow resistance (Pa) for the valve zone; $\Delta p_{\rm other}$ is the total flow resistance (Pa) for other zones, mainly including pipes and reactor internals.

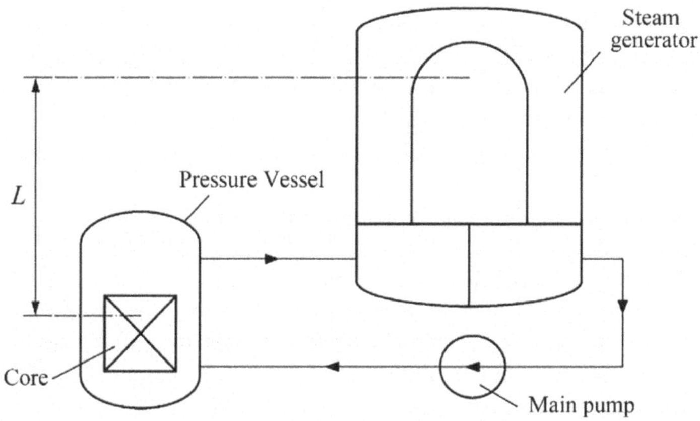

Fig. 2.21 Reactor system process

Figure 2.21 shows the flow loop of the reactor system. The driving pressure head of the natural circulation is formed by cold/hot core potential difference between the core and the steam generator as well as the density difference. A simplified calculation equation of the driving pressure head of natural circulation is

$$\Delta p_d = (\rho_c - \rho_h)gL \tag{2.76}$$

where, L is the cold/hot core potential difference of the system (m); ρ_{h} is the average density of the reactor core zone (kg/m^3); ρ_{c} is the average density of the steam generator zone (kg/m^3); g is the acceleration of gravity (m/s^2).

For methods to calculate the friction pressure drop and elevation pressure drop of system loops, refer to Sect. 2.3.3. It is worth mentioning that because the flow rate is relatively low under natural circulation condition, the frictional coefficient and form resistant coefficient shall be determined with actual type of channels and Reynolds numbers of fluid. See relevant manual on the calculation of hydraulic resistance for details.

In addition, during the design of marine reactors, the conversion between natural circulation condition and forced circulation condition will be encountered, and a major task of the thermo-hydraulic design is to realize the safe, quick and stable transition of such condition conversion. For the conversion from natural circulation condition to forced circulation condition, a large amount of fluid of the low temperature side flows into the core within a short time due to the start-up of the main pump, and thus a jump in the nuclear power occurs under the common action of moderator feedback and Doppler feedback of fuels caused by the dramatic decrease in the coolant temperature, which may trigger the reverse insertion resulting from reactor overpower or short-cycle shutdown. For the transition from forced circulation condition to natural circulation condition, the shutdown of main pump will lead to the quick decreasing of coolant flowrate, weakening the heat transfer between the primary and secondary sides of the steam generator, thus resulting in the reduction of steam pressure in transient process, which may not satisfy the steam demand. All the above problems need to be investigated through the calculation of reactor thermo-hydraulic transient and avoided by system design optimization.

2.3.4.8 Calculation of Reactor Transient Parameters

As required by its missions, the quick maneuverability is always regarded as one important technical feature of nuclear-powered ships. That means the reactor shall be able to rapidly increase or decrease its output power to ensure that the ship can change its movement quickly; In addition, other general requirements of ships and operation requirements of reactor system may lead to the conversion between forced and natural circulations, speed regulating operation of main pump, asymmetric loop operation and other transient conditions, and the quick and smooth transition between different conditions is an important manifestation of advanced technical performance of reactors. All the aforesaid conditions belong to normal transient conditions of reactors, and during the change of transient conditions, large fluctuations in coolant temperature, pressure and flowrate might introduce various problems for system pipes, components and structures, such as thermal stress, airtightness, mechanical shock and flow-induced vibration. Hence it is necessary to compute the parameters of thermo-hydraulic transients to get the variation range of relevant parameters of marine reactors in transient process.

2.3.4.9 Main Design and Analysis Software

As mentioned above, the reactor thermo-hydraulic design involves a lot of work like the verification of reactor main parameters, the thermo-hydraulic analysis of reactor core, the analysis of the reactor system, the calculation of reactor transient parameters, the design of natural circulation capability and the design of protection setting values, which generally depend on the specific programs and software. Compared with the commercial software developed for NPPs, the thermo-hydraulic design software for marine reactors is normally more targeted and has relatively simple and practical models.

2.3.5 Reactor Thermo-Hydraulic Test

For reactor thermo-hydraulic design, some design parameters regarding the thermal safety of reactor core, reactor structure design, main component design and reactor system operation as well as operational characteristics need to be studied further and validated through relevant tests due to the complexity of subjects investigated and the uncertainty of fluid flow and heat transfer phenomena. Some important reactor thermo-hydraulic phenomena need to be studied through experiments, while some design parameters are obtained through tests and some designs are verified and evaluated through tests.

In general, relevant reactor thermo-hydraulic tests include thermo-hydraulic mechanism test, thermo-hydraulic engineering test and thermo-hydraulic engineering verification test.

(1) Thermo-hydraulic mechanism test
Thermo-hydraulic mechanism test aim to study and explore the flow and heat-transfer mechanism characteristics of reactor-related thermo-hydraulic fluid. These tests provide intuitive and authentic data for scientific and technical personnel to learn about the thermo-hydraulic characteristics, guide the design and carry out research, which is the foundational work to conduct reactor research and development. They also provide engineering personnel with important basis for developing new theoretical models, writing specific computer programs and verifying relevant calculation methods.
(2) Thermo-hydraulic engineering test
In thermo-hydraulic engineering test, relevant design inputs need to be provided by carrying out certain specific tests, because such inputs are not yet available during reactor thermo-hydraulic schematic design and the design is based on assumed values or other reference values or relations. The input parameters provided by such tests are usually targeted for specific reactor structure or component type, and it is generally difficult to obtain their accurate values by theoretical means but they are critical to thermal safety of reactors.
(3) Thermo-hydraulic engineering verification test
It is normally a comprehensive test, mainly for verifying the overall performance and capability.

2.4 Fuel Assembly and Core Components

2.4.1 Fuel Assembly

(1) Functions

The nuclear fuel is a general term for fuel components of a nuclear reactor, which can either be a single rod or an assembly. The fuel assembly is a group of fuel elements assembled together and should not be disassembled during the fuel loading and unloading in the reactor, which is the smallest structurally independent component with nuclear fuels used as the major ingredient inside the reactor. Its functions are as follows: to contain fissile materials and thus to avoid the contact of fissile materials with the coolant, and to confine the fissile products in the fuel elements; to guarantee the proper spatial arrangement of fissile materials in the reactor; to ensure the discharge of fission heat, the proper flow path of the coolant and the capacities of fuel loading, unloading and refueling. Therefore, it is critical to the safe and reliable operation of the reactor.

(2) Analysis of application environment

Compared with the fuel assembly used in land nuclear power plants, the fuel assembly used in marine nuclear power plants requires special application environment, mainly including:

① Requirements for impact resistance

In addition to withstanding the high temperature, high pressure, vibration and other loads like those in NPPs, the fuel assembly in marine nuclear power plants have to withstand the tilting and the swing of the ship hull as well as the ship collision.

② Requirements for long lifetime (long refueling cycle)

In order to maximize the availability and the endurance of ships and avoid the impact on the service life of the ship hull, the core shall not be refueled frequently, and the ideal design objective is to achieve the same lifetime as the ship without refueling during the whole lifetime of the core. This requires large design margins for the selected material and the structure design of fuel assembly so as to meet the requirements for long lifetime and high reliability.

③ Requirement for power regulation

The core with its power regulated by control rods is commonly used in the marine nuclear power plants due to the limitation of accommodation space. The power regulation capacity is required depending on the service conditions. For example, evasive actions shall be taken in case of emergency, and the output power of the core is required to increase from low power level to high power level or full power level.

(3) Selection of materials

Materials for fuel elements shall be selected based on the features of the reactor and service environment.

The fissile materials potential for fuel elements are shown in Table 2.3 [24], and some common fuel materials will be introduced in the following parts.

① Featuring high melting point, good stability under irradiation, easy fabrication and low manufacturing costs, UO_2, as the most common fuel material, is widely used in the reactors such as the commercial nuclear power plants, nuclear power reactors, fast neutron reactors and high temperature gas cooled reactors.

② UAl_x and U_3Si_2 are mostly used in research reactors because their good stability under irradiation and good compatibility with aluminum material.

③ UC and UN are suitable for nuclear rockets, high-temperature reactors and the long lifetime reactor cores due to their high melting point, high thermal conductivity and large share of uranium in a single molecule.

④ UZr alloy was used in US submarine nuclear power reactors at the early stage, but abandoned quickly due to its significant irradiation swelling, which seriously affected the lifetime. Presently it is mainly used in the fast neutron reactors.

Table 2.3 Fissile materials potential for fuel elements

Compounds	Density/(g/cm^3)	Uranium volume per cubic centimeter in the compound/cm^3	Melting point/°C
U	18.9	1.0	1132
UAl_4	6.0	0.22	730
UO_2	10.96	0.51	2750
U_3O_8	7.3	0.33	Decomposition
UC	13.6	0.69	2535
UC_2	11.7	0.56	2450–2500
UN	14.3	0.71	2850
UZr_2	10.3	0.31	600①
UBe_{13}	4.37	0.15	2000
U_2Ti	15.22	0.73	890②
U_3Si	15.6	0.77	930
U_6Ni	17.6	0.89	790②
$MgO–UO_3$	7.2–7.5	0.29	1700–1800
$CaO–UO_3$	7.2–7.5	0.21	1700–1800
$BaO–UO_3$	7.2–7.5	0.35	1700–1800
UAl_2	8.1	0.35	1590
UAl_3	6.7	0.26	1320
U_6Fe	17.7	0.91	815③
UFe_2	13.2	0.48	1235
UPb	14.5	0.41	1280
U_3Si_3	12.2	0.59	1650
U_6Mn	17.8	0.91	726②
U_6Co	17.7	0.90	830②
U_5Sn_4	13.0	0.49	1500
USi	9.25	0.40	1600
USi_2	8.98	0.38	1600
UNi_5	11.31	0.27	1300
UCu_5	10.6	0.24	1052

Notes ①Disordering temperature; ②Peritectic temperature; ③Decomposition temperature

The metallic materials potential for the structural materials (cladding or matrix) of fuel elements are shown in Table 2.4 [24]. Zirconium alloy, aluminum alloy and stainless steel are commonly used.

① Featuring high melting point, small thermal neutron absorption cross section, strong resistance to high temperature water corrosion, and excellent mechanical property, zirconium alloy, as the most common cladding material, is widely used in various water-cooled reactors such as commercial nuclear power plants and

Table 2.4 Structural metallic materials potential for fuel elements (Cladding or Matrix)

Materials	Melting point/°C	Thermal neutron absorption cross section/(Barn)	Density at normal temperature/(g/cm^3)	Thermal conductivity at normal temperature/[W/(m K)]	Thermal expansion coefficient/(10^{-6} °C^{-1})	Specific heat capacity at normal temperature/[J/(kg K)]
Al	660	0.21–0.23 0.21–0.23	2.7	230	23.1	880
Zr	1855	0.18	6.45	16.7	5.89	280
Fe	1539	2.43–2.53 2.43– 2.53	7.87	95	12.5	460
Stainless steel	1400–1425 1400–1425	2.88	7.95	14.7	16.0	502
Ni	1455	4.5–4.8 4.5–4.8	8.90	67	13.7	470
Ti	1668	5.8	4.51	16.75	8.5	523
Be	1284	0.009	1.85	188.4	10.3–13.1 10.3–13.1	1800
Nb	2468	1.16	8.67	52.4	7.10	269
Ta	3000	21.0	16.6	54.5	6.50	139
Mo	2610	2.70	10.2	125.7 (204 °C) 125.7 (204 °C)	5.44	245
Pb	3380	19.2	19.3	129.9	4.59	134

marine nuclear power plants. However, it is not suitable for fast neutron reactors because its mechanic property degrades rapidly under high temperature and it is with poor resistance to metallic coolant corrosion.

② The aluminum alloy is commonly used in research reactors with low operation temperature and frequent refueling, for its small thermal neutron absorption cross-section and low cost. However, it is not suitable for power reactors with relatively high temperature due to its low melting point and poor resistance to high temperature water corrosion.

③ Though the stainless steel has large thermal neutron absorption cross section and is unfavorable in terms of the neutron economy of thermal reactors, it is widely used as the cladding material of fuel elements in fast neutron reactors for its high melting point, good resistance to metal coolant corrosion and excellent mechanical property.

A majority of worldwide marine nuclear power reactors adopt PWRs with thermal neutron spectrum, and mainly use fuel elements with UO_2 as the fuel material and zirconium alloy as the cladding. Rich experience in the R&D and application has been accumulated, and the operation performance and fuel cycle have been improved significantly through the continuous improvement in the design, manufacture and materials.

However, zirconium will react with water and produce hydrogen under severe accidents. If the hydrogen cannot be discharged in time, explosion would easily occur when the produced hydrogen and oxygen in the air reach a certain proportion. In the 2011 Fukushima accident, the zirconium alloy cladding had high temperature chemical reaction with the water steam under the loss of coolant condition, resulting in many times of hydrogen explosion, finally causing serous radiological hazard. This accident revealed the weakness of existing zirconium alloy claddings in severe accidents. The accident-tolerant fuel materials (such as SiC ceramic cladding) has been studied worldwide since then.

The SiC ceramic material is an ideal cladding material for nuclear fuel elements because it has small neutron absorption cross section, good heat conduction, excellent mechanical property under high temperature and excellent irradiation resistance, and it is not easy to react with water and will not produce hydrogen, which is mainly used for sphere fuel elements in high temperature gas cooled reactors. However, as a ceramic material, the single phase SiC material has the disadvantages of high brittleness and is liable to fracture, which restricts its application in the land NPPs and marine nuclear power plants with water cooled reactors. Therefore, in order to improve the toughness of SiC cladding materials, the three-layer composite SiC cladding has been studied in recent years, in which the inner layer is the high density single-phase SiC that contains and shields the fission products, the intermediate layer is the braided SiC fiber meeting the toughness and strength requirements, and the outer layer is the high density single-phase SiC resistant to corrosion. Such composite material not only possesses enhanced toughness, but also satisfies the air tightness and corrosion resistance requirements. However, the manufacturing process becomes complicated and the requirements on the process control are very strict. In addition, the key technique (the welding and sealing problems of the end plug) is still under development and it will take a long time to realize engineering applications.

(4) Type selection

Fuel elements vary with their applications, functions and service environments. Generally, two classification methods are used.

The first method is to classify the fuel elements by their shapes, mainly including rod, plate, tube, sphere and complicated shapes (see Fig. 2.22), which correspond to fuel rod, fuel plate, fuel tube and fuel sphere, respectively.

① Rod-shaped fuel elements

The rod-shaped fuel elements fall into two categories, i.e., smooth rod and ribbed rod, and they are simple in structure, easy to fabricate and thus have been widely used in nuclear power plant reactors.

② Plate-shaped fuel elements

Plate-shaped fuel elements can be straight or cambered and the thickness of plate can be constant or variable. Due to the excellent structural strength, large heating area, strong adaptability to quick power change, they are preferred in research reactors and power reactors.

③ Tube-shape fuel elements

Tube-shape fuel elements are similar to plate-shaped fuel elements. However the processing techniques for the former are more sophisticated. These elements are generally used in research reactors and fast neutron reactors.

④ Sphere-shaped fuel elements

They are simple in structure and easy to fabricate with large heating area, but are difficult to be located due to the looseness of the structure. They are commonly used in high temperature gas cooled reactors.

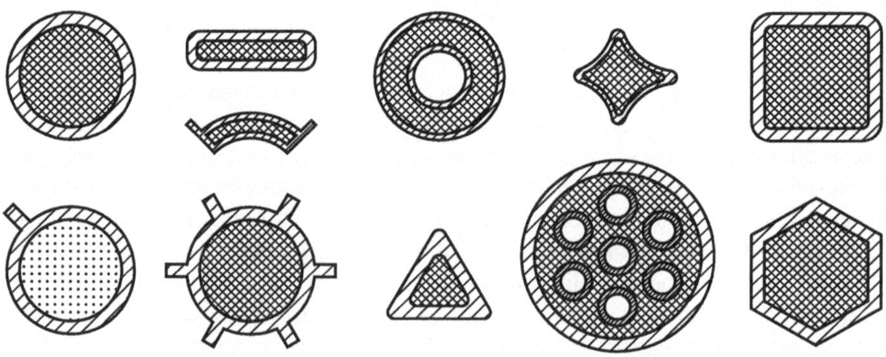

Fig. 2.22 Possible cross-section shapes of fuel elements

⑤ Fuel elements of complicated shapes

This category mainly includes prismatic, star-shaped and cross-shaped fuel elements. Generally they are designed for special purposes based on special requirements, and the processing technologies are relatively sophisticated. For example, the cross shaped rod fuel element was once used to increase the heating area and reduce the central temperature in Russia.

The second method is to classify the fuel elements based on fuel types, including metallic, ceramic and dispersion fuels.

① Metallic fuel elements

Metallic fuel elements adopt metallic uranium or uranium alloy as the fuel pellet, and the uranium alloy includes U–Zr, U–Mo, U–Nb, and U–Al alloys.

Their advantages include simple fabrication, high thermal conductivity and less thermal energy storage. They are suitable for small reactors with frequent refueling and power changes, such as research reactors and power reactors, yet have some disadvantages such as low operating temperature (e.g., U–Al alloy) and poor irradiation stability (e.g., metallic uranium and U–Zr alloy).

Metal fuels were often used in reactors at early stage. For example, metallic uranium and U–Al alloy were used for production reactors, U–Al alloy for research reactors, and U–Zr alloy for fast reactors and nuclear power reactors of U.S. navy. Such fuels have been gradually replaced by other types due to their disadvantages, and now only some of fast reactors are still using U–Zr metal fuels. For research reactors, because of the Reduced Enrichment for Research and Test Reactor (RERTR) Program in 1980s, the fuel materials have changed from U–AL alloy with low density of uranium to U_3Si_2 with high density of uranium in order to adapt to the change of enrichment. U–Mo alloy fuels with higher density are being developed, but it has not been applied in engineering. For power reactors, the nuclear reactor program of U.S. navy initially focused on U–Zr alloy, but for the significant irradiation swelling that seriously affected its lifetime, U.S. navy soon gave it up and then adopted UO_2–Zr dispersion fuels [25].

② Ceramic fuel elements

Ceramic fuel elements are mainly consisting of metallic claddings and ceramic fuel pellets. Ceramic fuels include uranium dioxide, uranium dioxide–plutonium dioxide (MOX), uranium carbide (UC), uranium nitride (UN) and thorium dioxide (ThO_2). These fuel elements are easy to fabricate and stable under irradiation, but have the disadvantage of low thermal conductivity and large thermal energy storage, and are not suitable for conditions experiencing rapid power change. Besides, their ability to withstand impact is poor and thus they are usually used in NPP reactors.

The light water reactors (PWR and BWR) and heavy water reactors (Candu reactor) often adopt ceramic fuel elements with uranium dioxide pellets and zirconium alloy claddings, and some fast reactors also use ceramic fuel elements consisting of the fuel pellets of uranium dioxide or uranium dioxide–plutonium dioxide, and stainless steel claddings.

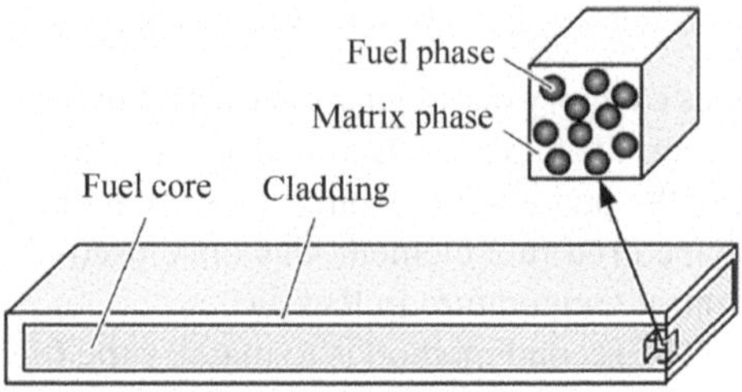

Fig. 2.23 Structure of dispersion fuel

③ Dispersion type fuel elements

Dispersion type fuel elements are formed by dispersing the fuel materials in other materials (matrix) in the form of fine particles and covered by cladding materials (see Fig. 2.23), such as UAl_x–Al, U_3Si_2–Al, UO_2–SS, UO_2–Zr. They are also named dispersion fuels.

The basic principle of dispersion fuels is to isolate the fuel particles from each other so that most part of the metallic matrix will not be damaged by fission fragments. Dispersion fuels have the advantages of good irradiation stability, high thermal conductivity, low fuel temperature, strong ability to maintain fission gas and higher burnup than large fuel, but the technologies used to fabricate them are sophisticated and the manufacturing cost is high. Because of high burnup and high thermal conductivity, dispersion fuels are suitable for small reactors featuring high burnup, high power density and frequent power change, such as research reactors and power reactors.

Internationally, research reactors mainly adopt Al-based dispersion fuels. The evolution of dispersion fuels can be divided into three main periods: U–Al alloy materials in 1950s; U_3O_8–Al and UAl_x–Al dispersion fuels in 1960s, the uranium density of which is 30–50% higher than that of U–Al alloy; and U_3Si_2–A1 dispersion fuels with higher uranium density in 1980s under the RERTR program, while the maximum core uranium density approved by IAEA is up to 4.8 gU/cm^3. Since U_3Si_2–Al dispersion fuels cannot satisfy the concentration reduction requirements from certain high-performance research reactors (e.g., ATR and HFIR) and R&D requirements of advanced research reactors with higher neutron flux, and the reprocessing of spent fuels is difficult, great efforts was made to develop UMo–Al dispersion fuels with higher density of uranium. The uranium density in the fuel core of such fuel can reach 8–9 gU/cm^3.

For power reactors, the nuclear reactor program of U.S. navy initially focused on U–Zr alloy, but later adopted UO_2–Zr dispersion fuels after short use of U–Zr alloy fuels at early stage [25].

During the selection of the type of marine nuclear power fuel elements, not only the special requirements on the application environment of marine nuclear power plants, but also the state of the art and the characteristics of the technologies and process possessed by the country shall be taken into account, and then appropriate technical routes can be established.

(5) Design criteria

① General requirements

The design of fuel assembly together with core structure, reactor control system, protection system and emergency cooling system ensure that:

i Fuel damage within the design lifetime is not expected during Condition I and Condition II events. Even if a few random failures occur, the coolant purification system shall be able to cope with the radioactive substances released so that the released amount is in line with reactor design basis.

ii Under Condition III, the failed fuel rods in the core shall be limited to a small portion of the total fuel rods.

iii Under Condition IV, the fuel assembly array geometry shall be maintained to enable the core cooling; the deformation or damage of the fuel assembly shall not impede the RCCA insertion to shut down the reactor and the reactor shall be maintained in the subcritical state; the radioactive substances released due to the failure of fuel elements shall not result in excessive hazards to human health and environment.

② Design criteria for fuel rod

The fuel rod shall meet the following criteria under Condition I and II events:

i During the initial period of reactor operation, the cladding tube of fuel rod must be self-supporting under the action of coolant pressure and temperature.

ii The creep collapse shall not occur to the fuel rod cladding tube within the design lifetime.

iii During the design lifetime, the internal gas pressure of fuel rod shall be lower than the coolant working pressure under the hot state.

iv The maximum temperature of fuel pellet shall be lower than the melting point of fuel. The effect of burnup, uncertainty and other factors on the maximum temperature limit shall be taken into account at the design stage.

v Within the design lifetime, the volumetric-average effective stress of cladding tube shall not exceed the yield strength of cladding tube material with the effect of temperature and neutron irradiation influence considered.

vi Compared with the state without irradiation, the total positive tensile creep strain of cladding tube shall be less than 1% for steady-state operation. For each transient event, the circumferential elasticity and plasticity strain of cladding tube shall not exceed 1% of the tensile strain under steady-state condition.

vii The accumulative strain fatigue of cladding tube of fuel rod shall be lower than the designed strain fatigue life, i.e.:

$$\sum_i \frac{n_i}{N_i} < 1 \tag{2.77}$$

where, n_i is the number of cycles for a given effective strain, ε_i; N_i is the allowable number of cycles for a given effective strain, ε_i.

viii At the end of design lifetime, the even equality corrosion depth of cladding tube shall be less than 10% of the wall thickness of cladding tube.

ix At the end of design lifetime, the maximum mass fraction of hydrogen of all cross-sections of the cladding tube averaged over the wall thickness shall be lower than 0.025%.

x The equivalent water content per cubic centimeter of free cold space shall be lower than 2 mg in the fuel rod. The free cold space includes axial gas cavity, annular space between pellet and cladding tube, and the open pore volume of the pellet.

③ Design criteria of fuel assembly

i Fuel elements shall be positioned in the fuel assembly and the fuel assembly shall be positioned in the core in a suitable manner, to construct and maintain the geometrical shape, and the axial and radial positions arrangement of fuels that meet the physical and thermo-hydraulic requirements under Condition I and II events.

ii The design of fuel assembly shall consider irradiation, temperature and other factors that may cause deformation and dimensional changes.

iii In the fuel assembly, proper space shall be provided for accommodating the fuel-associated components and the in-core measurement devices (if needed) and for creating the appropriate cooling conditions.

iv Fuel assembly shall be able to withstand various actions such as vibration, fretting, corrosion, lift force, pressure fluctuation and flow instability caused by fluids under Condition I and II events.

v The structural integrity of the fuel assembly shall be maintained and no deformation beyond the specified limits shall occur under Condition I and II events.

vi Under accident conditions and external impact loads, the deformation and destruction of fuel assembly shall not interfere with the insertion of control rods, and the array geometry shall be maintained to enable the core cooling.

vii During handling, transportation and storage, fuel assembly and their components shall maintain structural stability and integrity under extreme loads and shocks.

viii Under Condition I and II events, the strength design of each component of fuel assembly shall comply with the following provisions:

a. Austenitic steel components. The stress intensity of components shall be calculated according to the maximum shear stress theory. The values below, whichever is smallest, shall be adopted for the design stress intensity (Sm) of component materials:

(a) 1/3 of the minimum tensile strength or 2/3 of the minimum yield strength at room temperature; (b) 1/3 of tensile strength or 90% of yield strength at design temperature, but no more than 2/3 of the minimum yield strength stipulated at room temperature; (c) the limits of allowable stress strength, as shown in Table 2.5.

Table 2.5 Limits of allowable stress strength

Stress classification	Limits (Sm)
Overall primary membrane stress intensity	1.0
Local primary membrane stress intensity	1.5
Primary membrane stress and bending stress intensity	1.5
Total primary stress and secondary stress intensity	3.0

Table 2.6 Limits of allowable stress strength of austenitic steel components

Stress clarification	Limits
Primary membrane stress intensity	2.4Sm or 0.7Su①, whichever is smaller
Primary membrane stress and bending stress intensity	3.6Sm or 1.05Su, whichever is smaller

Note ①Su is the tensile strength for unirradiated materials at operating temperature

Table 2.7 Limits of allowable stress strength of zirconium alloy components

Stress classification	Limits
Primary membrane stress intensity	1.6Sy① or 0.7Su②, whichever is smaller
Primary membrane stress and bending stress intensity	2.4Sy or 1.05Su, whichever is smaller

Note ①Sy is the yield strength of unirradiated materials at operating temperature; ②Su is the tensile strength for unirradiated materials at operating temperature

b. Zirconium alloy component (fuel cladding excluded)

(a) The stress intensity of zirconium alloy components shall be computed according to the maximum principal stress theory; (b) The maximum principal stress shall not exceed the yield strength of un-irradiated zirconium alloy at operating temperature, or the design of zirconium alloy components is evaluated with the maximum shear stress theory.

⑨ Under Condition III and IV events, the limits for allowable stress strength of different components of fuel assembly are shown in Tables 2.6 and 2.7.
⑩ A grip shall be provided on each fuel assembly for handling, transport and loading into or unloading out of the core, and it shall be able to withstand the loads arising therefrom, and be compatible with relevant equipment.

The design of spring shall meet the following requirements:

a. During the design lifetime, the acting force of the spring in a fuel assembly in the core and the acting force from the spring (if any) of core components shall be compatible with the bearing capacity of the supporting structures at the upper part of the reactor core;

b. Under conditions I and II events, the fuel assembly shall not be separated from the bottom support plates of the core. However, under the specified action of impulse loads, the lift off of fuel assembly is allowed, but the characteristics of the spring shall not be changed and the spring shall be able to perform the compression function when recovered from lifting off.

The design of control rod channel of fuel assembly shall meet the following requirements:

a. The design of control rods and control rod drive mechanism shall be taken into account to ensure that the drop time of a control rod is within the limits.
b. The control rod channel shall be able to bear the abrasion caused by the action of control rods;
c. Together with the control rod design, the design of control rod channel shall ensure that no bulk boiling of coolants occurs in the control rod channel.

(6) Design analysis of fuel rod

① Thermal property

The fuel rod design shall ensure that the maximum temperature of the fuel pellets is lower than the melting point of fuels during the design lifetime. In the fuel rod, the heat released by the fission of UO_2 is transferred to coolant via pellets, gap, and cladding. The thermal properties of these materials are described below.

i Melting point of fuel

The melting point of UO_2 varies with the O/U ratio and trace impurities. In addition, because the O/U ratio gradually decreases with the oxygen release from UO_{2+x} at high temperature, it is difficult to measure the true melting point of UO_2. For this reason, the melting points measured by different researchers vary from each other, but the value is generally about 2800 °C [26]. After irradiation, the melting point of UO_2 decreases with the increasing of burn-up, reduced by 32 °C per 10 000 MW·d/tU on a conservative basis [4].

ii Thermal conductivity of the fuel

The thermal conductivity is a very important property of the nuclear fuel, and it determines the operating conditions of the fuel. The thermal conductivity of UO_2 pellet depends on temperature, porosity, burnup, O/U ratio and other factors.

The Lucuta model [27] considers the effects of temperature, burnup, fission products and porosity on the thermal conductivity of UO_2 pellets and it is suitable for high burnup conditions:

$$k_{UO_2} = k_0 \cdot f_d \cdot f_p \cdot f_{por} \cdot f_r \tag{2.78}$$

where, k_{UO_2} is the thermal conductivity of the UO_2 pellet [W/(m K)]; k_0 is the thermal conductivity of the unirradiated and fully dense UO_2 [W/(m K)]:

$$k_0 = \frac{1}{0.0375 + 2.165 \times 10^{-4}T} + \left[\frac{4.715 \times 10^9}{T^2}\right]\exp\left(-\frac{16361}{T}\right) \tag{2.79}$$

where, T is the temperature (K); f_d, f_p are the factors affecting the thermal conductivity by the dissolved and precipitated fission products respectively:

$$f_d = \left[\frac{1.09}{Bu^{3.265}} + 0.0643\sqrt{\frac{T}{Bu}}\right] \cdot \text{artan}\left[\frac{1}{\frac{1.09}{Bu^{3.265}} + 0.0643\sqrt{\frac{T}{Bu}}}\right] \tag{2.80}$$

$$f_p = 1 + \left(\frac{0.019Bu}{3 - 0.019Bu}\right)\left[\frac{1}{1 + \exp\left(\frac{1200-T}{100}\right)}\right] \tag{2.81}$$

where, *Bu* is the atomic percent burnup (%); f_{por} is the influence factor of porosity:

$$f_{por} = \frac{1-p}{1+(s-1)p} \tag{2.82}$$

where, *p* is the pellet porosity (%); *s* is the shape factor, and it is 1.5 for spherical hole; f_r is the influence factor of irradiation damage:

$$f_r = 1 - \frac{0.2}{1 + \exp\left(\frac{T-900}{80}\right)} \tag{2.83}$$

iii Thermal conductivity of the gap

Two cases are considered for thermal conductivity of the gap: the fuel pellet with and without contact with the cladding. Both can be described by one model [28]:

$$h_{gap} = h_g + h_s + h_r \tag{2.84}$$

where, h_{gap} is the total gap heat transfer coefficient [W/(m^2 K)]; h_g is the heat transfer coefficient of the interstitial gas [W/(m^2 K)]; h_s is the heat transfer coefficient [W/(m^2 K)] due to the contact of the pellet with the cladding; h_r is the radiation heat transfer coefficient [W/(m^2 K)].

The heat transfer coefficient h_g of the interstitial gas can be calculated by the model proposed by Ross and Stoute [29]:

$$h_g = \frac{k_g(T_g)}{d_g + C_r(r_1 + r_2) + g_1 + g_2} \tag{2.85}$$

where, k_g is the thermal conductivity of the interstitial gas [W/(m K)]; d_g is the gap width; C_r is the surface roughness coefficient; r_1 and r_2 are the roughness of external surface of pellet and internal surface of cladding respectively; g_1 and g_2 are the skip distances of external surface of pellet and internal surface of cladding respectively.

The heat transfer coefficient h_s due to the contact of pellets with cladding can also be calculated using the relational expression recommended by Ross and Stoute [29]:

$$h_s = C_s \frac{2k_f k_c}{k_f + k_c} \frac{P_c}{\delta^{1/2} H} \tag{2.86}$$

where, C_s is an empirical constant, and the recommended value is 10 $m^{-1/2}$; k_f and k_c are the thermal conductivity [W/(m K)] of fuels and cladding respectively; P_c is the contact pressure; δ is the average thickness of gas film [approximately 0.8 $(r_1 + r_2)$] (m); H is the Meyer hardness of flexible materials.

iv Thermal conductivity of the cladding

The thermal conductivity of zirconium alloys (Zr-2 and Zr-4) from room temperature to the melting point can be calculated using the following equation [30]:

$$k = 7.51 + 2.09 \times 10^{-2}T - 1.45 \times 10^{-5}T^2 + 7.67 \times 10^{-9}T^3 \tag{2.87}$$

where, k is the thermal conductivity coefficient [W/(m K)]; T is the temperature (K).

② Fission gas release

If the gaseous fission products (i.e., fission gases) resulted from fuel fission inside the pellets enter the cavity of fuel rod (including the gas cavity and gap between pellets and cladding), it will result in the increasing of the internal pressure of fuel rod. During the design lifetime, it shall be guaranteed that the gas pressure inside the fuel rod is lower than the working pressure of the reactor coolant in the hot state so that the fuel rod shall be protected from damage due to excessively high internal pressure.

The physical processes affecting the release of fission gases include the generation, recoiling and ejection of gas atoms, the diffusion of gas atoms in a lattice, the capture of fission products, irradiation re-dissolution and hot re-dissolution, thermal diffusion, and grain boundary diffusion, grain growth and cleaning of grain boundary, nucleation of bubbles and growth of stationary bubbles, migration and polymerization of bubbles, and saturation and connection of bubbles at the grain boundary.

As many physical processes are involved, the release model of fission gas usually only focuses on the physical processes playing a major role. According to the dependence of these physical processes on the fuel temperature, the release processes are generally divided into non-thermal release and thermal release.

i Non-thermal release

When the temperature is below 1,300 K, the mobility of fission gas atoms is too low and they can neither move to the releasing surface nor to the places where bubbles form. These fission gases freeze in the solid matrix, and only the gas generated near the outer surface of the pellet can release. Gas release is realized in the following three ways: firstly, when the gas atom is a high-energy fission fragment, it flies out directly from the fuel (recoiling); secondly, the fission fragment and the collision cascade collide with static gas atoms near the surface, leaking the gas atoms (ejection); thirdly, the fission fragment releases energy when passing through the matrix, creating a hot zone. The evaporation or emission of the hot zone results in the release of fission products contained therein. All the three mechanisms are independent of temperature, and the resultant release is not quite important at moderate burnup. When the rod burnup is below 45 MW d/kgU, the release caused by non-thermal mechanism accounts for less than 1%; when up to 60 MW d/kgU, such release increases up to around 3%.

ii Thermal release

When the temperature is between 1300 and 1900 K, the movement of gas atoms in the solid becomes important and the gas may be released by diffusing to the leaking surface. Bubbles may be formed within such temperature range, but the mobility is low and they cannot move too far away under the action of the temperature gradient. If the density of grain boundary bubbles is large enough to allow the connectivity of bubbles and thus provide access to the open hole, or sufficient to make the grain boundary pretty fragile and the stress within the fuels can cause the grain boundary to crack, then the gas gathered on the grain boundary can be released.

When the temperature is higher than 1900 K, the bubbles and the closed pores possess certain mobility and the movement distance within a few days or several months is almost the same as the grain size under the drive of the temperature gradient. When such bubbles or closed pores reach a crack or other surface in direct connectivity to the free space within the fuel rod, the fission gas is released.

The Benard-Tunbull model [31] considers both non-thermal and thermal releases of the gas. The later consists of two stages: in the first stage, gas atoms diffuse from the interior of the grain to the grain boundary, where they accumulate to a certain saturation (incubation) threshold. Gas atoms are brought back to the grains by the fission peak for dissolving, forming a thin layer at the grain boundary. The radiation-induced re-dissolution counteracts the diffusion flow and delays the beginning of release. In the second stage, when the grain boundary concentration exceeds a certain saturation threshold, the gas beyond this threshold will be immediately released. All phenomena that bring gas atoms from the grain boundary to the free space of fuel rods can be neglected, such as penetration, channel formation, and diffusion along grain boundaries, channels and fuel columns. Although these phenomena contribute to the release, they do not have significant effect on the total amount of gas released.

The thermal release of fission gas refers to the process that the fission gas atoms generated at random in the crystal are absorbed by grain boundary bubbles after arriving at the grain boundary by means of atom lattice diffusion, and the bubbles gradually grow up and connect with each other to form a serpentine channel connecting to the free space, from which some fission gases are released.

The process of fission gas diffusion in the crystal is expressed by the intracrystalline diffusion equation [see Eq. (2.88)]. Assuming that the crystal grain is spherical and its radius is a; $C\ (r,\ t)$ denotes the gas concentration at a radius of r for time t; β is the rate of gas production; D is the effective diffusion coefficient, then, the equation is expressed as follows:

$$\frac{\partial C}{\partial t} = \beta + \nabla \cdot \left(D\vec{\nabla} C\right) \tag{2.88}$$

The process from the growing-up of the grain boundary bubbles to their interconnection and release is described by the assumption of the grain boundary gas concentration threshold. This assumption assumes that the gases to be accommodated by the grain boundary are limited. When the concentration threshold is exceeded, a large number of grain boundary bubbles are interconnected due to excessive gases, and gas release will not stop until the concentration equals to the threshold.

Therefore, before the surface density of the grain boundary gas, *N*, reaches the saturation threshold N_S, the gas satisfies the following conservation equation, i.e., the gas production quantity is equal to the gas concentration in the crystal and the gas concentration at the grain boundary:

$$\int \left(C + \frac{3N}{2a}\right) dV = \int dt \int \beta dV \tag{2.89}$$

When the surface density of the grain boundary gas, *N*, reaches the saturation threshold, N_S, the gas concentration at the grain boundary will remain at N_S because all the gases reaching the grain boundary later will be released. The share of heat release can be obtained by formulas (2.90):

$$F = 1 - \frac{\int (C + \frac{3N_S}{2a}) dV}{\int dt \int \beta dV} \tag{2.90}$$

③ Densification and irradiation swelling of fuel pellets

After the fuel pellets are irradiated, two phenomena with opposite effects occur: densification and swelling.

The densification is the result of being irradiated and heated, mainly the gap annihilation. When fuel pellets are subjected to irradiation, gaps of small size (<1 μm) disappear immediately, while gaps of medium size (1–3 μm) gradually disappear, resulting in increased fuel density, volume reduction, and densification.

Densification of fuel pellets can be calculated with the ESCORE empirical model [32]:

$$\varepsilon_D = \Delta\rho_0 (e^{(\frac{Bu \ln(0.01)}{C_D Bu_D})} - 1) \tag{2.91}$$

where, ε_D is the densification strain; $\Delta\rho_0$ is the total potential densification; *Bu* is the burnup; Bu_D is the burnup under full densification; C_D is the coefficient, and it is 7.2–0.0086 (T − 25) below 750 °C and 1.0 above 750 °C, where T is the temperature (°C).

Irradiation of fuel pellets can lead to irradiation swelling, including the swelling of solid fission products and gas fission products.

After irradiation, the accumulation of solid fission products in the fuel matrix will cause solids to swell, resulting in an increase in the volume of pellets. Studies have shown that the swelling of solid fission products can be approximately described as a linear function of burnup [3U]:

$$\Delta\varepsilon_{sw-s} = 5.577 * 10^{-5}\rho\Delta Bu \quad (2.92)$$

where, $\Delta\varepsilon_{sw-s}$ is the solid swelling per unit volume; ΔBu is the burnup increment (% FIMA); ρ is the density (kg/m^3).

The contribution of transgranular and inter-granular bubbles formed by fission gas atoms to fuel swelling is called gas swelling. This phenomenon is very complex, involving the generation, growth, merging and connectivity of bubbles, and has an impact on the creep and deformation of $U0_2$. Fission gas swelling and fission gas release are closely related.

Fission product swelling can be calculated with the following semi-empirical model [30]:

$$\Delta\varepsilon_{sw-g} = 1.96 \times 10^{-31}\rho\Delta Bu(2800 - T)^{11.73}e^{-0.0162(2800-T)}e^{-0.0178\rho Bu} \quad (2.93)$$

where, $\Delta\varepsilon sw-g$ is the gaseous swelling increment per unit volume; Bu and ΔBu are the burnup and the burnup increment (% FIMA) respectively; ρ is the density (kg/m^3); T is the temperature (K).

④ Creep of the cladding

The creep of zirconium alloy cladding includes the thermal creep and the radiation creep.

The thermal creep can be calculated by Matsuo model [33]:

$$\dot{\varepsilon}_{th} = A\frac{E}{T}\left(\sinh\frac{a_i\sigma_{eff}}{E}\right)^2 e^{-Q/RT} \quad (2.94)$$

$$a_i = 650\left[1 - 0.56\left(1 - e^{\left(-1.4\times10^{-27}\Phi^{1.3}\right)}\right)\right] \quad ((2.95))$$

$$E = 1.148 \times 10^5 - 59.9T \quad (2.96)$$

where, $\dot{\varepsilon}_{th}$ is thermal creep rate (l/hr); A is 1.08×10^9 K/(MPa h); T is the temperature (K); Q is 201 kJ/mol; R is 8.314 J/(mol K); Φ is the fast neutron fluence rate (m^{-2} s); σ_{eff} is the effective stress (MPa).

Zirconium alloys will experience irradiation creep under stress and irradiation from fast neutrons, which can be described by the following relational expression [34]:

$$\dot{\varepsilon}_{ir} = C_0\Phi^{C_1}\sigma_m^{C_2} \quad (2.97)$$

where, $\dot{\varepsilon}_{ir}$ is the effective irradiation creep rate (1/s); Φ is the fast neutron fluence rate ($m^{-2}\,s^{-1}$); σ_m is the effective Mises stress (MPa); C_0, C_1, C_2 are material parameters.

⑤ Irradiation growth of the cladding

Due to the vacancy and interstitial atoms generated by fast neutrons, the material will grow freely under irradiation, showing an increase in volume. For the zirconium alloy material, according to the grain orientation, the positive growth occurs along the direction of maximum strain at the time of manufacture.

The irradiation growth of the zirconium alloy (Zr-2 and Zr-4) cladding can be expressed by the empirical correlation [33]:

$$\frac{\Delta L}{L} = A[\exp(240.8/T)](\phi \cdot t)^{1/2}(1 - 3f_x)(1 + 0.02CW) \qquad (2.98)$$

where, $A = 1.407 \times 10^{-16}(m^{-2})^{1/2}$; T is the temperature (K), with the applicable range of 310–630 K; ϕ is the fast neutron fluence rate per unit time ($m^{-2}\,s^{-1}$); f_s is the texture factor along the length; CW is the cold processing volume factor, i.e., the cross-sectional area reduction coefficient.

Free-growth phenomenon and the axial strain under actual condition will lead to the growth of fuel rods under irradiation in the reactor.

⑥ Water side corrosion of cladding

The corrosion reaction of zirconium alloy in water or water vapor can be expressed as follows:

$$Zr + 2H_2O \rightarrow ZrO_2 + 2H_2$$

The corrosion reaction of zirconium alloys falls into two stages:

i In the first stage, the metal surface will be covered by a layer of dense black zirconia. Under high stress within the oxide layer, the zirconia is of square structure, and the reaction depends on the diffusion rate of the O_2 anion. The dynamics of the protective effect of this oxidation film is inversely proportional to the thickness of the oxidation layer, and the corrosion rate is in relationship to the cube root of time.

ii In the second stage, the oxidation film will continue to grow until the cracking of dense oxidation film due to the increasing of stress, and then the oxide layer will be separated into the inner layer of a certain thickness and the outer layer substantially broken. The former is of square structure and a monoclinic structure, accelerating the migration of oxygen ions to the metal. The corrosion rate at this stage or "post transition" is almost linearly related to time.

The corrosion rate of staggered alloy before transition is determined by the cube of typical Arrhenius formula for oxide diffusion in zirconium alloy:

$$\frac{d\Delta\varepsilon^3}{dt} = K_{pre}\exp[-\frac{Q_{pre}}{RT_i}] \tag{2.99}$$

where, ε is the thickness of oxidation film; t is the time; K_{pre} is the kinetic constant before transition; Q_{pre} is the reaction activation energy; R is the ideal gas constant; T_{i} is the interface temperature of metal-oxidation film.

The transition time is related to the interface temperature of the metal-oxidation film:

$$\theta = C_{trans} T_i^{D_{trans}} \tag{2.100}$$

where, θ is the transition time; T_{i} is the interface temperature of the metal-oxidation film; C_{trans} and D_{trans} are the transition constants.

The rule after the transition follows the linear Arrhenius equation, indicating the diffusion of oxygen ions in zirconia. The model recommended by STEHLE [35] is as follows:

$$\frac{d\Delta\varepsilon}{dt} = K_{post}\exp[-\frac{Q_{post}}{RT_i}] \tag{2.101}$$

where, K_{post} is the kinetic constant after transition; Q_{post} is the reaction activation energy; R is the ideal gas constant; T_{i} is the interface temperature of metal-oxidation film; δ_{c} is the thickness of zirconia; t is the time.

⑦ Test verification of fuel assembly

When newly designed fuel assembly or existing operating experience cannot justify an important design change, the out-of-pile and in-pile tests of materials, components, and assembly shall be carried out to verify the reasonable performance of the design.

i Material performance tests

Such tests are usually carried out at the material development stage and include: ① mechanical property tests, covering conventional mechanical property, fatigue and creep; ② physical property tests, covering thermo-physical property, melting point and phase-transition temperatures; ③ studies on metallographic structure, covering grain size, uniformity, phase structure, texture; ④ corrosion performance tests, covering high temperature corrosion of hydrostatic and hydrodynamic water, hydrogen absorption analysis; ⑤ irradiation performance tests, covering irradiation damage, irradiation creep and irradiation growth, etc.

ii Component-level performance tests

In the R&D of fuel assembly, component-level performance tests may be required, which include: ① mechanical tests, covering the strength of various components, spring stiffness, cladding tube, internal pressure blasting and external pressure instability of cladding tube, failure of spacer grids; ② thermo-hydraulic tests, covering pressure drop of each component, critical heat flux of bundles, flow distribution of guide tubes; ③ mechanical connection structure tests.

iii Assembly-level and comprehensive performance tests

Such tests are all comprehensive design verification tests or engineering application tests, including ① thermo-hydraulic performance tests of fuel assembly, covering pressure drop, water power and buoyancy lift, flow distribution; ② flow-induced vibration tests of fuel assembly; ③ tensile, compressive, flexural, torsional stiffness tests of fuel assembly; ④ high temperature and high pressure dynamic water flushing tests of fuel assembly; ⑤ fuel assembly impact test; ⑥ anti-seismic performance tests of fuel assembly or control rod drive lines; ⑦ control rod drop test, covering buffer test of control rods, cold and hot test of drive lines; ⑧ tests of the handling of fuel assembly; ⑨ tests of the transporting of fuel assembly.

2.4.2 Core Components

(1) Design of rod cluster control assembly
① Functions of rod cluster control assembly

Safe and reliable operation is the primary requirements on nuclear reactors. The rod cluster control assembly (RCCA), the only movable part for the reactivity control, acts as the neutron absorber, and it is driven by the driving system and moves in the reactor, thus controlling the reactivity and realizing the reactor start-up, normal shutdown, regulation of reactor power, sustaining the reactor power and guaranteeing the reactor safety under accident condition.

② Design criteria of rod cluster control assembly

The design of RCCA shall satisfy the requirements of nuclear design, safety design, thermo-hydraulic design and thermo-mechanical design. The principal design criteria are as follows: i All the materials used for the RCCA shall satisfy related standards; ii The total reactivity worth at the time when all control rods are inserted into the core, i.e., the sub-critical depth under hot zero power condition or the shutdown margin, shall meet the nuclear design requirements; iii the rapid insertion time of control rod assembly or the reactor shutdown time shall meet the safety design requirements; iv The highest central temperature of the control rod neutronic absorber shall be lower than the phase-transition temperature or the melting point of the neutron absorber during operation; v Bulk boiling is not allowed in the circular coolant channel between the fuel assembly guide tube and the control rod; vi In cases of neutron irradiation, flow induced vibration, fretting, lift force and pressure fluctuation and control rod movement, earthquake or impact, and shutdown due to rod drop, the control rod assemblies shall maintain structural integrity and can move freely within the guide tube; while in the design lifetime, such assemblies can withstand the action of the specified steps and numbers of rapid insertions; vii Deformation of each components of control rod assemblies shall not affect the reactor scram under Condition III/IV.

③ Absorber material

i Ag–In–Cd alloy

Currently, the absorber materials used in the PWR control rod are usually Ag–In–Cd alloy with the respective mass percentages of 80, 15 and 5%. Firstly, this ternary alloy has excellent nuclear performance, in which the thermal neutron absorption cross section of Cd is very large, and Ag and In have strong epithermal resonance absorption peaks without the overlap of main resonance absorption peaks, giving it high neutron absorption efficiency. Secondly, it has appropriate mechanical properties. Although the mechanical properties of the alloy vary with the grain size, processing methods, annealing temperature and time, but in general, its strength and the creep resistance at high temperature are poor. However, the usage of cladding tubes resistant to high temperature water corrosion can offset this shortcoming. Thirdly, it has stable irradiation performance. The Ag–In–Cd alloy is a face-centered cubic solid solution, and the composition of the alloy changes after irradiation, and the second phase of hexagonal close packed structure appears. The geometric size increases just slightly and it has good radiation stability. Fourthly, this alloy has good moulding and processing properties, and it is easy to be smelted, extruded, drawn, rolled and cut. It can be made into various shapes and is insensitive to impurity content and deviation from components.

ii Boron carbide

B_4C has the following advantages as a reactivity control absorption material: a. large thermal neutron absorption cross section; b. relatively low cost; c. extensive in-reactor operating experience; d. excellent chemical compatibility with stainless steel cladding; e. sufficient supply due to abundant raw materials.

iii Hafnium (Hf)

As a reactivity control absorption material, Hf has excellent performance: a. large absorption cross sections for thermal neutrons and epithermal neutrons; b. excellent corrosion resistance performance in the PWR and it can contact with coolant directly without cladding; c. good mechanical properties and easy to be manufactured; d. good irradiation dimensional stability; e. little degradation in absorption performance along with the increasing of neutron fluence.

④ Design verification of RCCA

i Loads

The main loads covered by design analysis are as follows: gravity, buoyancy, reactor coolant system pressure, fluid-induced vibration, the in-reactor radiation, spring force of the absorber, the stepping-induced loads, loads in case of RCCA drop, manufacturing loads (such as those caused by the assembling of rod end plugs, threaded connection torque of finger tubes of the rod and connecting handle), handling and transporting loads.

ii Check of the dimensions

Dimensions to be checked include absorber volume, length of the control rod, total length of the RCCA, outer diameter of control rods, gap between the absorber and the cladding under hot state, and diameter and alignment interval of control rods.

iii Check of strength

The structure stress analysis under normal or accident conditions, transportation and handling during the lifetime, the stress analysis of each structural components and joints under design loads (stepping-induced loads) during normal operation, the impact stress analysis of parts bearing the impact when RCCA drops, the spring stress analysis of connecting handles, and the stress analysis of claddings and welds.

The structural integrity is checked by stress analyses and comparison of the stress levels of the components measured in the tests with those criteria specified in the ASME guidelines.

iv Test verifications

In addition to adequate theoretical calculations and analyses, necessary out-of-pile test verifications are also required and the main tests include: a. abrasion test of cladding tube; b. rod drop test of RCCA under cold condition; c. rod drop test of drive lines under hot condition.

(2) Design of burnable poison assembly

① Functions

The main functions of burnable poison assembly are to provide supplementary control over the excess reactivity of the core, and at the same time gradually release the core reactivity along with the burnup, partially compensate for the decrease in reactivity caused by burnup and the like. Reasonable layout of burnable poison assembly in the core can improve the neutron flux distribution and thus improve the core power distribution.

② Design criteria

In addition to meeting the nuclear design and thermo-hydraulic design requirements, the design of burnable poison assembly shall satisfy the following requirements:

i The materials used for burnable poison assembly shall comply with relevant standards.
ii Under the vibration, lift force and pressure fluctuations resulted from neutron irradiation and the fluid force, the burnable poison assembly shall maintain structural integrity and always maintain the specified relative axial position with the fuel assemblies.
iii During the entire lifetime, the cladding of burnable poison rods shall maintain their mechanical integrity when subjected to the external pressure of system and internal pressure due to the helium release.

iv After being exposed to radiation for a long term, the cladding shall not be subjected to the risk of collapsing due to circumferential creep and can maintain its structural integrity.

v During the operation of the reactor, the maximum temperature of the burnable poison absorbers shall be lower than the melting points of materials or the temperature at which the material undergoes significant deformation or creep; Bulk boiling shall not occur in the annular channel between the guide tube of fuel assembly and the burnable poison rods.

vi The dimensions and tolerances of all components shall meet the requirements of neutronics, thermodynamics, hydraulics, and mechanical strength, and be compatible with relevant components.

vii All structural materials shall have good mechanical properties, corrosion resistance, compatibility with the core environment and fuel storage pool under thermal and irradiation conditions, and the poison materials shall also meet the neutronics requirements.

③ Materials for burnable poisons

i Borosilicate glass tube (Pyrex)

Borosilicate glass tube is commonly selected as the poison material for NPP burnable poisons, and it is mainly composed of SiO_2 and B_2O_3 and has following advantages: a. good boron uniformity; b. low swelling rate at the design temperature c. no mechanical interaction with the cladding (the coefficient of thermal expansion and hardness is lower than those of the stainless steel cladding); d. easier to be manufactured at lower cost than other poison materials.

ii Aluminum oxide-boron carbide

Ceramics made by dispersing B_4C particles in Al_2O_3 matrix are also commonly used as burnable poison materials and have following advantages: a. low helium release rate; b. good chemical stability at high temperature; c. good chemical compatibility with the zirconium alloy cladding.

iii Gd_2O_3/Er_2O_3 poison

a. The integrated burnable poison design can be adopted in the PWR core design by dispersing Gd_2O_3/Er_2O_3 in the UO_2 pellets; b. gadolinium (Gd) has a very large thermal neutron absorption cross section, and there will be small amount of remaining poisons at the end of fuel cycle and less reactivity loss with reasonable design; c. erbium (Er) also has a larger heat absorption cross section, but it consumed slowly as the core burnup increases, and can gradually release the excess reactivity.

④ Design verification of burnable poisons

i Loads

a. Coolant pressure and temperature; b. internal pressure, as gaseous products released under radiation gradually increase with gas release; c. local circumferential expansion caused by installation.

ii Check of dimensions

Dimensions to be checked include: a. general dimension of the compression system; b. axial dimensions of the hold-down device (under hot and cold conditions); c. lateral dimensions of the hold-down device (under hot and cold conditions); d. rod position; e. rod dimensions, covering rod diameter, length, gap between the cladding and pellet under hot and cold conditions, and axial position of poison materials.

iii Check of strength

a. Stress analyses on compression system components under cold condition and at the beginning of lifetime (BOL) when the pressing pressure is the largest; b. Stress analyses on the joints like screws; c. Stress analyses on cladding and welding joints (including operation and manufacture); d. check of the non-collapsing of cladding tube.

The structural integrity is checked by stress analyses and comparison of the stress levels of the components measured in the tests with the criteria specified in the ASME guidelines.

(3) Design of neutron source assembly
① Functions of neutron source

The neutron source assembly is used to increase the core neutron flux to a certain level during the start-up of the reactor core to ensure that the neutron detectors are operational and responding to core multiplication neutrons, thus ensuring the safe start-up of the reactor. Generally there are two kinds of neutron assemblies applied in the reactor, i.e. primary and secondary neutron source assemblies, with the former working only during the first loading of the core and first start-up of the reactor, while the latter working during the subsequent reactor start-up.

② Design criteria of neutron source assembly

i The source intensity and source position shall meet the requirements for the core control measurement and thermo-hydraulic design, while the requirements for the structural design of the neutron source assembly is basically the same as those for burnable poison assembly. ii The position of the primary neutron source shall be consistent with that of the measurement instrument.

③ Materials for neutron source

^{252}Cf is commonly used as the primary neutron source material, which emits neutrons after spontaneous fission with a neutron yield of $2.32 \times 10^{12}/(/(\mathrm{s\ g})$, and the a-decay half-life of ^{252}Cf is 2.64 years. The Po-Be source is also adopted, but it has a shorter half-life (about 138 days) and the requirements for the accuracy of critical time is rather strict in the engineering construction.

The Sb and Be powder mixture is often used as the secondary neutron source material for PWRs. When it is subjected to neutron irradiation, ^{123}Sb will transmute to ^{124}Sb and emit γ-ray, and the γ-ray will react with beryllium (γ, n) and produce radiation neutrons to start up the reactor again.

The reaction equation is:

$$^{123}_{51}\mathrm{Sb} + ^{1}_{0}n \rightarrow ^{124}_{51}\mathrm{Sb} \longrightarrow [\mathrm{Tr}_2 = 60d]^{124}_{52}\mathrm{Te} + ^{0}_{-1}\beta + \gamma(e \succ 1.6\mathrm{MeV})$$

$$^{9}_{4}\mathrm{Be} + \gamma \rightarrow ^{8}_{4}\mathrm{Be} + ^{1}_{0}n$$

The neutron yield of Sb–Be material is about 1.3×10^6/s.

④ Design verification of neutron source assembly

Loads covered by design analysis: a. temperature and pressure of the coolant; b. internal pressure by the gass release of the Sb–Be pellets; c. different thermal expansion among various components of neutron source rod; d. localized circumferential expansion caused by assembling; e. cooling conditions for neutron source rod in the guide thimble. Dimension and strength verifications of neutron source assembly are the same with that of the burnable poison assembly.

2.5 Reactor Pressure Vessel

2.5.1 Overview

A reactor pressure vessel (RPV) refers to a closed vessel used to contain the reactor coolant under pressure and to accommodate the components such as the core and the reactor internals. RPV is one of the key components of a reactor.

Generally, RPV provides such functions as containing the radioactive substance, maintaining the integrity of the pressure boundary, supporting (locating) the reactor internals, and meeting the requirements for alignment.

The marine RPVs are installed on various kinds of ships (such as icebreaker, merchant ship). The work environment thereof is significantly different from that of the RPVs installed on land (such as RPVs of NPPs). More severe conditions such as tilting and swing of ship hull caused by sea waves and the effects of marine environment on materials shall be considered.

Compared with the common pressure vessels in other industries (such as petrochemical industry), the marine RPVs shall have more special functions and withstand more severe work environment (such as irradiation). Therefore, the requirements for design, manufacture, inspection, transportation, installation, operation surveillance and maintenance of the marine RPVs are much more rigorous.

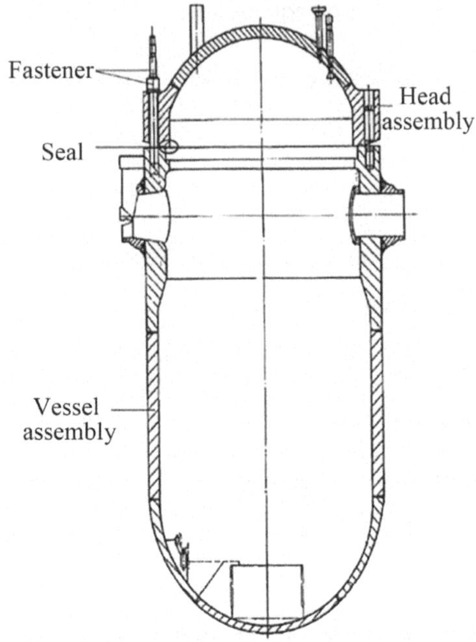

Fig. 2.24 Example of structure of reactor pressure vessel

2.5.2 *A Brief Introduction to Structure*

A RPV is usually composed of head assembly, vessel assembly, seals and fasteners (see Fig. 2.24). The head assembly and the vessel assembly are integrated as a monoblock by fasteners and effectively sealed.

(1) Head Assembly

The head assembly (see Fig. 2.25) includes closure head, lifting lugs, various kinds of holes for instrumentation penetrations or device channels (such as control rod drive mechanism adapters, flux instrumentation adapters, temperature measuring adapters), etc.

A closure head is classified into flat closure head (namely a flat cover, a circular flat panel), spherical closure head (namely a hemispherical or a quasi hemispherical shell with flanges), spherical cap-shaped closure head (namely a spherical cap-shaped shell with flanges), ellipsoidal closure head (namely a semi-ellipsoidal head flanged with a straight cylinder section), etc. When the reactor power and the RPV diameter are both relatively small, the effects of the type of the closure head on stress, deformation, sealing and bolt stress are not obvious and flat closure heads that are simple in structure and easy to process are commonly used. When the reactor power or the RPV diameter is relatively large, the effects of the type of closure head on stress, deformation, sealing and bolt stress are relatively large, and non-flat closure heads that can bear more uniform load are commonly used, or the flat closure heads combined with the counter flanges for cylinder section are adopted at the same time to reduce the diameter of the closure head.

Fig. 2.25 Example of head assembly

The lifting lugs welded on a closure head shall be used during the hoisting of the head assembly. The lifting lugs are used for hoisting the head assembly only. When relatively heavy objects are installed on the head assembly or the head assembly is connected to other relatively heavy objects, using the lifting lugs for hoisting operation is prohibited.

The adaptors of the control rod drive mechanism, of which the position accuracy and perpendicularity are effectively controlled, are usually installed at the preset opening of the closure head. The adapters and closure head are usually connected through the fillet weld at the inner wall of the closure head or connected in other ways such as mechanical connection structure.

Various kinds of measurement and process adapters (such as temperature measuring adapter, flux instrumentation adapter, vent pipe) are also installed at the preset openings of the closure head. The adaptors and closure heads are usually connected through the fillet weld at the inner wall of the closure head.

(2) Vessel Assembly

A vessel assembly (see Fig. 2.26) is also called a reactor vessel, which is the main body of RPV, is used to contain the reactor core and reactor internals. The vessel assembly is usually formed by welding the vessel flanges (matching with the closure head), nozzle shell, core shell, inlet and outlet nozzles, transition ring, bottom head and other relevant components together. A vessel assembly is usually divided into cylindrical shell sections mentioned above. Certain cylindrical shell sections can be eliminated or combined depending on the reactor layout and other requirements. With the development of manufacturing process and technology, the development trend of cylindrical forgings is to combine and integrate various cylindrical shell

Fig. 2.26 Example of vessel assembly

sections. The integrated large complex cylinder forgings can be used to effectively reduce the amount of cylinder welds and improve the reliability of equipment.

The upper part of vessel assembly (namely vessel flange) is provided with the threaded holes for closure studs, which matches with the through holes of studs on the closure head to install closure studs for sealing of RPV. The vessel flanges are usually provided with a leak detection channel connected to a leak detection system to detect the leakage of RPV leakproof structures at any time to avoid accident. The barrel assembly of reactor internals is usually mounted on the counterbore steps of the vessel flange, which are compressed through the closure head and positioned through the keys or grooves provided inside the vessel flange.

The cylindrical shell where the inlet and outlet nozzles or other nozzles (such as the nozzles in the integral reactor used for supporting the main pump) locate is called the nozzle shell, which shall be provided with large openings or flanging lug boss to which the inlet and outlet nozzles are installed and welded. Due to the opening reinforcement, the thickness of the nozzle shell is usually large, which is only less than that of the vessel flanges in vessel assembly. In both the separately arranged reactors and the compact reactors, the inlet and outlet nozzles are usually provided as the channel for coolant entering and leaving the reactor. One end of inlet and outlet nozzles is welded to RPV nozzle shell, and the other end is connected with the main pipe (or other equipment, such as steam generators). In order to make a reliable connection of heterogeneous metals between the nozzle and the connected system pipe, the nozzle end is usually welded with a safe end corresponding to the pipe material in advance. In the integrated reactor, different nozzles will be provided depending on the actual layout. For example, the main pump nozzles are provided in order to accommodate the main pump at the outside steps of RPV; the steam transportation nozzles are provided in order to install the steam generator in RPV.

The shell section directly facing the reactor core is called the core shell (namely the cylindrical ring corresponding to the length of the fuel assembly placed in the vessel). For RPV, the core shell bears the maximum neutron fluence, therefore the requirements for the material performance and the structural design of the core shell are extremely rigorous. For example, the content of elements such as P, S, Cu and Ni is strictly controlled in the base metal of the core shell. The reference nil ductility transition temperature of the base metal is low enough. The longitudinal weld is prohibited in the core shell and the circumferential welds at both ends shall stay as far away from the core area as possible. The wall thickness of the core shell shall be constant, and the discontinuous zone shall be avoided.

The bottom end (namely the bottom head) of the vessel assembly is similar to the closure head. They have the same types of structure and show similar development trend. The flat bottom head is adopted at the early stage, and then the spherical or ellipsoidal bottom head is adopted by almost all the pressurized water reactors. The structure of the inner wall of the bottom head shall match with the bottom reactor internals as much as possible to optimize the flow distribution at the reactor core inlet. When there is no transition ring between the bottom head and the core shell, a straight section shall also be provided for the bottom head to prevent the weld joint between the bottom head and the core shell from locating at a discontinuous zone with high stress.

The transition ring is a transition shell between the bottom head and the core shell, which includes a part of spherical shell or the ellipsoidal shell of the bottom head and a part of the straight-sided cylindrical shell section. The transition ring is set to reduce the manufacturing difficulty of the bottom head. However, the setting of transition ring increases the number of welds and the inspection work in later period.

(3) Fasteners and Seals

Fasteners are used to connect the head assembly and the vessel assembly and to provide enough pre-tightening force to seal RPVs. Fasteners mainly include closure studs, closure nuts and gaskets.

Seals include sealing elements and their accessories. Sealing elements (see Fig. 2.27) usually include O-rings [1], C-rings or flat gaskets. The accessories of sealing elements include a series of fastening screws and spacers, which are mainly used to fix the seals.

(4) Other Equipment

The equipment closely related to RPVs includes irradiation surveillance capsules and thermal insulation.

RPV base metal would present an irradiation embrittlement effect [36] such as increasing of strength and decreasing of ductility after being irradiated by neutrons. Irradiation surveillance capsules are usually set in RPV in order to monitor the irradiation embrittlement of materials in the strong irradiation zone of RPV and modify the safe operation limit of RPV according to the degree of embrittlement to ensure the safe operation of RPV. Irradiation surveillance capsule is a pressurized

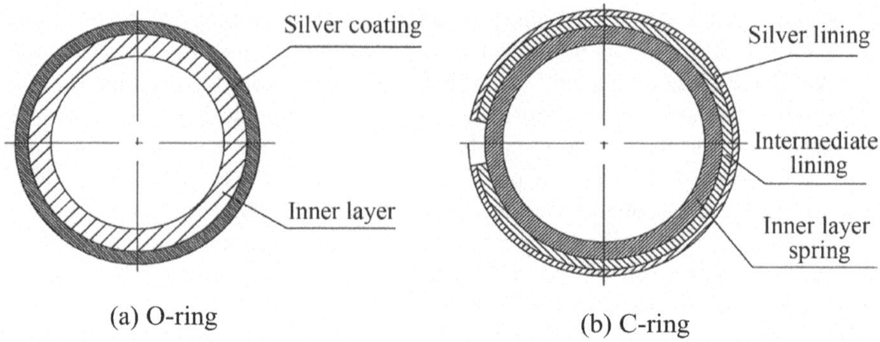

Fig. 2.27 Example of sealing rings

sealing pipe (shell) assembly, in which an irradiation monitoring specimen is mounted to monitor the effect of irradiation on the performance of RPV materials, and the temperature and dose detectors are also included.

The outer of RPV shall be provided with thermal insulation in order to effectively reduce the heat loss of the reactor, improve the environmental conditions of the compartment and reduce the temperature difference of RPV wall surface, further to reduce the stress of temperature difference and thus play an important role in RPV safe operation. Currently, the thermal insulation is mainly classified into three categories based on materials and structural characteristics.

The first category is non-metallic insulation made of mineral wool, the material of which is generally irradiation-resistant glass fiber wool or felt. The non-metallic insulation made of mineral wool is installed at the external wall of RPV. The advantages are lightweight, easy to install and can adapt to various kinds of complex structures, while the disadvantages include the generation of mineral wool dust during installation and disassembly (causing certain hazards to personnel and environment) and the degradation in long term application (requiring a regular replacement of local parts).

The second category is metal reflective insulation with whole stainless steel metal structure (see Fig. 2.28), which is usually made of 304 or 316 material. The insulation performance of the metal reflective insulation is equivalent to that of non-metallic insulation made of mineral wool. It is reusable, easy to assemble and disassemble, in corrosive to the insulated equipment, and shows higher irradiation stability. Its disadvantages include slightly heavier in weight than the non-metallic insulation made of mineral wool, relatively high requirements for manufacturing process and poor economic efficiency.

The third category is composite insulation combining metallic and non-metallic materials, which possesses the advantages of metal reflective insulation and non-metallic insulations as well as a better insulation performance. This composite insulation can drastically reduce the thickness of the insulation and also possesses additional functions such as a certain degree of shielding, which is an optional direction for the future development of insulation.

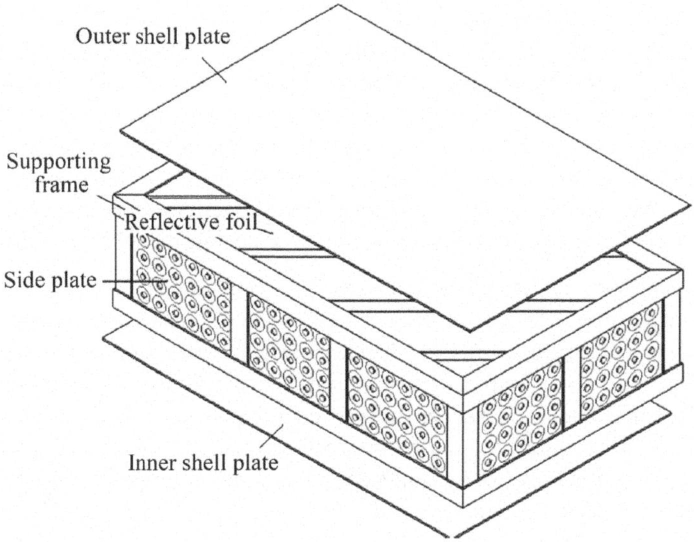

Fig. 2.28 Metal reflective insulation

2.5.3 Materials

(1) Principles for Selection

The selection of materials for RPV is mainly based on following principles.

① The materials shall possess enough purity, compactness and uniformity, appropriate strength, enough ductility and toughness, low sensitivity to neutron irradiation embrittlement and aging embrittlement, and good weldability as well as excellent cold and heat processing performance.

② The compatibility of physical, chemical and mechanical properties of materials of the pressure vessel shall be taken into consideration in material selection to reduce the effect of constraint stress and corrosion.

③ The materials contacting the reactor coolant shall possess good corrosion resistance and are compatible with the coolant to avoid the corrosion of the pressure vessel as well as the effect of the corrosion products on the overall performance of the reactor.

④ The cobalt content of material shall be strictly controlled to improve the safety during maintenance and inspection.

⑤ The well-developed materials that have been used before shall be given preference. The newly selected materials shall be qualified by tests.

⑥ In selecting welding materials, the deposited metal of welds shall match with base metal in strength, ductility and toughness in addition to the principles mentioned above. Furthermore, the effect of welding with large thickness and multiple welding beads on the weld quality shall also be considered.

(2) Development of Materials for RPV

RPV works in the environment of high temperature, high pressure, and strong irradiation. As the main pressure-bearing component, the base metal of RPV usually adopts low alloy steel with a relatively high strength to reduce wall thickness and reactor weight. In order to prevent the coolant from corroding the base metal, the corrosion-resistant stainless steel is usually bead welded or lined at the inner wall of the low alloy steel. Other components contacting the reactor coolant are usually made of stainless steel or nickel base alloy.

① Base Metals of Main Pressure-Bearing Components

The development of the base metal used for making the main pressure-bearing component of RPV began nearly 70 years ago. At the beginning, there was no available RPV material. Based on massive screening, C–Mn steel that had been successfully used for petrochemical vessel was used as the base metals for the main pressure-bearing components of RPV. For example, the RPVs of Saxton Reactor and Indian Point Unit −1 of America were made of C–Mn steels such as A212B, A105 and A350 in ASTM standards.

Due to the low strength and the low toughness of thick section, the previous C–Mn steel could not satisfy the requirements for the material of RPV with large thick wall. Based on the low strength C–Mn steel, the improved Mn–Mo steel plate with a higher strength through adding Mo and the improved Mn–Ni steel forgings with a higher strength through adding Ni were developed. In order to further improve the hardenability and property uniformity of the thick section steel to satisfy the requirements for manufacturing RPV with a larger wall thickness, the improved Mn–Ni–Mo steel through adding Ni or Ni–Mo were developed consecutively. In the early 1960s, France had developed the improved Mn–Ni–Mo steel forgings through adding Ni and decreasing carbon based on Mn–Ni–Mo steel. The Mn–Ni–Mo steel forgings had lower welding crack sensibility and were used to make RPV of Chooz NPP. The Mn–Ni–Mo steel forgings were later named as A508-3 steel and further improved and developed in France, Japan and Federal Germany. The harmful impurity elements (such as sulfur, phosphorus and copper) and the microelements (such as arsenic, tin and antimony) are strictly controlled in specifications and production. The smelting, ingot forging and heat treatment process were improved through decreasing carbon and chrome, and increasing manganese, which endowed A508-3 steel with excellent comprehensive performance. A508-3 steel came to be the most ideal and mature forging materials for large and super large RPV in the contemporary era. For most of the marine RPVs in the above countries, A508-3 steel is used as the base metal for main pressure-bearing components of RPVs. Russia has developed the improved Cr–Mo–V steel through adding V and the improved Cr–Ni–Mo–V steel through adding Ni–V based on Cr–Mo steel (such as 15X2HMOA-A steel[42] adopted by BB3P_1000). For most marine RPVs in Russia, Cr–Mo–V or Cr–Ni–Mo–V steel is used as the base metal of main pressure-bearing components of RPV. We can expect that with the development of larger size RPV which can bear super high pressure, the base metals for main pressure-bearing components of next generation RPVs with better strength, hardenability, ductility and toughness are still the key points in the future development.

② Stainless Steel
Due to the limit of smelting process and manufacturing cost, the austenitic stainless steel materials used for RPV at preliminary stage have relatively high carbon content and are relatively sensitive to intergranular corrosion. The main measure is to add Ti (or Mo) to stabilize carbon and thus to improve the ability of resisting intergranular corrosion. However, this measure increases the content of carbide inclusion and aggravates the tendency of pitting corrosion. Therefore, the austenitic stainless steel is not an ideal anti-corrosive material. With the development of smelting technique, the super-low-carbon stainless steel is developed, which fundamentally solves the intergranular corrosion resistant problem of the austenitic stainless steel. The super-low-carbon austenitic stainless steel is commonly used for the stainless steel parts of modern RPVs.

③ Nickel Base Alloy
Inconel 600 alloy was generally used to make components such as RPV penetrations in the world at early stage. In September, 1991, many cracks (including SCC) were detected on the penetrations of French PWR Bugey 3. The penetrations in NPPs in Belgium, Brazil, Japan, South Africa, Spain, Sweden, Switzerland and the United States were inspected, and about 2% of the penetrations had cracks. In order to improve the capability of SCC resisting, Inconel 690 alloy has generally been used as the material of penetrations as well as the weld connecting material of penetrations and vessels of nuclear reactors built after 1990s. Inconel 690 alloy is a kind of austenitic nickel-based anti-corrosive alloy containing about 30% of chrome. In the solution containing chloride, this alloy has better SCC resistance than Inconel 600, Incoloy 800 and 304 stainless steel, and is featured with high strength, good metallurgical stability and good processing characteristics, especially in various kinds of high temperature water. Inconel 690 alloy shows low corrosion rate and excellent SCC resistance.

④ Materials for Fasteners
The closure studs used for RPV at early stage were made of high strength low alloy steel (such as Cr–Mo steel and Cr–Ni–W steel adopted in RPVs of Yankee and Dresdner NPPs in the United States). This kind of steel is suitable for manufacturing the studs with small diameter. In order to get a higher hardenability to satisfy the requirements for the uniform performance of studs with large diameter, Cr–Ni–Mo steel was adopted in studs at intermediate stage. With the further enhancement of requirements, the improved Cr–Ni–Mo steel (such as A540 B24) with better properties is generally adopted at later stage. Since this kind of steel has relatively high manganese

and molybdenum content and relatively low sulfur and phosphorus content, it has better performance and is suitable for making the fastener studs of modern large RPVs. Subsequently, in order to further refine the structure and grain of the steel and improve the strength and toughness of the steel to satisfy the requirements for manufacturing the studs with larger diameter, the improved Cr–Ni–Mo–V steel (such as A540 B24V) is developed by adding vanadium. This kind of steel has higher hardenability, finer structure and grain, higher strength as well as excellent toughness. Therefore, it has become the commonly used material for the fastener of modern large and super large RPVs.

(3) Material Performance Requirements and Test Methods
According to the above principles for the material selection as well as the state of the art of RPV material, the following tests shall be conducted to verify the consistence of the produced and manufactured material with the design expectation in order to select the suitable RPV material:

① Analysis and examination such as chemical composition analysis and metallographic examination implemented for verifying the material purity;
② Tests such as tensile test, impact test, hardness test, drop weight test and fracture toughness test implemented for verifying the mechanical property;
③ Corrosion tests such as those for intergranular corrosion, pitting corrosion, crack corrosion, uniform corrosion and stress corrosion implemented for verifying corrosion resistance (generally, only first or first two corrosion tests shall be implemented in engineering phase; all the corrosion tests shall be implemented in material development phase);
④ Tests such as bending test, flattening test and expansion test implemented for verifying cold processing performance;
⑤ Weldability test implemented for verifying welding performance.

(4) A Brief Introduction to Manufacturing Process of Forging
The manufacturing of A508-3 forging from charging to delivery usually includes the following process: melting—ingot casting—forging—rough machining—non-destructive examination—heat treatment—non-destructive examination— mechanical properties test—finish machining—non-destructive examination—delivery.
The following is a brief introduction to the main process.

① Melting and Ingot Casting
The manufacturer formulates strict chemical composition control measures based on the requirements for the shape and the performance of forgings, improves the purity of molten steel through the measures such as careful selection of raw material, baking of furnace burden at high temperature and purification of molten steel, and exercises strict control over the elements sensitive to irradiation embrittlement, the elements of residual impurities and the content of non-metallic inclusion. The melting is usually performed in two steps. The electric arc furnace is adopted for the primary refining to

make molten steel sufficiently and highly active in the boiling period, so the gas (especially H_2) can escape in the intense boiling period; and the temperature of molten steel is controlled, so that the impurities and gases can be sufficiently transported and discharged with the help of slag to reduce the content of impurities and gases in the molten steel. The secondary refining (external refining) refers to the processes such as desulfuration, deoxidization, inclusion removal, alloying composition adjustment and homogenization for molten steel. There are multiple processes for the external refining, such as vacuum carbon deoxidization (VCD), electro-slag remelting (ESR) and ladle refining (LRF).

The process of pouring molten steel into ingot is usually classified into single ladle pouring, pouring of two ladles or pouring of multiple ladles at the same time. The process of pouring multiple ladles at the same time refers to the smelting of molten steel by adopting multiple electric arc furnaces. Steel ladle is subjected to vacuum degassing at the time of tapping, and then molten steel is discharged for external refining till the next furnace of molten steel; at the beginning of pouring, intermediate casting ladles are set above large vacuum chambers to implement a successive and continuous pouring.

② Forging and Pressing

The basic purpose of forging and pressing is to shape the billet of forgings and reduce or eliminate various defects in steel ingot through thermal deformation and heat treatment in forging and pressing, such as segregation band, slag inclusion, crack and porosity. Enough forging pressure and forging ratio are usually required, so the steel ingot can deform thoroughly and complete plastic deformation under reasonable stress state, to bridge, disperse or weld together most of the defects under high temperature and pressure, and thus to make the composition compact.

③ Heat Treatment

A508-3 steel forgings are subjected to the following heat treatments in the manufacturing phase.

i Dehydrogenation and annealing treatment: to remove the residual hydrogen in the forgings and prevent micro cracks.

ii Preliminary normalizing treatment: to refine and homogenize the grains, to improve the penetrability of ultrasonic flaw detection, and be structurally ready for the subsequent quenching and tempering heat treatment.

iii Quenching and tempering heat treatment: to conduct a rapid quenching and cooling, and then high temperature tempering, thus to obtain excellent comprehensive performance.

The heat treatments are usually implemented in heat treatment furnaces. The furnaces shall be provided with uniform heating function to ensure the uniformity of the heating temperature of forgings. Enough thermocouples shall be arranged at different positions of forgings, to accurately measure and record the actual temperature of various parts of forgings, and to control the uniformity of heating temperature in the heat treatment of forgings, and it shall prevent the deformation of forgings in the heat treatment.

④ Mechanical Properties Test and Non-Destructive Examination.
Mechanical properties test mainly includes tensile tests at room temperature and high temperature, impact test, drop-weight test, and the evaluations of nonmetallic inclusion and grain sizes. As forgings will be subjected to post-welding heat treatment in the subsequent RPV manufacturing process, which has a certain influence on the properties of forgings, the properties test after simulated stress-relieving heat treatment is also required at the mechanical properties test of forgings.
Non-destructive examination mainly includes ultrasonic examination (UT, for inspecting the internal defects of forgings), magnetic particle examination (MT, for inspecting the defects on the surface or near the surface of forgings), liquid penetration examination (PT, for inspecting the defects at the surface of forgings) and visual examination (VT, for inspecting surface quality). Forgings are usually subjected to a non-destructive examination after finish machining as well as heat treatment for mechanical properties.

2.5.4 *Design Analysis and Verification*

During the structural design of RPV, RPV strength calculation and stress analysis are usually implemented according to the requirements stipulated in related standards. Generally, RPV strength calculation is implemented by adopting the maximum principal stress theory, RPV stress analysis is implemented by adopting the maximum shear stress theory and the rationality of RPV structure is evaluated according to the limit values for different working conditions.

RPV is usually subjected to hydrostatic test after manufacturing to verify its overall strength and reliability, and comprehensively examine the manufacturing quality of RPV, the strength and rigidity of various pressurized components, the sealing performance of welded joints and various joint faces.

2.6 Control Rod Drive Mechanism

2.6.1 *Overview*

Control rod drive mechanism (CRDM) refers to the mechanism to drive the reactor control rods. CRDM is the actuator of the reactor control system and the protection system, it drives the control rod assembly in accordance with particular instructions to accomplish reactor startup, reactivity adjustment, maintaining power, shutdown, scram, etc.

CRDM is an indispensable component of the nuclear reactor. The work environment of marine CRDM is quite different from that of onshore reactors. It is necessary to take into account the effects of the tilting and swing of ship hull induced by waves and marine environment (high humidity, condensation, salt mist, mold, etc.) on materials, and the effects of frequent power changes in service on the abrasive resistance and fatigue performance of moving components of CRDM.

2.6.2 A Brief Introduction to Structure

There are different types of CRDM that can be used in pressurized water reactors, including magnetic lifting type CRDM which directly impel the control rods to move up and down, hydraulic and mechanical combined CRDM, rack-wheel type CRDM which converts motor rotation into lifting motion through rack and wheel, and reluctance motor type CRDM which converts motor rotation into lifting motion through roller nut screw drive pair. Usually CRDMs are mounted on the closure head of the pressure vessel.

As the reluctance motor type CRDM and the rack-wheel type CRDM are more suitable to the special working conditions such as tilting and swing of marine reactors, they are widely used on the marine reactors in various countries at present. West Germany had studied the adoption of hydraulic transmission type CRDM on Otto Hahn Nuclear-powered Merchant Ship built in 1970s. This particular type of CRDM cannot satisfy the operating requirements of the marine reactor due to its inherent defects, therefore, there is no report on the application of hydraulic transmission type CRDM in marine reactors all over the world.

Reluctance motor type CRDM is mainly composed of rod position detector, reluctance motor, pressure housing, and transmission shaft. Reluctance motor type CRDM can convert the rotation of the rotor of the reluctance motor into the up-and-down motion of the transmission shaft, and drive the control rod up and down through the roller nut screw drive pair, to accomplish startup and reactivity adjustment; the stator field disappears after switching off the reluctance motor, and then the transmission shaft drives the control rod to quickly insert into the core to realize a safe shutdown.

Rack-wheel type CRDM is mainly composed of DC drive motor, reduction gearbox, universal joint, clutch, position indicator, gearbox, and rack. Rack-wheel type CRDM can convert the circular motion of the drive motor into the linear motion of the rack through the rack-wheel drive pair to drive the control rod which is coupled with the rack to move up and down.

Hydraulic transmission type CRDM is mainly composed of water tank, magnet, piston, and chuck. The static drive mode is adopted and the energy storage equipment is provided for accelerating the rod drop in case of the tilting and swinging of ships. West Germany had developed the prototype and implemented tests of its performance and swing bench. The performance of the prototype was greatly improved compared with the dynamic pressure hydraulic stepping driving mechanism developed by the United States in 1950s.

2.6.3 Materials

In light of the working condition of CRDM which is of high temperature, high pressure, high irradiation, high humidity and salt mist, the materials contacting the reactor coolant shall be compatible with the coolant, and the cobalt content shall be strictly controlled. All the materials contacting the environment of the reactor compartment shall be resistant to salt mist, damp heat and corrosion.

The materials of components at the reactor pressure boundary of CRDM shall possess the toughness, strength, corrosion resistance and weldability required by high pressure vessels. If the component is a part of the magnetic circuit of the drive motor, the materials shall also have modest magnetic conductivity.

The insulating materials of electrical components of CRDM shall be resistant to high temperature, irradiation, hydrolysis and mold, and shall be compatible with each other.

The deposited metal of welds shall match with the strength of base metal. The welding materials shall be resistant to welding crack and intergranular corrosion, and can be welded multiple times. Referring to the materials to be welded, it is necessary to control the α phase area content and C/Ti content ratio to prevent defects such as cracks and intergranular corrosion after welding.

2.6.4 Design Analysis and Verification

(1) Design analysis

The component at the pressure boundary of CRDM is a part of primary circuit pressure boundary. It yields primary stress and secondary stress. It is necessary to conduct stress analysis and assessment, fatigue analysis and anti-fatigue verification test.

Electromagnetic calculation of CRDM shall be implemented by combining the traditional magnetic circuit analysis and finite element analysis. The torque characteristics and operation of CRDM motor are mainly analyzed to study the relationship between motor output torque and specific structure, the relationship between the maximum/average output torque and electric current and the operational magnetic field distribution and operational magnetic circuit saturation. The main factors to be considered include working load, operating speed, temperature, pressure and environmental conditions.

(2) Design verification

It is required to conduct design verifications for key components of CRDM which generally include the verifications of the fatigue life, the hydraulic bursting strength, the pressure fluctuation fatigue, the lifetime, and the anti-irradiation.

The overall performance design verification of CRDM mainly includes the cold hydrostatic test intended to inspect the performance parameters, hot hydrostatic test intended to verify the useful life and conducted at the condition similar to that of real reactor which is of high temperature and high pressure, and the impact test intended to verify the shock resistance of CRDM.

2.7 Reactor Internals

2.7.1 Overview

Reactor internals are installed inside the reactor pressure vessel and are one of the key equipment in the reactor. Reactor internals can provide accurate positioning, reliable support and axial compression for fuel assemblies, reliable guiding for the control rod assembly, reasonable flow path for the coolant, the channel and protection for in-core temperature and flux measurement, and the shielding of the neutron and γ irradiation to reduce irradiation damages to the inner wall of the reactor pressure vessel.

The marine reactor internals operate at the water environment with high temperature and high pressure, therefore, they need to withstand relatively intense neutron and γ irradiation, as well as the load caused by flow-induced vibration and mechanical vibration in multiple states (such as tilting and swinging) of ship hull.

2.7.2 A Brief Introduction to Structure

The marine reactor internals generally include barrel assembly, compression assembly and compression spring assembly (see Fig. 2.29). The flanged suspension structure is generally adopted.

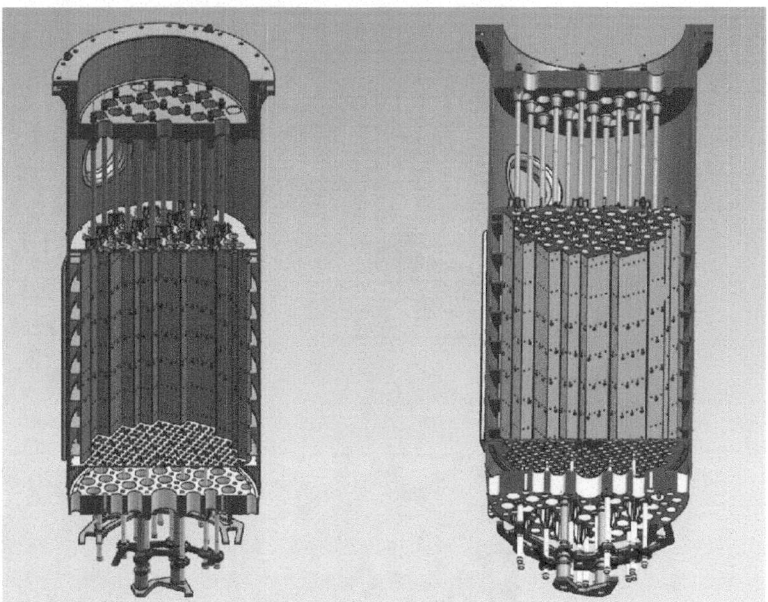

Fig. 2.29 Structure of marine reactor internals

The barrel assembly is generally composed of barrel shell, lower core plate, brattice-forming plate assembly, and outlet nozzle, as shown in Fig. 2.30. The annular cavity between the outer wall of the barrel shell and the inner wall of the pressure vessel cylinder forms the main flow channel for the reactor coolant after entering the pressure vessel; the openings at the lower core plate corresponding to each group of fuel assemblies provide the coolant flow channels for the fuel assemblies; The locating pins are set at the position on the upper surface of the lower core plate. The circumferential and radial positioning of the fuel assemblies are realized by the connection of the locating pins and the corresponding fuel assemblies; the baffe-forming plate assembly constitutes the outline of the core, to form the core flow channel, and shield the core neutron and γ irradiation together with the barrel shell and water layer therein to reduce the irradiation damages to the inner wall of the pressure vessel. The sealing between the outlet nozzle of the barrel shell and the outlet nozzle of the reactor pressure vessel is generally realized through the difference of thermal expansion coefficients of various structural materials.

The compression assembly generally includes the compression barrel, upper core plate, support plate and guiding assembly, as shown in Fig. 2.31. The locating pins are set at the position on the lower surface of upper core plate. The circumferential and radial positioning of the fuel assemblies are realized by the connection of the

Fig. 2.30 Example of structure of barrel assembly

locating pins and the corresponding fuel assemblies; under the effect of preload of the compression spring assembly, upper core plate is used to compress the plate spring of fuel assemblies to realize the axial pressing and position limiting of fuel assemblies; the center-line of the corresponding openings of the support plate and upper core plate can be used as a baseline for installing the guiding assembly of the control rod; the guiding assembly is used to guide the control rod assembly to ensure a smooth lifting and inserting of the control rod assembly. Generally, the compression assembly is equipped with the in-core instrumentation mechanical structure to provide the channel and protection for the measurement system.

After the reactor internals installed in place, the compression spring assembly deforms under the gravity of the pressure vessel head and the pretightening force of the main bolt to generate a relatively large pressing force, and firmly press the compression assembly and the barrel assembly to avoid the axial movement and influence on the safe operation of the reactor.

2.7.3 Materials

The special service conditions, complex welding structure and manufacturing process as well as high safety requirements of reactor internals determine the adoption of materials with excellent comprehensive performance for reactor internals to satisfy the requirements for service conditions, manufacturing process and safety functions. Generally, the main materials of reactor internals shall possess the following capacities.

① The materials shall possess excellent metallurgical quality, that is, the materials shall possess enough purity, density and homogeneity.

② The materials shall possess low sensitivity to neutron irradiation embrittlement and aging embrittlement; the mass fraction of elements with large activation cross-section, long half-life and easy irradiation embrittlement shall be as low as possible.

③ The materials shall possess excellent weldability as well as cold and hot processing performance.

④ The welding materials adopted by the welding structure shall make the strength and plasticity of the deposited metal in compliance with that of the base metal.

⑤ The materials shall possess excellent corrosion resistance.

The experience from domestic and international engineering projects shows that the austenitic stainless steel is generally adopted for the main materials of reactor internals. The matched welding materials generally shall fit the base metal. The chemical composition of welding materials and deposited metals is equivalent to that of the base metal. The mechanical property of weld metal and welding joint fits the base metal.

(1) Development overview

The austenitic stainless steel used for manufacturing reactor internals has a development history of more than 60 years. Due to extremely adverse service

Fig. 2.31 Example of structure of compression assembly

environment for reactor internals and the extremely rigorous requirements for the materials, there is a long way from research and development to engineering application. Materials such as 0Cr18Ni9, 1Cr18Ni9Ti and 1Cr18Ni9Nb are commonly used as the materials of reactor internals at early stage. It becomes difficult to forge this kind of austenitic stainless steel due to the inclusions in the microstructure steel, which degrades the corrosion resistance of the steel. Moreover, the heat treatment is usually required after welding to improve the weld microstructure and the mechanical property, causing a series of engineering problems such as the difficulty manufacture and the complex process ocurr. Recently, in order to improve the strength and the corrosion resistance, the nuclear grade 304 series austenitic stainless steel is more frequently adopted in some countries, such as 304 steel of the United States and Z2CN19-10 steel and Z3CN18-10 steel of France. Compared to the austenitic stainless steel at early stage, this kind of austenitic stainless steel has a greatly improved local corrosion resistance, equivalent plasticity and toughness as well as a processing performance and production process similar to the corresponding steel without nitrogen, which makes it easy for material producing and equipment manufacturing.

With the development of reactors with higher comprehensive performance, new materials of reactor internals that can withstand higher temperature and higher irradiation dose shall be developed.

(2) Material performance and test requirements

Generally, the materials of reactor internals shall be subjected to destructive tests and nondestructive tests for investigating the chemical composition, mechanical properties, corrosion resistance and structure form. The test indexes shall meet the design requirements. The test methods and means specified in corresponding standards are usually adopted. The fatigue resistance and irradiation resistance of materials are usually assessed through the vibration endurance test for reactor internals and the irradiation test for materials.

(3) A brief introduction to material manufacturing

① Smelting process

Limited by the process level, the early ingot of austenitic stainless steel is usually smelted by using the alkaline electric arc furnace refining method and treated by using the vacuum degassing method. With the development of science and technology, the combination of electric furnace, AOD, and electroslag remelting process or other more advanced smelting methods are preferably adopted.

② Heat treatment

The section material of austenitic stainless steel for reacor internals shall be delivered under solution heat treatment condition and a part of section material shall be subjected to acid pickling and passivation. The section materials with different thickness have different thermal insulation time for solution heat treatment.

2.7.4 Design Analysis and Verification

(1) Design analysis

During the design of the reactor internal structure, the strength and stress of components are usually analyzed based on the specifications. During the mechanical analysis of reactor internals, the design conditions and different operation conditions shall be subjected to appropriate load combinations and evaluated by combining with the basic allowable stress intensity. Generally, the calculation of the structural strength of reactor internals shall be assessed by using overall primary membrane stress intensity or overall primary membrane stress intensity plus primary bending stress intensity.

(2) Design verification

After designing the reactor internals, the related design verification (which is usually test verification) shall be implemented. The test verification mainly includes flow-induced vibration test for reactor internals, thermo-hydraulic test for the reactor, buffer performance test for the buffer structure implemented and guiding performance test for guiding assembly by combining the relevant tests on control rod drive lines.

① Flow-induced vibration test for reactor internals

The purpose of flow-induced vibration test for reactor internals is to understand the flow-induced vibration behavior of reactor internals, confirm that the vibration caused by liquid under typical operation conditions is similar to the predicted vibration and does not cause an unexpected violent vibration or the vibration damaging the structure, ensure the structural integrity of reactor internals of test reactor, finally provide a basis for safety review. The tests usually include vibration characteristic test in air and still water, dynamic water response test, vibration endurance test, theoretical analysis and calculation of flow-induced vibration response of reactor internals in dynamic water.

② Reactor thermo-hydraulic test
As for new reactor internals, the thermo-hydraulic test for the reactor must be implemented in order to provide the necessary test basis for thermo-hydraulic design as well as structural design of reactor internals. Reasonable thermal channel, bypass flow distribution and leakage flow limit shall be confirmed through the tests.

③ Relevant tests of control rod drive lines
When the reactor internals are required to provide buffering for the moving parts of control rod drive lines at the end of rod drop, the reactor internals shall be designed with buffering structures. After determining the reasonable form of buffering structure based on the input conditions, in consideration of the buffering type selection test for control rod drive lines, the performance parameters such as the rod drop time of moving parts of control rod drive lines, impact force and the rebound height of the control rods for different buffering structure should be measured. Through the evaluation of the test results, the optimal buffering structure shall be chosen and adopted for subsequent cold state and hot state verification tests of control rod drive lines.

2.8 Reactor Support and Shielding

2.8.1 Overview

Reactor support is the bearing structure of the reactor and mainly has the following functions: to support the weight of the reactor body and bear the specified external load; to bear the load of the reactor under various operation conditions and pass it onto the ship hull.

During the operation, the reactor will generate plenty of energy through atomic nucleus fission and also will generate plenty of fission products radiating around the reactor, such as fast neutron, thermal neutron and γ ray, which resulting in different degrees of influence on equipment and human body; therefore, the space around reactor shall be set with shields to reduce the irradiation and protect equipment and staffs in the reactor compartment. The shielding include the reactor shielding and biological shielding depending on their locations.

The reactor shielding refers to the shielding around the reactor, which mainly has the following functions:

To stop and weaken various irradiation from the reactor and its components ensuring the irradiation in the reactor compartment below the specified limit and protecting the equipment in the reactor compartment from intense irradiation; during the reactor shutdown, to shield and keep the delayed and induced irradiation in allowable limit, so that staffs can enter the reactor compartment at the specified time after reactor shutdown to inspect and repair the equipment around reactor.

The biological shielding refers to the shielding arranged at the outside of the reactor compartment, which is the last barrier to protect human body and equipment.

2.8.2 A Brief Introduction to Structure

Reactor support and primary shielding structures are mainly composed of reactor support, lateral shielding, top shielding and bottom shielding structures.

Reactor support is a very important force-bearing component in the reactor structure, which not only fastens the reactor body and bear the load under the specified impact, but also required to ensure that no irradiation fracture will occur under irradiation within its lifetime. Reactor primary shielding embraces the reactor, which is used to shield the radiation dose from reactor to protect human and equipment from the damage by irradiation. Reactor support and shielding structures (lateral shielding) are shown in Fig. 2.32.

Top shielding refers to the shielding structure to shield the upward irradiation from the reactor core; lateral shielding refers to the shielding structure embracing the side of the active region of the reactor core, which is mainly composed of water tank (refers to the water tank structure consisted of primary shielding structural components, which is responsible for absorbing and stopping the neutron) and lateral shielding structure.

2.8.3 Reactor Shielding Design

The design objective of radiation shielding for the reactor compartment is to design the technically feasible, small sized and light weighted shield under the premise of ensuring the safety of staffs and equipment.

In order to achieve the design objectives above, the following measures can be adopted in the shield design:

① Treat the reactor shielding and the ship reactor compartment shielding as a whole and provide appropriate configuration.
② Reasonably allocate the thickness of shield.
③ Adopt different design dose limits based on the duration of the staffs staying at different positions outside the reactor compartment.
④ Take a full consideration of the shielding effects of the reactor body as well as the pipes and equipment at the primary loop.

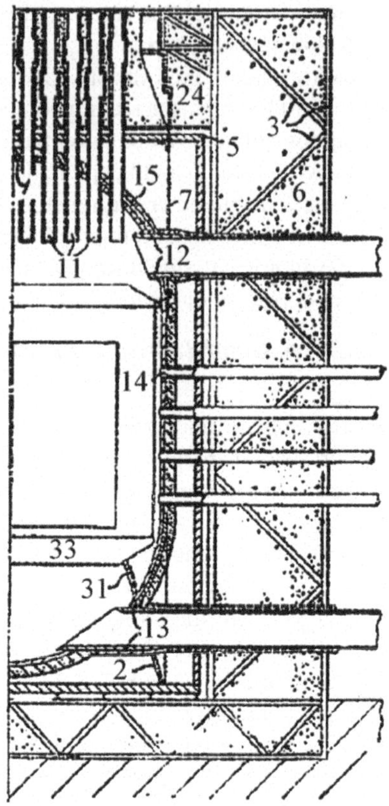

Fig. 2.32 Reactor support and shielding structures (Lateral Shielding)

2.8.4 *Shielding Materials*

(1) Effect of shielding materials

① Effect on neutrons.
Within a certain shielding thickness range and a certain energy range, the attenuation of fast neutrons in materials can be approximately described by using the following equation according to the segmented exponential weakening rule:

$$H = H_0 e^{-T/\lambda_{eff}} \tag{2.102}$$

where, H refers to the reading of instrument with shielding materials; H_0 refers to the reading of instrument without shielding materials; T refers to the thickness of shielding materials (cm); λ_{eff} refers to the relaxation length (cm) of fast neutrons in shielding materials (this parameter is related to neutron energy as well as characteristics and thickness of shielding materials, and is a constant within a certain shielding thickness and a certain energy range).

When fast neutrons pass through the shielding layer, they will be scattered or captured by the atomic nucleus of various material elements in shielding materials, and the degree of scattering or capturing depends on neutron cross sections of various elements. The neutron cross section of most of the elements is a function of neutron energy. When fast neutrons pass through shielding materials, the following effects will occur:

i The elastic and inelastic scattering with heavy element (such as iron and lead) can reduce the energy. In most cases, the neutrons with the energy less than 1 MeV mainly belong to elastic scattering. When the energy exceeds 1 MeV, the probability for the inelastic scattering will increase with the increasing of the neutron energy. The neutrons with the energy more than 2 MeV are moderated to 1 MeV or even lower by inelastic scatting.

ii The collision with hydrogen nuclei can reduce the neutron energy and the total cross section of hydrogen will increase rapidly with the declining of neutron energy, therefore, in the low energy range, the collision of hydrogen nuclei with neutrons is an important process for neutrons to be moderated to thermal neutrons.

iii Finally, the neutrons are lost due to the capturing of neutrons by hydrogen nuclei or other elements with a large thermal neutron capture cross-section.

Simply speaking, the interaction process between fast neutrons and materials is as follows: the inelastic collision and elastic collision with the nuclei of heavy elements in the high energy range will reduce the energy, while in the low energy range, the neutrons mainly collide with hydrogen nuclei and are moderated to thermal neutrons, and finally absorbed by the elements with a large thermal neutron capture cross section.

② Effect on γ ray

The degree of attenuation and absorption of γ photon during the interaction with substance can be expressed by the linear attenuation coefficient.

Under the narrow-beam geometric conditions, the attenuation of γ ray collimated beam after passing through materials shall be in accordance with the exponential attenuation law, i.e.

$$H = H_0 e^{-\mu T} \tag{2.103}$$

where, H refers to the reading of instrument with materials; H_0 refers to the reading of instrument without materials; T refers to the thickness of materials (cm); μ refers to the linear attenuation coefficient of materials (cm^{-1}).

(2) Common shielding elements

① Water

Water is abundant and favorable in price, containing rich hydrogen. Therefore, it is an extremely effective neutron moderating material. Water does not generate micro clearance, has stable chemical properties and is an effective coolant. However, water has some disadvantages; for example, complete waterproof measures are required; in addition, the turbidity will occur in case of long term sit. Problems such as the container, temperature control, corrosion and purification (softening) shall be considered when water is used as the shielding material.

② Graphite

Graphite is a very good neutron moderating and reflecting material, which has high temperature resistance characteristics and has been widely used in reactor design. Graphite can also be used as thermal shielding or primary shielding besides the reflector and moderator.

③ Boron

Boron is a valuable element, with the thermal neutron capture cross section up to 750 barns (b). The radiation energy generated after capturing thermal neutrons is low and can be shielded easily. Its half-life is short and the post treatment after decommissioning is convenient, therefore, the boron-containing material is an excellent thermal neutron radiation shielding material for reactors. In the design and research of high boron steel, the hardest problem encountered is the poor ductility and toughness of material, i.e. the occurrence of so-called "boron embrittlement" phenomenon. There are many factors for this phenomenon, which not only relate to the high boron content in the steel, but also the existing forms of boron in steel (solid dissolved boron, oxides and nitrides of boron) and the mutual transformation of the forms. Especially this phenomenon is closely related to the amount, the shape, the distribution variation, etc. of the boron phase.

④ Hydrogen

Hydrogen is the most harmful element resulting in unallowable cracking or pitting in the steel, and also the deadly element that causes welding cold cracks, therefore, the hydrogen in steel shall be removed through vacuum degassing or hydrogen diffusion heat treatment. However, the shielding material containing hydrogen can be used as an effective moderator.

⑤ Lead

Lead is the main γ radiation shielding material. As the lead has excellent photoelectric absorption property and relatively large cross section for electron pair effect, it can provide effective shielding of both low energy γ radiation and high energy γ radiation. As the lead is with high density, a relatively small volume for shielding material can be obtained when it is used as the shielding material, and is applicable to nuclear powered ships. In addition, lead is a relatively ideal γ radiation shielding material, as it is easy to be processed, and with small amount of impurities, stable property under irradiation and the capability to resist damage. However, lead also has disadvantages. When the temperature exceeds 65 °C, its structural strength will obviously degrade. If the temperature exceeds 260 °C, obvious creep will occur. Therefore, the application temperature shall be limited when lead is used as the shielding material. When lead and steel are immersed in water at the same time, electrochemical corrosion reaction is likely to occur, therefore, lead shall be arranged at a confined space without water. During the manufacturing of lead, the internal defects shall be strictly controlled and the density shall be guaranteed.

⑥ Concrete

Concrete is a mixture of various materials, with optimal structure and shielding property through proper preparation and formation. Concrete is usually classified into common concrete and heavy concrete. The density of common (silicate) concrete is 2.2–2.4 g/cm^3, which depends on the aggregate adopted (granite, sandstone and limestone). This kind of concrete contains plenty of oxygen, silicon or calcium. The density of heavy concrete is 3–6 g/cm^3. The aggregates for heavy concrete are usually iron or iron ore, barite (barium sulfate), steel particles, steel impact blocks, reinforcement cut into pieces or other metallic packing. Concrete is usually used as the shielding material for stationary reactor systems.

⑦ Tungsten

Tungsten is a kind of refractory metal with a high melting point, which can be used to increase the strength, hardness and brittle transition temperature of steel. The density is up to 19.3 g/cm^3. Tungsten is added into shielding materials through powder metallurgy or smelting methods to increase the density of shielding materials and improve the shielding effect.

⑧ Polyethylene

As the hydrogen content in the polyethylene is relatively high, polyethylene is a very good neutron shielding material. The polyethylene is non-toxic, odorless, reluctant to react with acid and base, and easy to be processed, and it is with a certain mechanical strength and applicable as the shielding with a relatively complex local shape. However, when polyethylene is adopted, the following problems shall be considered in structural design: the coefficient of linear-expansion of polyethylene is large (which is almost 12 times of that of iron) and the softening temperature is low.

⑨ Composite shielding material.

The composite shielding material manufactured with multiple shielding components is usually adopted as the reactor shielding material. The composite shielding material can comprehensively shield the neutron and γ ray, and achieve a relatively good shielding effect within a limited space.

(3) Shielding material selection principle

The selection principles for shielding materials are as follows:

① The materials with high hydrogen content shall be selected in preference as the materials for shielding fast neutrons, the materials with a large density for shielding γ ray, and the materials containing boron for shielding thermal neutrons.

② The shielding materials shall be with excellent shielding property, stable chemical property, small expansion rate and shrinkage rate, non-toxic, without special odor, and reluctant to decompose and evaporate, and shall not be a comburent. Meanwhile, the shielding materials shall possess excellent irradiation stability and the secondary effect resulted from irradiation shall be as low as possible.

③ The shielding materials shall possess relatively good corrosion resistance.

2.8.5 Reactor Support Materials

(1) Material selection principles

The selection principles for the materials used as load-bearing structure are as follows:

① The materials shall possess excellent metallurgical quality, enough purity, compactness and uniformity.

② The materials shall possess excellent mechanical properties, welding properties, and cold & hot processing properties.
The materials shall possess relatively good corrosion resistance.

③ In addition, the fast-neutron irradiation test data for this material shall be available, and no brittle failure occurs in the material after irradiated by the fast neutron fluence within the design lifetime of the equipment.

(2) Irradiation effect

As an important structural component for the reactor, the reactor support will be irradiated by fast neutrons within its lifetime, and the irradiation embrittlement effect will occur in the material (the origin for irradiation embrittlement is that the fast neutrons "bombardment" cause the displacement of the atoms in the metallic crystal structure, resulting in the defects such as dislocation and vacancy). The degree of irradiation embrittlement is evaluated by the variable (increasing) ΔRT_{NDT} of reference nil-ductility transition temperature RT_{NDT} as well as the decreased value ΔE of the upper shelf energy (USE) of the material. The material shall possess low sensitivity to fast neutron irradiation embrittlement (namely low irradiation embrittlement effect) to maintain the stable property of materials within the whole lifetime and thus to mitigate crack propagation and to decrease brittle fracture probability. The factors affecting irradiation embrittlement effect mainly include material composition (including purity), steel type, irradiation temperature, neutron fluence, neutron energy spectrum, and material state. The sensitivity to irradiation embrittlement can be obtained through corresponding tests.

(3) Reactor support welding

① Welding materials.

The basic requirements for the welding materials are as follows:

i The welding materials have the mechanical properties and welding properties compatible with the base metal; ii The welding materials used for the reactor support shall satisfy the irradiation requirements for materials and have the high temperature characteristics that are basically compatible with the material of the pressure vessel. iii The content of harmful elements such as sulfur, phosphorus and copper in the welding materials shall be strictly controlled, and the technical specifications of relevant welding materials shall be satisfied.

② Factors affecting welding.

Weldability is a kind of process property of metallic materials. This property will be affected by the property of materials themselves as well as the process conditions, the structural conditions and the application conditions.

i Material factors.

Materials include base metal and welding materials. Under the same welding conditions, the main factors determining the weldability of base metal are its physical and chemical properties. As for steel, these factors mainly include chemical composition of steel, smelting and rolling states, heat treatment conditions, structural state, and mechanical property. The chemical composition (including the distribution of impurities) is the main influencing factor, and can determine the hardening tendency and embrittlement at heat affected zone as well as the sensitivity to cracking. Meanwhile, in the welding process, the chemical composition of welds also will be affected due to the participation of the base metal in the metallurgic reaction in the molten pool. The elements that have a relatively large effect on welding mainly include carbon, sulfur, phosphorus, oxygen, hydrogen, and nitrogen. They may result in the defects in welding process and degrade the welding joints. In addition, the factors such as smelting and rolling states, heat treatment conditions, and structural states of steel will also affect welding. The correct selection of welding materials is also an important condition for obtaining high quality welding joints.

Weldability is the adaptability of the materials to welding and mainly refers to the degree of difficulty and easiness in obtaining high quality welding joints under certain process conditions. Specifically, the weldability is represented by the sensitivity to welding defects and the adaptability the joint to operating requirements (i.e. the property).

The weldability of steel can be assessed by using the sensitivity to cracking and the hardening tendency at heat affected zone. There are many factors that affect the weldability. From the aspect of materials, the weldability mainly depends on the chemical composition and impurity content (i.e. purity) of materials. The carbon content is the most important factor that affects the weldability. International Institute of Welding (IIW) recommends the following carbon content be used to assess the hardening tendency at heat affected zone of low-alloy high-strength steel:

$$C_E(IIW) = C + \frac{\mathrm{Mn}}{6} + \frac{\mathrm{Cu+Ni}}{15} + \frac{\mathrm{Cr+Mo}+V}{5}\ (\%) \tag{2.104}$$

where, the symbol of each element respectively represents its content. When $C_E \le 0.4\%$, the materials basically have no hardening tendency and have excellent weldability, therefore, the measures such as the preheating before welding or the strict control of weld heat input are not required.

When $C_E = 0.4–0.6\%$, the hardening tendency becomes obvious. The steel with low hardening tendency shall be properly preheated, and the welding heat input shall be controlled. The process measures shall be adopted based on the type and thickness of joints to avoid cold cracks.

The formula of carbon equivalent created by Nippon Steel Corporation is the one that is widely used currently and has a relatively high precision:

$$CEN = C + A(c)\left(\frac{\mathrm{Si}}{24} + \frac{\mathrm{Mn}}{16} + \frac{\mathrm{Cu}}{15} + \frac{\mathrm{Ni}}{20} + \frac{\mathrm{Cr+Mo}+V+\mathrm{Nb}}{5} + 5B\right)\ (\%) \tag{2.105}$$

The formula is applicable when C = 0.034–0.254%, in which A(c) is the accommodation coefficient of carbon. For the relationship between A(c) and carbon content, please refer to *Welding Data Manual* and other relevant data.

The formula specified by JIS:

$$Ceq = C + \frac{Mn}{6} + \frac{\mathrm{Si}}{24} + \frac{\mathrm{Ni}}{40} + \frac{\mathrm{Cr}}{5} + \frac{\mathrm{Mo}}{4} + \frac{V}{14}(\%) \tag{2.106}$$

As for the quenched and tempered steel with low carbon, low alloy, high strength and relatively high yield strength, the following formula can be adopted:

$$Ceq = C + \frac{\mathrm{Mn}}{9} + \frac{\mathrm{Ni}}{40} + \frac{\mathrm{Cr}}{20} + \frac{\mathrm{Mo}}{8} + \frac{V}{10}\ (\%) \tag{2.107}$$

The following formula is used to speculate the weldability through chemical composition.

The exponential experimental formula for the root crack:

$$Pc = Pcm + \frac{[H]}{60} + \frac{\delta}{600} \tag{2.108}$$

where, [H] refers to the content of diffused hydrogen (ml/100 g); δ refers to the thickness of test plate (mm).

The exponential formula for the sensitivity to cold crack:

$$Pcm = C + \frac{\text{Si}}{30} + \frac{\text{Mn} + \text{Cu} + \text{Cr}}{20} + \frac{\text{Ni}}{60} + \frac{\text{Mo}}{15} + \frac{V}{10} + 5B\ (\%) \tag{2.109}$$

The experimental formula for the predicted reheating crack:

$$\Delta G = \text{Cr} + 3.3\text{Mo} + 8.1V - 2\ (\%) \tag{2.110}$$

When $\Delta G > 0$, the reheating crack will occur.

$$P_{SR} = \text{Cr} + \text{Cu} + 2\text{Mo} + 5\text{Ti} + 7\text{Nb} + 10V - 2\ (\%) \tag{2.111}$$

When $P_{SR} > 0$, the reheating crack will occur.

It can be seen that the strict control of the alloying elements, especially decreasing the content of the harmful residual impurity elements to reduce the sensitivity to cracks is the key factor to improve the weldability of materials. In addition, the control of thermal parameters (heat input, preheating temperature, interlayer temperature, heating and cooling speed, post-heating, etc.) shall be attached great importance so as to mitigate the constraint stress of the joint structure and eradicate the source of hydrogen, thus to guarantee the quality of the joint.

For load-bearing welds, the welding materials shall have enough toughness and the strength equal to or slightly higher than that of the base metal. However, the match of welding materials with the base metal depends on the strength of the base steel. The overmatch (i.e. the strength of welding materials is slightly higher than that of the base metal) is generally adopted for medium and low strength steels or medium and high strength steels, while the under match is adopted for ultra-high strength steel (i.e. the strength of welding materials is slightly lower than that of the base metal). In addition, the effect of high neutron fluence on the sensitivity to material irradiation shall also be considered. For the load-bearing welds, the toughness shall be attached greater importance, as the strength of weld metal is too high and generally deteriorates its crack resistance, ductility and toughness. For the weld of low alloy steel, the content ratio of the manganese and silicon in the weld metal is the main factor for determining its toughness. When the manganese and silicon content ratio is greater than 2, the toughness can be easily guaranteed.

ii Process factors.

Process factors mainly include welding methods, welding parameters, preheating, post-heating and post-weld heat treatment. Welding methods and welding parameters determine the range of welds and heat affected zones, structural changes, and potential for defects. Preheating, slow cooling, post-heating and post-weld heat treatment are adopted to reduce the hardening tendency and the sensitivity to cold cracks. The selection of reasonable welding sequence can improve the constraint and stress of structures.

iii Structural factors.

Structural factors mainly refer to the effects of design forms of welding structures and welding joints (such as structural shape, dimension, thickness, joint groove type, weld arrangement and its cross-section shape) on welding. Different plate thickness, joint types or groove types will cause different heat transfer directions and speeds, and thus affect the crystallization direction and grain growth in the molten pool. Structural shape, plate thickness, and weld arrangement determine the stiffness and restraint intensity of joints, and thus affect the stress state of joints. Severe stress concentration and excessive welding stress are the basic conditions for the occurrence of weld cracks. The important measures for improving the weldability in the design process include the measures to reducing the joint stiffness and the cross welds, avoiding dense welds and reducing various factors causing stress concentration.

iv Service conditions.

Service conditions refer to the operating temperature, loading types (such as static load, dynamic load, and alternating load) and operating environment of welding structures. For example, creep may occur when working under high temperature; brittle failure may occur when work under low temperature or impact load; the possibility of corrosion damages to welding joint shall be considered when working in corrosive medium.

2.8.6 Design Analysis

(1) Design principles

① Reactor support and shielding water tanks shall have the same lifetime with that of the reactor pressure vessel.

② Shielding materials shall be able to adapt to environmental characteristics, and the material failure shall not occur due to creep and other reasons. Structural materials and shielding materials shall be able to withstand the failure and degradation of structural materials resulted from irradiation embrittlement. The design temperature of the reactor shielding water tanks shall be determined based on the heat-resistance temperature in which shielding materials work for a long time and the temperature limited by specific environment.

③ Composite shielding materials are selected in preference to reduce the shielding volume and weight. All the structural and welding materials used for manufacturing shields must conform to the requirements of relevant technical specifications, and possess excellent shielding performance and stable chemical properties and non-toxicity.

④ Under the premise of satisfying strength requirements, the materials with relatively good ductility and toughness shall be selected for the load-bearing reactor support, and the chemical elements that may cause irradiation embrittlement shall be strictly controlled.

(2) Mechanical analysis

The analysis of equipment load shall usually include:

① internal pressure or external pressure; ② the dead weight of the shield, including the weight of the medium contained; ③ the equipment weight supported by the shield, the weight of the medium contained in the equipment and the load caused by liquid flow; ④ the thermal expansion load of structures; ⑤the load generated by the vibration of the system; ⑥ environmental load; ⑦ all the external load transferred to the shield.

The stress analysis of equipment and support shall usually include:

① mechanical load stress analysis; ② thermal stress analysis; ③ fatigue analysis; ④ thermal ratchet analysis. The results of stress analysis shall satisfy the requirements for the limits under different operating conditions.

References

1. She Zhongsheng, and Zhang Shaohong. 2000. *Physical Theories and Calculation Methods for Nuclear Reactor*. Xi'an: Xi'an Jiaotong University Press.
2. Stacey, W.M. 2001. *Nuclear Reactor Physics*. New York: Wiley.
3. James, J., and L.J.H. Duclerstaclt. 1976. *Nuclear Reactor Analysis*. New York: Wiley.
4. Cacuci, D.G. 2010. *Handbook of Nuclear Engineering*. Springer.
5. Briesmeister, J.F. 1997. *MCNP—A General Monte Carlo N-Particle Transport Cocle, LA-12625—M*. Los Alamos National Laboratory.
6. Petrie, L.M., and N.F. Landers. 2000. *Keno: An Improved Monte Carlo Criticality Program with Supergrouping, NUREG/CR-0200 Revision 6 Volume 2, Section F11 (ORNL/NUREG/CSD-2/R6)*.
7. She Ding, Wang Kan, and Yu Ganglin. 2012. Development of Burnup Calculation Function in Reactor Monte Carlo Code. *Nuclear Power Engineering* 33 (3): 1–5.
8. Kelly, D.J., T.M. Sutton, and S.C. Wilson. 2012. MC21 Analysis of the Nuclear Energy Agency Monte Carlo Performance Benchmark Problem. In *Proceedings of ANS Topical Meeting on Reactor Physics (PHYSOR 2012)*.
9. Gang, Li, Zhang Baoyin, Li Rui, et al. 2013. Design and Implementation of Pointwise Continuous Energy Cross Section Physics Module of JMCT Code. *Atomic Energy Science and Technology* 47 (supplement): 453–457.
10. Romano, P.K., and B. Forget. 2013. The OpenMC Monte Carlo Particle Transport Code. *Annals of Nuclear Energy* 51: 274–281.
11. Qiang Shenglong, Qin Dong, Chai Xiaomin, et al. 2014. Research on Calculation Error of Fuel Consumption Based on Monte Carlo Method. *Atomic Energy Science and Technology* 48 (12).
12. Jones, R.D. 1990. *Model Development for Plant Vogtle Utilizing ALPHA/PHOENIX—P/ANC Methodology*. Washington, DC: Amercian Nuclear Society (CANS) Writer Meeting.
13. SCIENCE V2 Nuclear Code PackagQualification Report. 2004. Rev A, NFPSD/DC/89. FRAMATOME ANP.
14. Joo, H.G., J.Y. Cho, and K.S. Kim. Methods and Performance of a Three-dimensional Whole-Core Transport Code DeCART. *PHYSOR 2004*, Chicago, IL, USA. American Nuclear Society.
15. Jung, Y.S., C.B. Shim, and C.H. Lim. 2013. Practical Numerical Reactor Employing Direct Whole Core Neutron Transport and Subchannel Thermal/Hydraulic Solvers. *Annals of Nuclear Energy* 62: 357–374.

16. Kochunas, B., B. Collins, and D. Jabaay. 2013. Overview of Development and Design of MPACT: Michiganparallel Characteristics Transport Code. In *International Conference on Mathematics and Computational Methods Applied to Nuclear Science Engineering(M&-C 2013)*, Sun Valley, ID, USA. American Nuclear Society.
17. Wenbin, Wu, Li Qing, and Wang Kan. 2014. Tiger-3D: 2D/1D Coupled Whole-Core Transport Code Based on Large-Scale Parallel Computation. *Nuclear Power Engineering* 35 (S2): 135–139.
18. Yu Pingan, Zhu Ruian, Yu Zhenwan, et al. 2002. *Nuclear Reactor Thermal Analysis (Revision 3)*. Shanghai: Shanghai Jiao Tong University Press.
19. Jiyang, Yu., and Jia Baoshan. 2003. *Nuclear Reactor Thermohydraulics*. Beijing: Tsinghua University Press.
20. Guanghui, Su, Qiu Suizheng, Tian Wenxi, et al. 2013. *Nuclear Power System Thermo-Hydraulic Calculation Methods*. Beijing: Tsinghua University Press.
21. Jijun, Xu, and Jia Dounan. 2001. *Boiling Heat Transfer and Vapor-Liquid Two-Phase Flow*. Beijing: Atomic Energy Press.
22. Shiming, Yang. 1987. *Heat Transfer*. Beijing: Higher Education Press.
23. Tang Liangsun, Weisman J. Pressurized Water Reactor Thermal Analysis [M]. Beijing: Atomic Energy Press, 1983.
24. СамойловАГ,КаштановАИ,ВолковВС.Дисперсионныетвэлы:В2 хт.Т.1.Материалыитехнология.М.:Энергоиздат, 1982.
25. Frost BRT. 1997. *Nuclear Materials (Part I), Materials Science and Technology Series*, vol. 10A. Translated by Zhou Bangxin. Beijing: Science Press.
26. Masayoshi, Hasegawa, and Mishima Yoshitsugu. 1987. *Nuclear Reactor Material Handbook*. Translated by Sun Shouren. Beijing: Atomic Energy Press.
27. Lucuta, P.G., H.J. Matzke, and I.J. Hastings. 1996. A pragmatic approach to modelling thermal conductivity of irradiated UO_2 fuel: review and recommendations. *Nuclear Materials* 232: 166–180.
28. Hales, J.D., S.R. Novascone, and G. Pastore et al. 2013. BISON Theory Manual The Equations Behind Nuclear Fuel Analysis. INL/EXT-13–29930, Idaho National Laboratory.
29. Ross, A.M., and R.L. Stoute. 1962. Heat Transfer Coefficient Between UO_2 ancl Zircaloy-2. Technical Report AECL-1552, Atomic Energy of Canada Limited.
30. Scdap/Relap5—3D. 2003. Code Manual Volume 4: Matpro A Library of Materials Properties for Light-Water-Reactor Acciclent Analysis. Idaho INEEL/EXT—02–00589, Volume 4, Revision 2. 2, National Engineering and Environmental Laboratory.
31. Bernard, L.C., and J.L. Jacoud. 2002. An Efficient Model for the Analysis of Fission Gas Release. *Journal of Nuclear Materials* 302: 125–134.
32. Rashid, Y., R. Dunham, and R. Montgomery. 2004. Fuel Analysis and Licensing Code: FALCON MOD01. Technical Report EPRI 1011308, Electric Power Research Institute, Dec 2004.
33. Matsuo, Y. 1987. Thermal Creep of Zircaloy-4 Cladding Under Internal Pressure. *Journal of Nuclear Science and Technology* 24 (2): 111–119.
34. Hoppe N.E. 1991. Engineering Model for Zircaloy Creep and Growth. In *Proceedings of the ANSENS International Topical Meeting on LWR Fuel Performance*, Avignon, France, 21–24 Apr 1991, 157–172.
35. Stehle, H., W. Kaden, and R. Hanzel. 1975. External Corrosion of Cladding in PWRs. *Nuclear Engineering and Design* 33: 155–169.
36. Dimitriev Sergey. *Main Equipment for Nuclear Power Plant with Thermal Neutron Shell Reactor*, 201. Chengdu: Sichuan University Press.

Chapter 3
Reactor Coolant System (RCS)

3.1 Overview

3.1.1 Functions

The reactor coolant system (RCSs) can:

① Transfer the heat from the reactor core to the secondary feedwater via steam generators (SG), in which steam is generated;
② Provide the coolant as the neutron moderator and the reflector within the reactor;
③ Remove the residual heat from the core and the sensible heat from the equipment during normal shutdown or scram of the reactor;
④ Provide precondition for core cooling in case of loss of coolant accident (LOCA);
⑤ Form a barrier for preventing radioactive substances from releasing to the environment of the compartments, as part of the reactor coolant pressure boundary.

3.1.2 System Composition

In general, the RCS consists of two or more loops, and each loop consists of SG(s), reactor coolant pumps (RCPs), and main pipes connecting the equipment mentioned above. For a marine nuclear power plant, the form of pipes connecting the equipment depends on the arrangement of the RCS (as detailed in Sect. 3.3) due to limited installation space. When the separated arrangement is adopted, large-diameter main pipes are often used to connect all the equipment. When compact arrangement is adopted, short pipes or the parts with special structure are used for firm connection. When integrated arrangement is adopted, the main pipes are cancelled.

J. Yu, *Marine Nuclear Power Technology*,
https://doi.org/10.1007/978-981-15-2894-1_3

3.1.3 System Process

The RCSs for all kinds of marine nuclear power plants are basically the same, but with distinctive features.

Figure 3.1 shows a typical RCS for marine reactors. Figure 3.2 shows the separated arrangement of primary circuit system. It has two loops, and consists of one reactor pressure vessel, two SGs, two RCPs and other items.

Reactor coolant is supplied by reactor coolant pumps to reactor, where the coolant is heated by fuel elements and is carried to the primary-side of the heat transfer tubes of SGs by forced circulation. The heated coolant transfers its heat to the secondary water and then returns to the RCPs. The secondary water in SGs is heated by coolant in heat transfer tubes and circulates in a natural way and becomes steam.

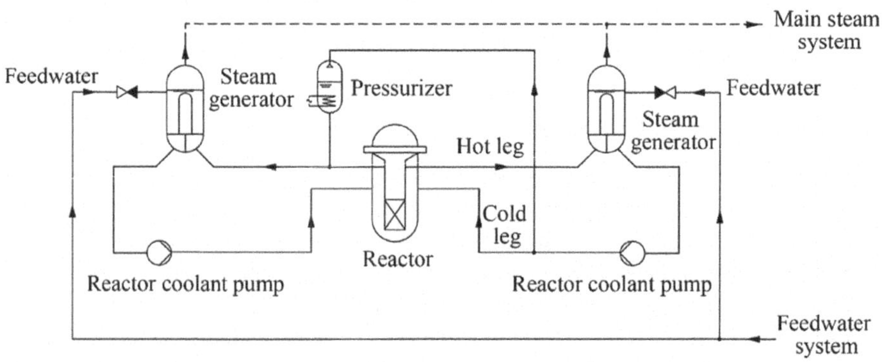

Fig. 3.1 Schematic diagram of reactor coolant system

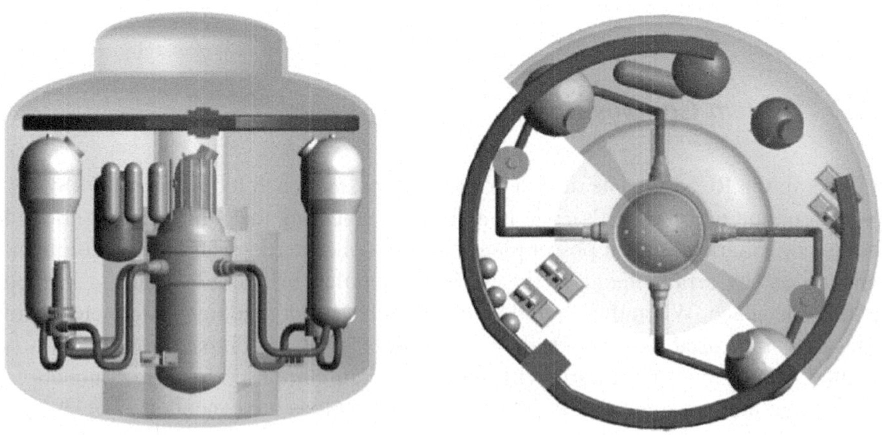

Fig. 3.2 Primary circuit system inside containment of reactor with separated arrangement

Fig. 3.3 Reactor coolant system with a compact arrangement

For the RCS with compact arrangement as shown in Fig. 3.3, the coolant flow of each loop is generally as follows:

After the heat transfer in the steam generator, the reactor coolant is driven by the main pump installed on the steam generator to the bottom of the SG, and then enters the downcomer annulus through the outer annulus, which is connected to the SG and the pressure vessel. It is mixed in the lower plenum, flows upwards, and completes the heat transfer in the core. After that, it flows to the inner annulus, which is connected to the SG and the pressure vessel, and enters the SG for heat transfering. Thus, the coolant circulation is completed.

For the integrated PWR as shown in Fig. 3.4, the reactor coolant is subject to closed circulation in the pressure vessel and the main pumps. After being heated in the reactor core, the reactor coolant enters the SG inlet annulus via the top of the barrel, and flows to the primary side of the SG heat transfer tube bundles, where it transfers its heat to the secondary feedwater, with the secondary water becoming slightly superheated steam for the secondary loop. The reactor coolant cooled by the secondary feedwater enters the lower annulus and the lower head of the pressure vessel via the main pumps, and then flows upwards through the core, where it is heated again. It then leaves the core, completing the constant closed circulation.

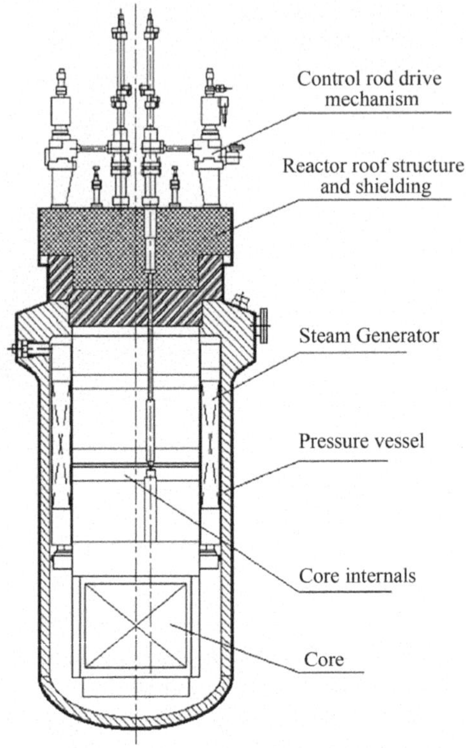

Fig. 3.4 A kind of integrated pressurized water reactor (PWR)

3.2 Design Requirements

As the main system of the primary loop, RCS determines the form, the weight and the overall size of the nuclear steam generator, and to a large extent, it also determines the reliability, the safety and the technical advancement of the plant. Therefore, besides meeting the functional requirements of the reactor and the nuclear power plant, the RCS shall be designed in such a way that the main equipment should be rationally configured, the structural type and performance parameters of the main equipment should be selected rationally, and the overall arrangement of the RCS should be reasonably carried out based on the current technical advancement and industrial manufacturing capability.

To ensure the system functions, the main performance parameters of the RCS shall meet the requirements of the technical indicators of the nuclear power plant and the overall performance of the ship, and the RCS shall also be designed in compliance with the applicable standards and regulations. The equipment contained in the RCS shall be designed appropriately according to its safety-related functions, classifications, and the design criteria of the primary loop and its equipment so as to meet the requirements of the integrity of RCS boundary and ensure the reliability of the equipment and system.

The general design requirements of the RCS are as follows:

① The design of the RCS shall be based on the rated thermal power transferred from the reactor to the secondary loop, with its design parameters, operational characteristics and operation modes meeting the requirements of the technical indicators and the overall performance of the nuclear power plant.
② During normal operation and general accident conditions, the RCS shall meet the requirements of core cooling, and ensure that the relevant parameters of the fuel elements and reactor coolant pressure boundary do not exceed the design limits. In case of severe accident conditions, the RCS shall provide core cooling to ensure that any damage to the fuel elements does not exceed the design limits. In case of extreme accident conditions, the RCS shall provide core cooling to ensure that the amount of radioactive substances released to the environment is within the defined limit.
③ The RCS shall have certain capacity of natural circulation, so that the natural circulation of the coolant can remove a certain amount of heat from the core in case the main pumps stop. To obtain a certain capacity of natural circulation, in the limited compartment the position of the steam generator shall be as much higher than that of the reactor pressure vessel as possible, and the flow resistance of the system shall be minimized.
④ The requirements on the integrity of the RCS boundary shall be met. The RCS is designed in full consideration of the operation performance of its components and the possible adverse change in the material performance within the service lifetime, with sufficient margin reserved. The design of the RCS shall be able to withstand all kinds of static and dynamic loads imposed on any parts of the system due to normal design transients or accident conditions. Under any normal transient or accident condition, the design of the reactor coolant system should ensure the integrity of the coolant pressure boundary.
⑤ The RCS shall be with double operation loops, and each loop shall be able to operate independently to increase the availability of the device.
⑥ The RCS layout should take full consideration of the accessibility and operability of the equipment, pipes, valves and support parts during installation, operation, maintenance, in-service inspection (ISI) and refueling, while minimizing the size of the compartment.

3.3 System Arrangement

The arrangement of a RCS has direct impact on the performance of the primary loop system and thus the overall performance of the ship, and different arrangements have different demands on the design, construction, operation and management of the nuclear power plant. In the development history of the marine PWR units, there are three arrangement types of reactor structures and systems, namely, separated type, compact type and integrated type. Those that adopted the separate type arrangement cover the US "Savannah" ship, the Japanese nuclear-powered merchant ship "Mutsu" and the Soviet "Lenin" icebreaker. Those that adopted the compact type arrangement include the French CAS reactor and the Soviet "Arktika" icebreaker. The French CAP reactor series and the German FDR reactor for "Otto Hahn" nuclear ore carrier [1] adopted the integrated type arrangement.

3.3.1 Separated Arrangement

The so-called separated arrangement means that the reactor, pressurizer, SGs, main pumps and other items are separate from each other and connected to each other by reactor coolant piping, forming loop type arrangement.

Most of the primary loops for marine reactors are of separate type arrangement. The RCS with separated arrangement has the advantages of good accessibility during maintenance and ISI and relatively lower demands on equipment reliability, as the pieces of the main equipment are independent. However, it has certain disadvantages: for example, it has large demand on installation space and shielding range due to the complexity of the system, the long pipes and the large number of valves, resulting in significant increase in the weight of the power plant; besides, the support systems for main equipment and main pipe are more complex, which is unfavorable for bearing the external dynamic loads. The displacement of the separate type loop caused by the expansion or shrinkage can be compensated by the flexible configuration of the main pipes and the movable support of the main equipment.

Figure 3.2 shows the separated arrangement of the primary loop.

3.3.2 Compact Arrangement

The so-called compact arrangement means that the reactor, SGs, main pumps and other items are firmly connected to each other by short pipes or special structures.

Compared to the separated arrangement, the compact arrangement is featured by compactness as well as smaller volume, weight and resistance, favorable for improving the capacity of natural circulation for the reactor and avoiding large break accident. However, challenges will be brought to the shielding arrangement due to the compact arrangement of the SGs and reactor pressure vessel.

Figure 3.3 shows a typical compact type arrangement

3.3.3 *Integrated Arrangement*

The integrated arrangement refers to integrating the equipment such as SGs into reactor pressure vessel or connecting them directly to the reactor pressure vessel, forming an integral structure.

The integrated PWR is popular due to its merits such as simplified system, compact structure, small size, light weight and high safety. It is known that the integrated PWRs have been adopted by or under the consideration of some countries such as Russia, France, the USA and Japan. By laying SGs into the pressure vessel or connecting them directly to the pressure vessel, the integrated reactor has the weakness of complex internal structure of the pressure vessel, as the main equipment of the RCS is highly integrated, and it makes the ISI, maintenance and replacement of the equipment more difficult. Thus, the demands on the reliability of the equipment operation are very high, and the quality of the pipe materials, welding, secondary water and other items shall be controlled strictly.

The main concerns for the design of integrated reactor include: the manufacture capacity of the reactor pressure vessel, the design of the compact and high efficiency SG, the simple and robust control rod drive technology and convenient maintenance and replacement means for the equipment (such as how to do the post-break repair of the SG heat transfer tubes).

Figure 3.4 shows a kind of an integrated reactor.

It shall be noted that, the three arrangement types mentioned above have their own characteristics and applications, and there are no absolute advantages or disadvantages. The system arrangement of the primary loop depends on various factors such as compartment space, industrial base, R&D technology level, technical requirements for operation conditions and equipment types. The countries may choose the specific arrangement type based on their own circumstances and actual conditions.

3.4 Characteristic Design

The functional relationship between the main parameters of the RCS and the nuclear steam supply system (NSSS) is called the steady-state characteristics of the RCS, and sometimes may be called the static characteristics of the NSSS. The steady-state characteristics of the RCS not only reflect the characteristics under rated operating conditions at full power, but also reflect the characteristics under operating conditions at partial power. The main parameters of the RCS include the system coolant flow, reactor inlet and outlet coolant temperature, reactor coolant average temperature, steam pressure and temperature, etc.

The steady-state characteristics of the RCS are closely related to the main parameters, the selection of the structure type of the main equipment and the design of the reactor control system. Reasonable selection of the steady-state characteristics of the RCS is a key part in the scheme design of the nuclear power plant. The guideline for the selection of the steady-state characteristics of the RCSs varies with the purposes and application of the nuclear power plants [2].

3.4.1 Operation Scheme with Constant Average Coolant Temperature

Figure 3.5 shows the operation scheme with constant average coolant temperature. The characteristic of the operation scheme is that the average temperature of the reactor coolant will not change with the plant load. For the primary loop, with the change in the reactor power, it is not necessary to use the control rods to compensate for the reactivity change caused by reactor coolant temperature change; the size of the pressurizer in the pressure control system can be minimized because the fluctuation of the load has the least effect on the volume of the coolant in the primary loop in this scheme; and the change in thermal stress in the plant is small, the response to the load is fast, and the time required for returning to the setpoint after the load fluctuation is short.

Nevertheless, the scheme has the major disadvantage that the parameters of the secondary steam vary significantly with the output power, especially during the low-power operation when the steam pressure is high. This imposes high demands on the pressure-bearing capacities of the secondary steam pipes, valves, turbines and other items. To meet the requirements of the manoeuverability, the marine nuclear power plant is subject to frequent change in operation conditions, significant change in power and long-time low-load operation. To address the disadvantage brought by such an option, automatic steam pressure control valves shall be provided on the SG outlet pipes.

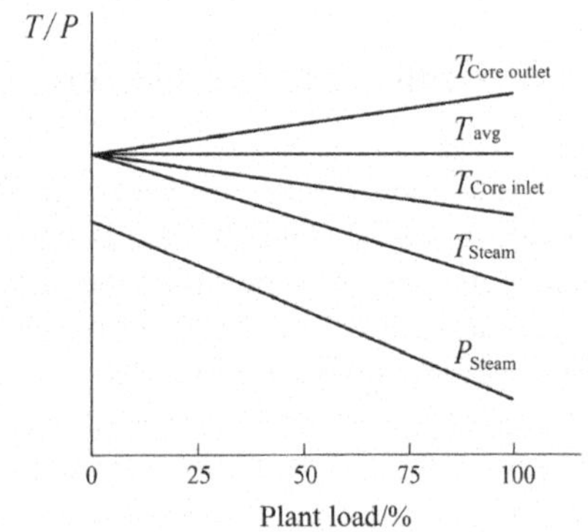

Fig. 3.5 Operation scheme with constant average coolant temperature

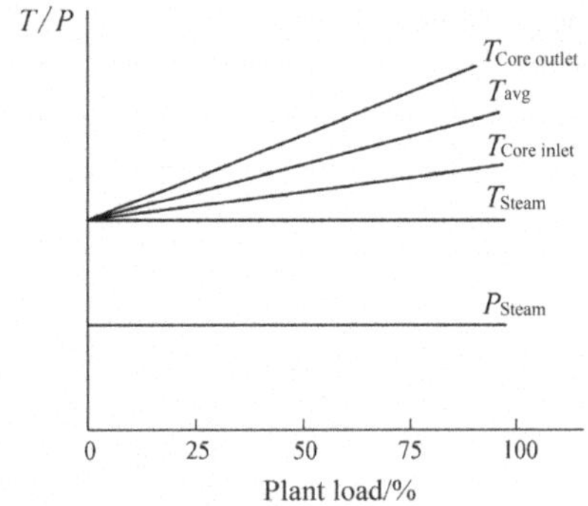

Fig. 3.6 Constant steam pressure option

3.4.2 *Operation Scheme with Constant Steam Pressure*

The operation scheme with constant steam pressure is a scheme that the steam pressure will not change with the load. Figure 3.6 shows some of the load characteristics of this scheme.

The scheme provides significant convenience for the design of the secondary loop, due to the unchanged steam parameters. For example, no automatic control valve is required to stabilize the secondary steam pressure and no special requirements are imposed on the turbine, still the feedwater pump can have the similar characteristics to that of the conventional steam-powered unit, without imposing any special requirements. But the change in the reactivity is significant due to the significant change in average temperature of the primary loop, the pressurizer is required to have larger volumetric compensation capacity, and the control rods will have larger displacement at the same load as compared with the previous scheme. It brings certain challenges to the design and operation of the primary loop.

3.4.3 *Compromised Solution*

The two schemes mentioned above have their own distinct advantages and disadvantages, thus having adverse impact on the operation of the primary loop or the secondary loop. In order to change this situation, a compromised solution can be adopted, namely, the approach of adjusting to the power regulation based on the output. With this approach, the constant steam pressure scheme is adopted for the low-power zone, and the constant average temperature scheme is adopted for the high-power zone, as shown in Fig. 3.7. It not only prevents excessively high steam pressure within the low-power zone, but also prevents significant change in the average temperature of the primary loop within the high-power zone.

It shall be noted that, there are various types of compromised solutions, and each type has its own advantages and disadvantages. Though the compromised solution can, to some extent, improve the operational performance of the marine nuclear

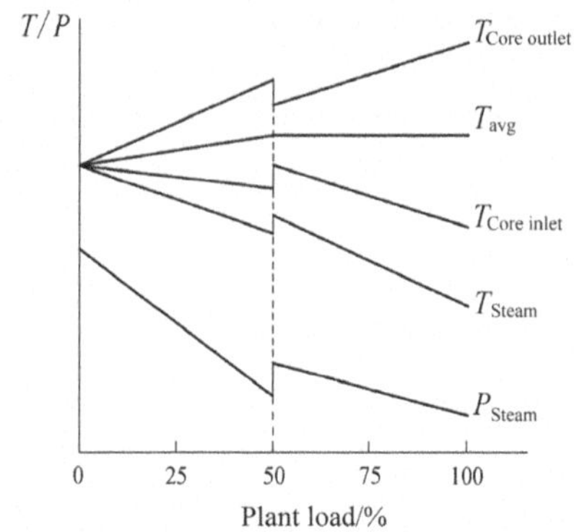

Fig. 3.7 Compromised solution

power plant and mitigate the operational challenges caused by a single operation scheme, namely the operation scheme with constant average coolant temperature or the operation scheme with constant secondary steam pressure, it does not solve the fundamental problems of the two basic schemes, and the design and operation challenges are just shared by the primary and secondary loops. That means, the overall operation performance of the nuclear power plant is not improved essentially If the average temperature of the primary coolant and secondary steam pressure are simultaneously kept constant during the steady-state operation of the nuclear power plant, the ideal steady-state operation with "double constant" can be achieved, which can keep all the merits of the two basic schemes to the maximum extent, so that the operation and regulation challenges to the primary and secondary loops can be mitigated effectively, the design and manufacturing conditions for many equipment of the nucelar power plant can be improved, and the overall operation performance of the plant can be enhanced significantly [3].

3.4.4 Static Characteristics of Once-Through Steam Generator

The several schemes discussed above are specific to the natural circulation steam generators. However, some PWR units, such as those for the Soviet "Lenin" and "Arktika" icebreakers, are provided with once-through steam generators. Any steam pressure change pattern can be selected, as the load characteristics depend on the change in heat transfer condition or the redistribution of the heat transfer area in each heat transfer zone of the SG. The most reasonable option is to keep the steam pressure constant or make it having a linear function with the turbine load.

Figure 3.8 shows the static characteristics of the once-through steam generator, reflecting the change patterns of the reactor outlet coolant temperature and the temperature of the superheated steam at SG outlet as a function of the plant load.

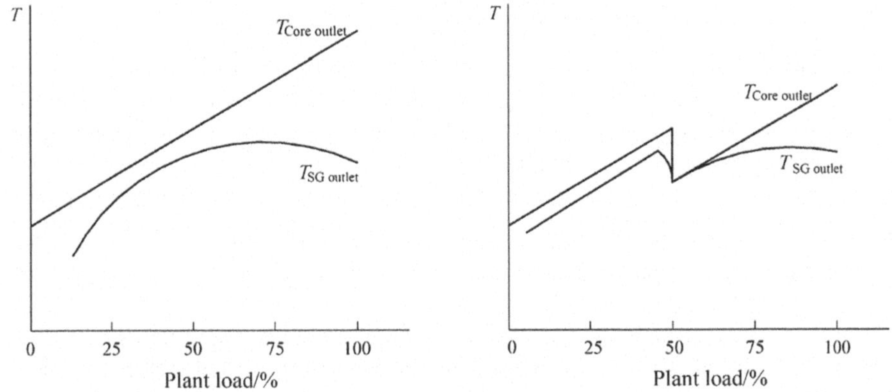

Fig. 3.8 Static characteristics of once-through steam generator **a** Constant coolant flow; **b** Step change in coolant flow

Nevertheless, the once-through SG, with small water inventory and poor transient behavior, has high demands on the design of the control system.

3.5 Brief Introduction to Main Equipment

Main equipment of the RCSs generally includes the SGs, RCPs and main pipes.

3.5.1 Steam Generator

SG is the key equipment of the primary loop system and plays a key role in the connection between the primary and secondary coolant circuits. Its main functions are as follows:

① During power operation, transfer the heat from the reactor coolant to the secondary water which will be heated to saturated temperature and turned into acceptable saturated steam;
② Remove the residual heat from the reactor and the sensible heat from the equipment during normal shutdown or emergency shutdown;
③ Form an integrated part of the reactor coolant pressure boundary and thus a part of the second barrier for nuclear safety, and play a key role in preventing radioactive substances from leaking into the reactor compartment and the secondary coolant circuit and in ensuring the safety of the nuclear power plant.

(1) Basic technical requirements and classification for SGs

As SG is a hub for the primary and secondary coolant circuits, it operates under the severe conditions. As part of the second barrier for nuclear safety, the SG is subject to higher requirements on its performance, reliability and maintainability. The following basic technical requirements shall be included in the design of a SG based on current status of SG operation and application in the nuclear power plant:

① When designing the SG and its components, the amount of the steam required and the specified parameters to the nuclear power plant under any operation conditions shall be ensured.
② Appropriate measures in all aspects shall be adopted to prevent the corrosion of the SG heat transfer tubes, including the selection of the heat transfer tube materials with good corrosion resistance, the specification of proper water chemistry indexes, the improvement of manufacturing and installation technology for the tubes, the provision of the blowdown devices and slurry collection devices, and the improvement of the structural design for adequate thermo-hydraulic behavior at tube root;
③ The pressure boundary including the primary header shall be safe and reliable, without any crack or failure;
④ For the natural-circulation SG, effective and compact moisture separators shall be selected to ensure the acceptable moisture content of the SG outlet steam and to reduce the size of the SG;
⑤ The resistance at primary side shall be minimized to ensure high natural circulation capacity of the primary loop system;
⑥ The SG shall be of simple and compact design, easy for installation and maintenance; and
⑦ The SG shall be easy for operation and control.

SGs are the most critical heat transfer device in the nuclear power plant. There are various types of SGs, with varied advantages and disadvantages, to meet the demand of the operation and application of nuclear power plants. SGs can be divided into several different types, based on their structural types and operational characteristics.

① SGs can be divided into natural-circulation SGs and once-through SGs by the flow pattern of the medium;
② SGs can be divided into vertical SGs and horizontal SGs by their installation orientation;
③ SGs can be divided into U-tube type, straight tube type, sleeve type, helical tube type and other types by the shape of the heat transfer tubes; and
④ SGs can be divided into modular SGs and independent SGs based on how they are integrated with other primary equipment.

(2) Natural-circulation SG

Natural-circulation SGs may also be subdivided into vertical natural- circulation SGs, which are widely adopted for marine nuclear power plants, and horizontal natural-circulation SGs.

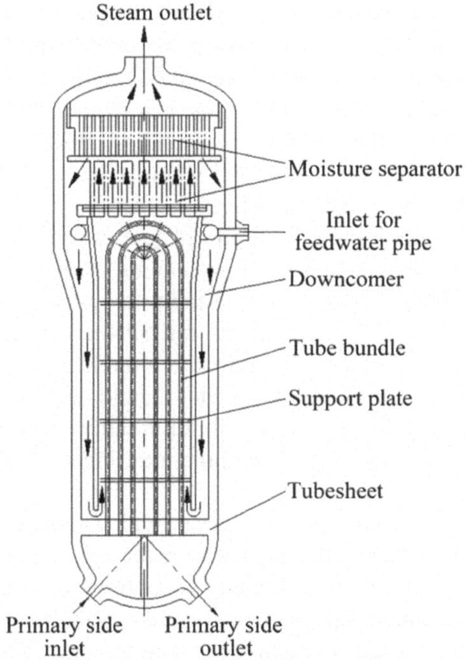

Fig. 3.9 Typical vertical natural-circulation steam generator

Figure 3.9 shows the typical vertical natural-circulation SG, mainly consisting of lower head, tubesheet, U-tube bundles, moisture separators and shell assembly.

① Lower head assembly.

The lower head assembly consists of lower head, splitter plate, nozzles for coolant inlet and outlet and manhole devices.

The lower head is generally forged with low-alloy steel. In order to reduce the workload of ISI and ensure the reliability of pressure boundary, the nozzles for primary coolant inlet and outlet, manhole base and spherical shell of the lower head are forged as a whole. The lower head and tubesheet are welded together. Nozzles for primary coolant inlet and outlet are provided with stainless steel (SS) safe ends, so as to weld with the main pipes.

The lower head is divided by the splitter plate into the inlet and outlet plenums, which are provided with manholes for ISI of the heat transfer tubes, tube plugging and other inspection and maintenance operations. The inner surface of the lower head is overlaid with extra-low-carbon stainless steel by welding.

② Tube bundle and tubesheet

The tube bundle assembly consists of tubesheet, heat transfer tubes, flow distribution plates, tube support plates, spacer rods, corridor stoppers, blowdown pipes, bundle sleeves and other items.

The tubesheet is low-alloy steel forging, and its primary side is overlaid with nickel-based alloy. The tubesheet is provided with flanges at top and bottom in order to weld with secondary shell and lower head.

The heat transfer tubes are of square or triangular arrangement, and are locked to the tubesheet by full-depth hydraulic expansion and sealing welding. A distribution baffle is provided above the secondary side surface of the tubesheet and within the bundle sleeve. A large hole is provided at the center of the distribution baffle so that most of the secondary coolant will flow through it. This can increase the transverse scouring velocity of the secondary side surface coolant flow and reduce the slurry deposits on the tubesheet.

Stainless steel support plates are provided evenly for the straight part of the tube bundle to keep proper gap among the heat transfer tubes and prevent the mechanical vibration and hydraulic-induced vibration of the tubes. Spacer rods are used for keeping longitudinal spacing between the support plates. Holes are broached for the support plates, through which the heat transfer tubes can penetrate. The structural design of the support plates should take into account the flow capacity at secondary side, flow resistance, vibration induced by restricted flow, and the concentration of chemicals in the tube-hole gap. Figure 3.10 shows the improved trefoil broached holes.

Anti-vibration bars are provided at the U-tube bend zone of the heat transfer tube bundles, and the gap between the anti-vibration bars and the heat transfer tubes shall be controlled reasonably so as to reduce the fretting wear to the tubes. To remove the slurry deposits from the tubesheet, blowdown pipes are provided at the center of the corridor to control the water quality. The bundle sleeve is fixed in the shell and is arranged around the bundles. It is used to separate the downward channel and the upward channel. The sleeve is positioned by locking block to ensure the size of the annular channel between the sleeve and shell.

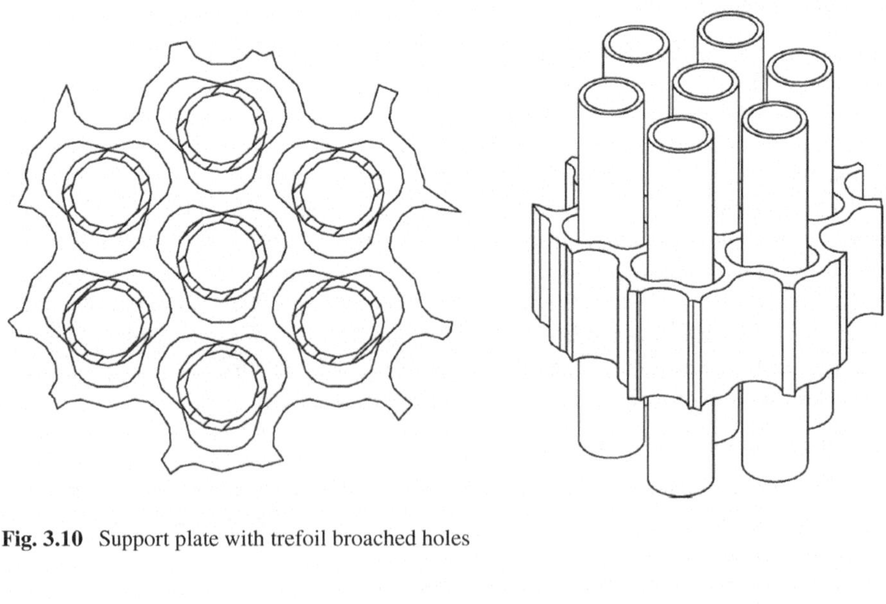

Fig. 3.10 Support plate with trefoil broached holes

③ Moisture separator

Natural circulation steam generator produces saturated steam. If the humidity of the steam sent to the secondary consumers is too high, the water droplets mixed in the steam and the salt dissolved in the water will cause the deposition of the salt in the flow passage of the steam turbine, reduce the efficiency of the steam turbine, and affect the reliability of the steam turbine. Thus, moisture separator shall be provided to remove the moisture from the steam. It is located in the upper shell of the SG, generally including the primary separator and secondary separator.

There are diversified types of moisture separators. The mostly widely used primary separator is swirl vane separator, which is for the primary separation of the moisture-steam mixture. Figure 3.11 shows the structure of a swirl vane separator.

A set of stationary helical blades is provided inside the separator bowl. These helical blades will change the motion of the mixture from linear pattern to rotary pattern as the mixture passes through the blades upwards along the separator bowl. In this process, the moisture droplets are removed from the steam by centrifugal force, to form an annular water film on the surface of the inner wall of the separator

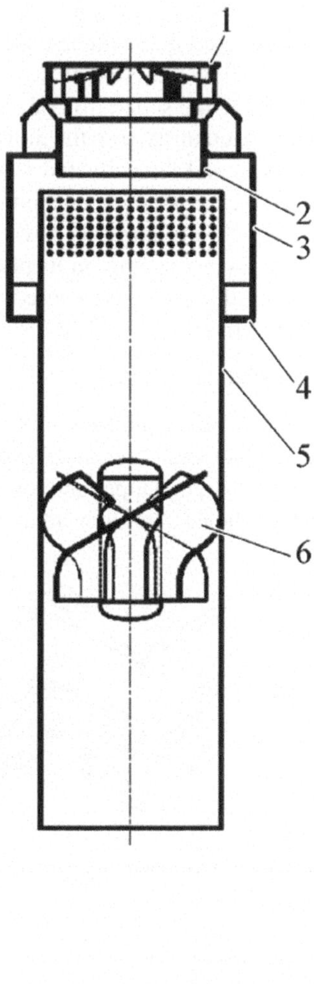

Fig. 3.11 Structure of moisture separator . 1-Cover; 2-Outlet sleeve 3-Sleeve; 4-Restrictor ring; 5-Separator bowl; 6-Helical blade

bowl, which flows upwards along with the steam. The steam reaches the top of the separator bowl, and flows out from the steam outlet pipe, which is coaxial with the bowl. The separated water flows upwards to the drain ring in a helical manner along the wall, and then is baffled to the drain channel formed by the separator bowl and the outer sleeve, and then enters the water space.

The corrugated plate separator is most widely used as the secondary separator for further separation of the residual moisture from the steam from the primary separator. After two-stage separation, the steam with specified quality will gather at the top of the SG and flow from the steam outlet pipe to the turbine via the secondary main steam pipeline.

The corrugated plate separators can be divided into horizontal type and vertical type by arrangements. The horizontal corrugated plate separator is laid horizontally, with the steam flow direction opposite against the drain water flow direction. As the drain water and the steam flow are opposite, effective separation is possible only when the velocity of inlet steam is very low. The vertical corrugated plate separator is laid vertically, with the steam flow direction normal perpendicular to the drain water flow direction. This significantly improves the conditions for water drain, allowing higher velocity for the inlet steam. Crooked corrugated plate separators are novel secondary separators inspired by vertical corrugated plate separators, and have been widely used in the nuclear power plants. Figure 3.12 shows the operating principles.

The flow direction of steam is changed for several times within the zigzag channel to enable the separation of the droplets carried. The crooks on the corrugated plate, large in quantity, trap the moisture in the steam and collect the water film on the plate surface, and the collected water flows to the drain device along the grooves. In this way, the crooked corrugated plate separator allows higher velocity for the fluid and higher elevation as compared to the conventional vertical corrugated plate separator, since the water film on its plate will not become thicker and thicker and will not be susceptible to breaks caused by the blow of the steam like that on the conventional separator.

④ Shell assembly

The secondary shell assembly consists of top head, upper shell section, conical section and lower shell section. The shell has various pipe nozzles, such as water level measuring nozzle. Upper and lower shell sections are provided with sufficient manholes, access holes and other items to allow ISI and maintenance of the SG.

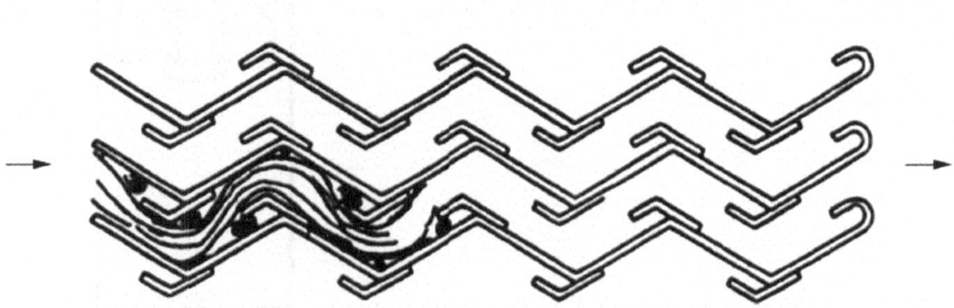

Fig. 3.12 Operating principles of crooked corrugated plate separator

The shell is forged or rolled with low-alloy steel, and the part adjacent to the tubesheet is thickened so that it can be attached to the tubesheet at short straight sides. The thickened zone is provided with manhole and access hole.

The top head is of standard ellipsoid shape, at the center of which is provided with the steam outlet nozzle. The steam outlet nozzle is provided with stainless steel safe end, to which the main steam pipeline is welded.

To restrain the release of substances and energy after the failure of the main steam pipeline and to reduce the load on the internals of the SG, a flow restrictor can be provided in the steam outlet nozzle.

Figure 3.9 shows the operating principles of the vertical natural-circulation steam generator.

At the primary side, the heated coolant from the reactor core, with high temperature and high pressure, enters the hot-side water chamber of the SG via the primary hot leg, and then flows into the U type heat transfer tubes. As it flows through the heat transfer tubes, it transfers its heat to the secondary water continuously, with its temperature lowered gradually. Then, the coolant goes from the tubes to the cold-side water chamber, and finally returns to the reactor for reheating by the main pump via the primary cold leg.

At the secondary side, the secondary feedwater enters the SG and mixes with the separated water from the moisture separator. It flows downwards along the downcomer in the annular space between the sleeve and the shell and turns by 90° at the lower part of the sleeve to flow into the gap between the tube bundles horizontally, and flows upwards. It is heated by the primary coolant in this process and becomes moisture-steam mixture. The mixture enters the moisture separator(s), and most of the moisture will be removed. The separated water enters the downcomer via the drain channel, forming water circulation at the secondary side. The steam, after leaving the separator, enters the dryer for further dehumidification so that the outlet steam can meet the design requirements of steam quality, and can be supplied for the operation of the turbine.

The vertical natural-circulation steam generator has the following features:

① The water inventory at the secondary side is large and can play a buffering role, with low requirement for the automatic control of feedwater and steam; ② The water can be treated and blown down within the secondary side of the steam generator so that requirements on materials of the heat transfer tubes and quality of the secondary water can be lowered to some extent, thus the auxiliary system can be simplified and the safety and reliability of the equipment can be improved.

The vertical natural-circulation SG has the following disadvantages:

① Only saturated steam is generated, and thus moisture separator is required to ensure the steam quality, resulting in complexity of the SG structure; ② The steam pressure varies significantly in the process from zero to full load, bringing certain challenges to the design, operation and management of the secondary steam system and the steam consumers.

(3) Horizontal SG

The horizontal natural-circulation SGs are one of the main types adopted for land PWR nuclear power plants and marine nuclear power plants in early stage. They are widely used in Russian PWR nuclear power plants. The Water-Water Energetic Reactors (VVER) are developed by Russia and are sold to the previous German Democratic Republic, Bulgaria, Hungary, Czech Republic, Finland and other countries. A kind of horizontal natural-circulation steam generator is shown in Fig. 3.13.

The horizontal natural-circulation SG has the following structural features:

① The thermo-hydraulic performance at the root of heat transfer tube bundle is good. The U-tubes and the vertical cylindrical header are both made of 1Cr18Ni9Ti SS. Blast expansion is adopted, and the U-tubes are welded to the inner wall of the header with TIG welding. Stagnation zone will not be formed on the surface of the header, and there is coolant flow with certain flowrate at the root of heat transfer tubes, preventing impurities from deposition and concentration. There are blowdown pipes at the bottom of the shell, so the slurry deposits will not contact with the heat transfer tubes, preventing the corrosion and failure of the junctions between the heat transfer tubes and the header.

② The circulation rate is large. The tube bundle is designed as three separate assemblies. Each assembly is divided into two layers vertically. The space between the assemblies forms the vertical downcomers. The heat transfer tubes are of square arrangement. Supports are used to ensure the proper spacing between the tubes and to bear the total weight of the tube bundles. While natural circulation loop is not significant, the circulation rate is large as the tube bundles are at low position.

③ Provisions are made to ensure uniform steam load per unit area. Feedwater enters the hot side along the feedwater pipes. Some steam is condensed by the feedwater, so that the quantity of the hot-side steam is reduced. In such a way, the steam flowrate is even over the hot-side and cool-side evaporation surface.

④ Simple moisture separators are used. With large steam space, simple moisture separators are sufficient to meet the requirements of steam quality, as the load per unit surface area is smaller than that of the vertical SG. The primary separation is gravity separation, and the moisture-steam mixture is efficiently separated by gravity, after the bubbles come out from the tube layers. The secondary separation is fine separation by means of corrugated plate separator, and the saturated steam is further dried, with the water droplets removed.

⑤ Austenitic SS heat transfer tubes are adopted, resulting in a low consumption of nickel and low cost of SG.

⑥ With large evaporation surface area and large water inventory, the SG can remove more heat from the reactor than a vertical SG, and enhance the natural circulation capacity of the reactor coolant.

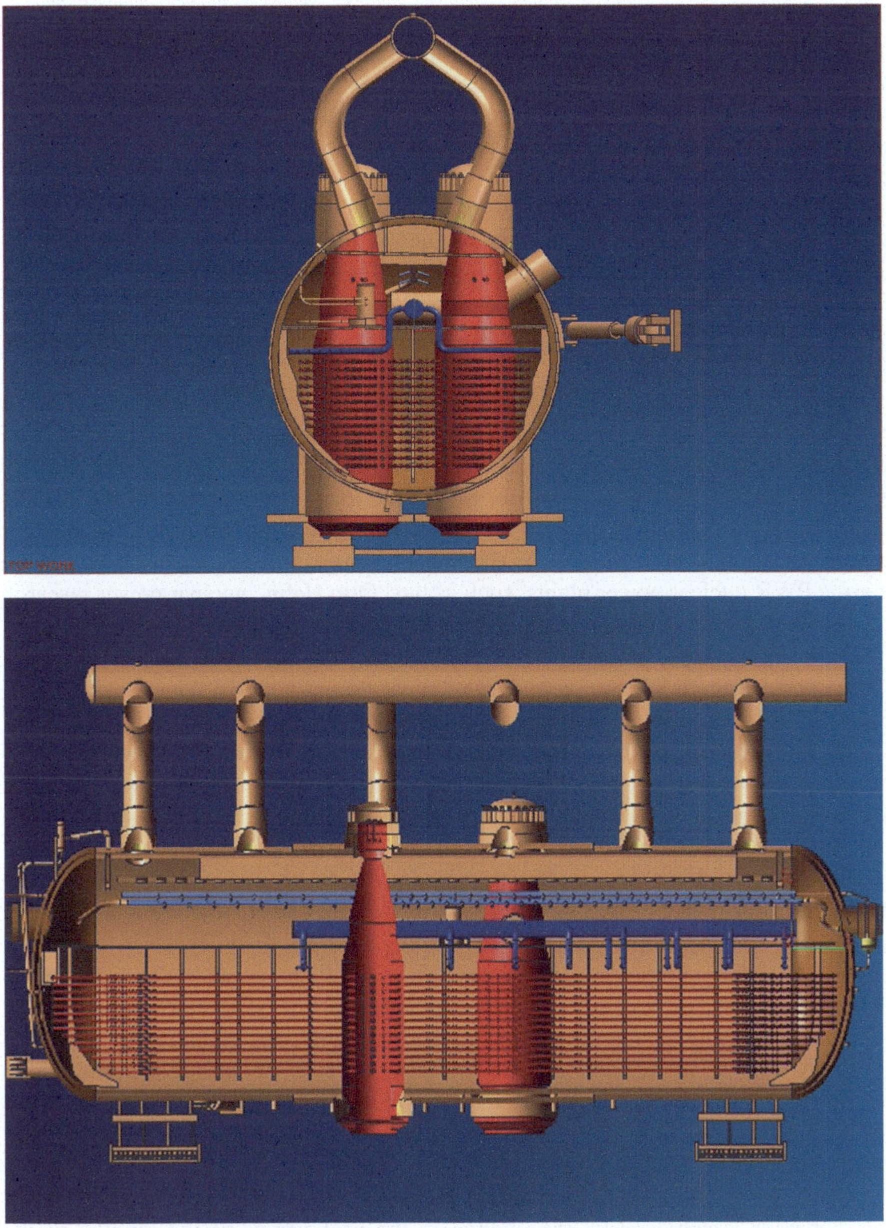

Fig. 3.13 Structure of a kind of horizontal natural-circulation steam generator

Nevertheless, it has such major disadvantages as the sensitivity of the humidity of the outlet steam to the water level fluctuation, thus imposing high demand on water level control. In addition, large volume, large installation space and large weight result in inconvenience for rail transportation and adverse effect on the configuration within the reactor compartment.

(4) Once-through steam generator

For once-through steam generator, the secondary medium flows to the heat transfer tubes in the manner of forced circulation, with the pressure head provided by the feedwater pump. It flows to the preheating section, evaporating section and superheating section in turn, and becomes superheated steam. Therefore, the circulation rate in the once-through SG is 1.

Once-through SGs are divided into outside-tube type and inside-tube type. For the outside-tube type, the secondary medium flows outside the heat transfer tubes, and the primary coolant flows inside the tubes. For the inside-tube type, the secondary medium flows inside the heat transfer tubes, and the primary coolant flows outside the tubes. The outside-tube type once-through SGs are adopted at some PWR nuclear power plants, while the inside-tube type once-through SGs are mostly used for the marine PWR nuclear power plants, such as the "Otto Hahn" nuclear power research reactor, the "Lenin" icebreaker and the "Arktika" icebreaker.

Figure 3.14 shows the structure of the once-through steam generator.

For once-through SG, the secondary medium flow is not driven by density difference as for the natural circulation, but by the pressure head of the feedwater pump. Under that pressure head, the feedwater flows over the heating surfaces of the preheating section, evaporating section and superheating section. As it flows over such surfaces, it is heated, evaporated and superheated, and finally, the steam so generated reaches the required temperature (superheated steam with certain degree of superheat). As the medium flow is driven by the pressure head provided by the pump in the once-through SG, the medium will move in a forced way over the heating surfaces. As the medium passes over the heating surfaces all at once, it is fully evaporated and there is no natural circulation. Any change in operation condition of the once-through SG (such as the change in feedwater quantity) will cause change in the medium specifications at all the points in the moisture-water channel.

This will result in the change in the lengths of the heating surface sections, which is different from that of the natural-circulation SG.

During operation of the once-through SG, when load varies over certain power range,it is feasible to keep a constant average temperature of the primary coolant and a constant inlet steam pressure of the turbine control valve. During the load variation, this behavior is achieved by the change in the proportions of the lengths of the evaporating section and superheating section without any change in the temperature difference between the primary and secondary loops. The feedwater flow is regulated so as to keep constant the inlet pressure of the turbine control valve, unlike the natural-circulation SG, the feedwater flow of which is regulated so as to keep the water level constant.

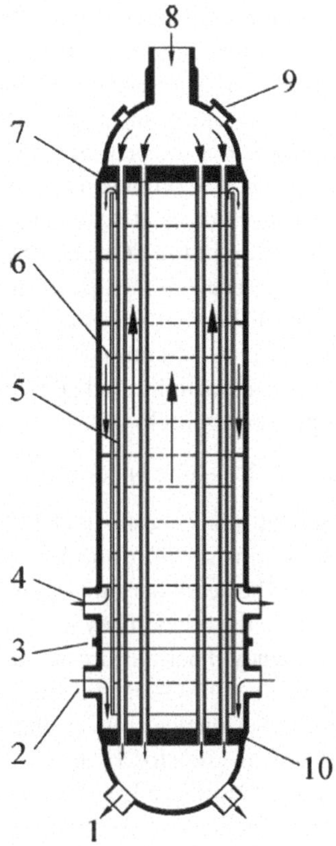

Fig. 3.14 Once-through steam generator. 1-Coolant outlet;2-Feedwater inlet; 3-Emergency feedwater inlet; 4-Steam outlet; 5-Heat transfer tube; 6-Sleeve; 7-Upper tubesheet; 8-Coolant inlet; 9-Manhole; 10-Lower tubesheet

As there is two-phase flow in the once-through SG, the flow instability may occur. This will result in periodical fluctuation of the steam pressure, steam flow and primary coolant outlet temperature. The fluctuationis related to the resistance between the feedwater head and the tube bundles. It can be eliminated or mitigated by adjusting the resistance at the tube bundle inlets. Therefore, throttling devices are often provided at the secondary inlets of the heat transfer tube bundles in the design of once-through SGs to provide the required resistance at the inlets, so as to improve the operational stability of the SGs.

Once-through SGs have the following features:

① They can generate superheated steam with stable pressure and require no moisture separation, with improved thermal efficiency of the nuclear power plant; ② Without complex moisture separators and large-diameter drums, they have simple and compact structure, smaller size, lighter weight, and are easier to be installed in the compartments with limited space, as compared with the natural-circulation SGs with the same capacity; ③ Heat and water inventories are both small at the secondary side, enabling faster startup and shutdown and thus improving the manoeuverability.

Once-through SGs have the following weakness:

① They cannot keep the quality of the water for the boilers as the natural-circulation SGs (which rely on blowdown) during operation due to the circulation rate of 1, imposing high demands on the quality of feedwater and the corrosion resistance of the heat transfer tube materials and thus requiring periodical chemical cleaning; ② Complex automatic control systems are required due to small water and heat inventories at secondary side; ③ Due to inherent features, the two-phase flow instability in the SGs shall be handled, and design and operation measures shall be taken to prevent the operation in the unstable flow zone.

With the merits of compact structure, light weight and small size, the once-through SGs can be of compact arrangement or integrated arrangement to fit for the compartments with limited space for the marine nuclear power plants, and become a hot topic for research.

① Helical-tube once-through steam generator.

Helical-tube once-through SGs are of inside-tube type. Each of their steam generation units consists of a number of heat transfer tubes bent into helical shape in some way. The secondary medium flows inside the tubes, while the reactor coolant flows outside the tubes. They are characterized by compact structure and high heat transfer efficiency. Thermal stress is minimized due to the free expansion of the helical tubes. This type of SGs can be used for PWRs, gas cooled reactors, sodium cooled reactors and other nuclear power plant, and can be either integrated with the reactor or laid outside the reactor. Figure 3.15 shows the helical-tube once-through steam generator for the "Northern Line" icebreaker, with compact arrangement. Figure 3.16 shows a kind ofhelical-tube once-through steam generatorwith integrated arrangement.

Helical-tube once-through SGs can be subdivided into large-scale type and small-scale type, based on the geometry of the helical tubes. The heat transfer tubes for large-scale type are bent and installed in the whole annular space between the pressure vessel and the core barrel, with compact but complex structure and higher utilization of the space. This type is used for "Otto Hahn" merchant ship.

The heat transfer tubes of small-scale type are bent and installed in a sector zone of the annular space between the pressure vessel and core barrel, generally in helical shape. This type is suitable for modular and standardized SG design, and the quantity of SGs is based on the reactor power. Each module consists of central cylindrical tube support, lower head for feedwater, upper head for steam and enclosure. Figure 3.17 shows the small-scale in-reactor once-through steam generator.

② Sleeve type once-through steam generator.

The sleeve type once-through SGs are provided with steam generation units that have sleeve structure. Each SG is composed of several steam generation units, and is of a box structure in a whole, thus called high efficiency box type SG. The box type SGs consist of steam generation units, headers, feedwater pipes, steam lines and other items, and are located in the reactor pressure vessel. Compared with the conventional SGs, these SGs not only have the merits of compact structure and high

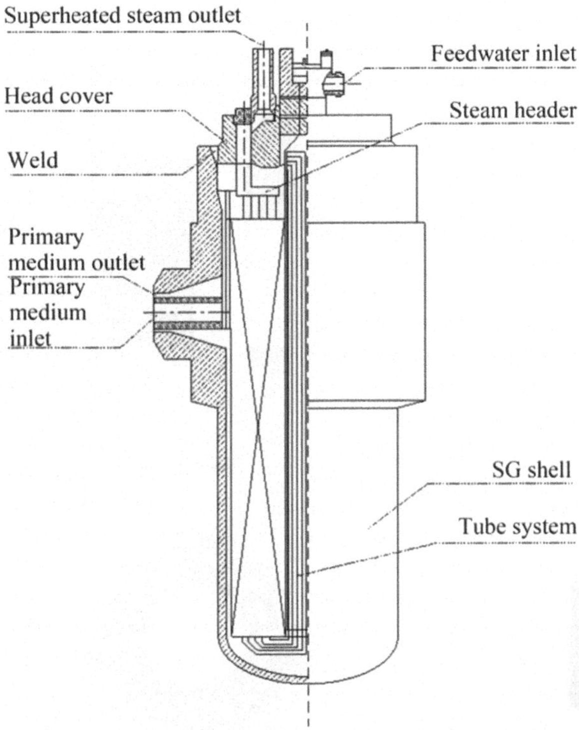

Fig. 3.15 A kind of helical-tube once-through steam generator

safety and reliability, but also provide convenience for modular design, manufacturing and installation. The Russian fourth-generation marine PWRs adopt integrated arrangement and are provided with the high efficiency box type SGs.

Taking the once-through SG forVPBER-600 reactor as an example (see Fig. 3.18 [5]), this SG is made of titanium alloy, which has excellent performance and improved corrosion resistance. It is of modular design and can operate by isolating the failed part in case of any break of heat transfer tube, allowing the identification and plugging of the leaking part after shutdown. It is provided with modular and standardized parts and components, which are manufactured by concurrent engineering process to ensure the manufacturing quality [1].

③ Printed circuit Steam Generator.

To enhance the power density of the integrated reactors, printed circuit heat exchangers (PCHEs) with more compact structure are proposed [5].

As a kind of compact heat exchangers, they can enhance the power density and reduce the pressure drop and maintenance demand. It is shown from the steady-state performance calculation that by adopting a PCHE, the diameter of a reactor pressure vessel can be further reduced as compared with that using the helical-tube once-through SG.

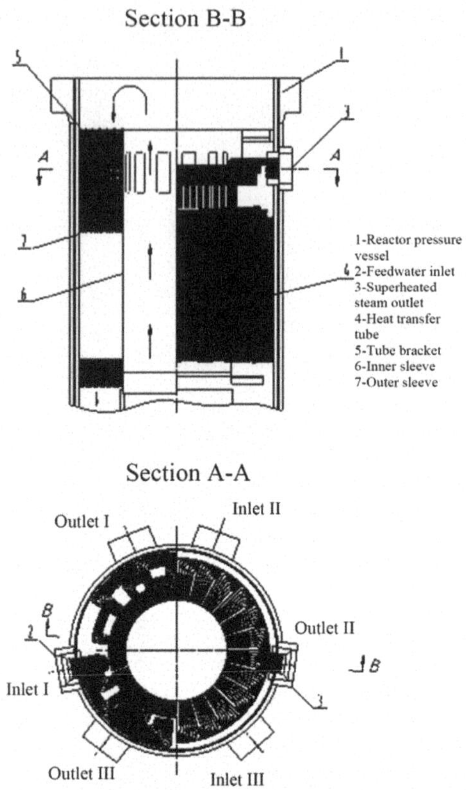

Fig. 3.16 Helical-Tube Once-Through Steam Generator with compact arrangement

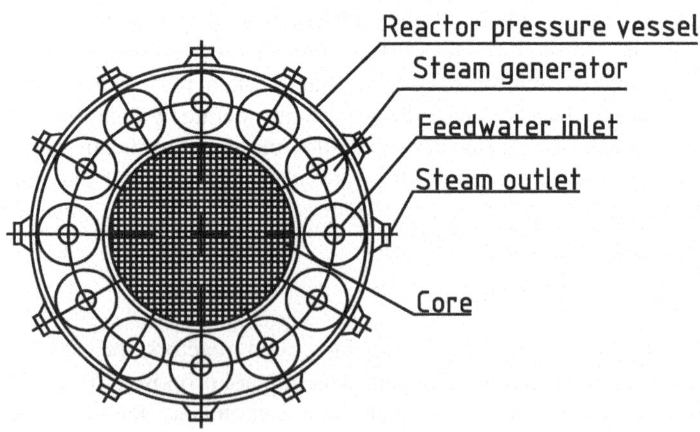

Fig. 3.17 Small-scale helical-tube in-reactor once-through steam generator

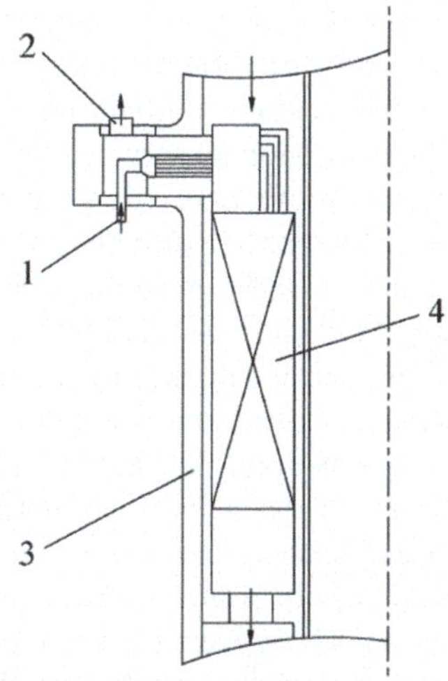

Fig. 3.18 High efficiency box type once-through steam generator. 1-Feedwater pipe; 2-Steam line; 3-Pressure vessel; 4-Heat transfer tube bundle

Figure 3.19 shows the structure comparison of Printed Circuit Steam Generator and helical coiled Steam Generator.

The PCHE can increase the power density and reduce the pressure drop and, to a larger extent, the space demand. PCHEs have not been applied in engineering yet, as further research is needed for two-phase flow behaviors, plugging between plates, inspection, cleansing and manufacturing technologies.

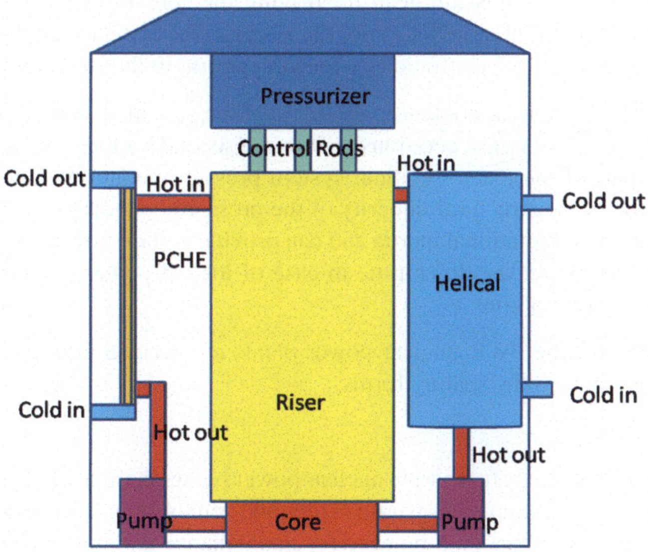

Fig. 3.19 Structure comparison of printed circuit steam generator and helical coiled steam generator

As SGs are taken as one of the key components in nuclear power plants, their operation performance has significant effect on the safe and reliable operation of the whole plant. With the development of science and technology, the nuclear power plants with longer service lifetime and higher safety, reliability and maintenability are developed, to reduce the probability of unexpected shutdown and the workload of ISI or maintenance caused by SGs, and to enhance the availability of the nuclear power plants. In addition to the satisfaction of the performance indexes, the volume and weight of the SGs shall be minimized to facilitate the arrangement and fabrication. Therefore, the natural-circulation SGs with compact structure, reliable performance, and easy operation and maintenance are one of the future development trends.

International leading players in nuclear power field focus on the basic technical R&D for the natural-circulation SGs, such as the design of overall structure, the research on effective moisture separation approaches, thermo-hydraulic analysis, R&D of novel materials for heat transfer tubes and other extensive tests and verification. This covers the key elements such as moisture separation approaches, slurry management approaches, R&D of novel materials for heat transfer tubes, thermo-hydraulic design approaches, heat transfer tube inspection, cleansing, and plugging approaches.

3.5.2 Reactor Coolant Pump

Reactor coolant pumps, also called the main coolant pumps or main pumps, are intended to provide the pressure head for driving the coolant in the primary loop system in the forced circulation. RCPs for the nuclear power plants are the only type of components that operate continuously for long period in the RCS, and,

① They enable the reactor coolant to circulate at the specified flow to remove the heat from the reactor core and transfer it to the secondary loop via the SGs;
② They are part of the reactor coolant system pressure boundary, thus providing the support for the structural integrity of the pressure boundary;
③ They have certain rotational inertia and can provide certain flow coast dowm and maintain specified coast down time in case of loss of power, favorable for the safety of the reactor core.

Main pumps for the PWR nuclear power plants are divided into canned pumps and shaft sealed pumps, by sealing forms.

(1) Canned Pump

The reactor coolant pumps for marine nuclear power plants shall satisfy the following requirements: ① The pump can operate safely and reliably for long period without maintenance; ② The pump shall be easy for maintenance and the auxiliary system is simple; ③ The rotatory parts of the pump can provide sufficient rotational inertia; ④ Austenitic SS or other equivalent anti-corrosion materials shall be adopted as the surface material for the parts contacting the coolant; and ⑤ No leakage of the radioactive coolant.

Based on the requirements mentioned above, the main pump sets for the marine nuclear power plants are generally vertical motor-driven single-stage centrifugal pumps or mixed flow pumps of shielded type. Vertical canned pump sets, usually driven by shielded motors, consist of canned motors and pumps. The shielding jackets are welded to the inner surface of motor stators and the outer surface of rotors for isolating the medium conveyed from the stator cores, rotor cores and windings. The space between them communicates to the pump chambers and withstands the pressure from the RCS. The auxiliary systems required for the canned pumps are simple, and only cooling water systems are required for cooling the stator windings.

Figure 3.20 shows the typical structure of the RCP. Its main components include: 1-Pump blade; 2-outlet guide vane; 3-serpentine cooler; 4-high-speed stator winding; 5-shielding jacket; 6-speed sensor; 7-impeller; 8-upper guide and thrust bearings; 9/12-component cooling water inlet and outlet; 10-low-speed stator winding; 11-rotor; 13-heat screen; 14-lower guide bearing; 15-water chamber

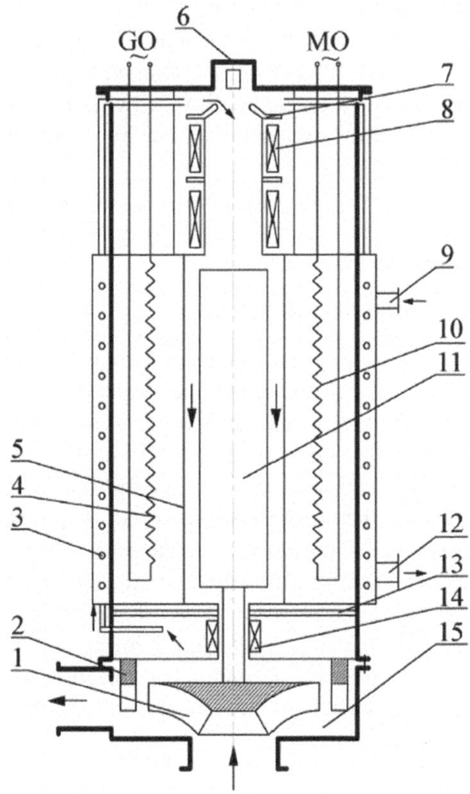

Fig. 3.20 Primary circulating pump

① Bearing

The RCPs are generally provided with bearings, including radial bearings and double direction thrust bearings, which are all located at the motor side and are of water lubricated type.

The thrust bearings are intended to bear the residue axial loads from the rotors, while the radial bearings to bear the radial loads from the rotors and restrict the rotors from displacement.

② Shielding jacket

Shielding jackets are welded to the inner surface of the motor stators and the outer surface of the rotors to isolate the stator windings and rotors from the primary coolant, and they are made of materials with high strength, plasticity and corrosion resistance. The space between the stator shielding jackets and the rotor shielding jackets are filled with medium during the operation of the pumps.

③ Stator winding and cooling structure

Due to the presence of the shielding jackets of the canned motors and the friction between the rotors and water, the canned motors have great attrition and large heat generation, and the stator windings are enclosed in a limited space, with poor heat dissipation. Therefore, the canned motors shall be with high grade insulation and heat resistance, and be provided with the circulation circuits for the medium to cool the windings and remove heat by means of heat exchange with the outer cooling water.

Moreover, the heat barriers are provided between the pumps and motors to decrease the heat transferred from the hot medium to the motors during the operation of the pumps.

④ Hydraulic components

The hydraulic components of the canned pumps mainly include the impellers, guide vanes and pump casings. The impellers are directly attached to the motor shafts, and are provided with balance holes to mitigate the axial loads.

With the technical advancement of the marine nuclear power plants, the reactor power is higher and the service lifetime is longer, hence large-capacity, long-life and highly reliable RCPs are required. Application of variable frequency technology in the pumps of the marine nuclear power plant invokes the attention in the nuclear fields, as it can mitigate the impact on the electric grid due to the direct startup or operation switchover of high-power main pumps and the disturbance to the thermo-hydraulic parameters of the reactor core, improve the safety for the change of operation mode of the plant, and smoothen the variation of the parameters. Indeed, this application requires sufficient analysis and demonstration in advance to address the technical problems concerned.

(2) Shaft sealed pump

Due to the increasing of the power of the nuclear power plants, the increased power of the canned pumps is required and brings about great problems. However, the technology of the shaft seals is constantly improved and optimized, becoming more technically mature. In light of the combined considerations of technology, cost, operation and maintenance, the shaft sealed pumps have been widely used as the main pump sets of the plants in operation and under construction internationally, with and show excellent operating performance. The canned pumps are still mainly used for the marine nuclear power plants.

The shaft sealed pumps are generally vertical motor-driven single-stage single-suction mixed flow pumps or axial flow pumps, and are often provided with and driven by drip-proof squirrel cage induction motors. The auxiliary systems for the shaft sealed pumps are complex, generally including the component cooling water systems for cooling the pump and motor components, the chemical and volume control systems for providing water injection to the seals, the nitrogen supply systems for providing nitrogen to the shutdown seals and the drain and venting systems for venting the exhaust from the main pumps and collecting gas from the seal leak offs. The shaft sealed pumps generally consist of motors, inner bearings, main shafts, hydraulic components, heat screens and shaft seal assemblies.

The main components of the shaft sealed pumps usually include:

① Motor

The three-phase drip-proof squirrel cage induction motor is generally adopted, and it is excluded from the pressure boundary, as the shaft seal is adopted for the pump. The main structural components of the motor are not necessarily made of anti-corrosive materials, and thus their manufacturing costs are low. In addition, the space between the stator and the rotor of the main pump motor is filled with air, requiring no anti-corrosive shielding jacket,hence no shielding jacket mechanical loss. Therefore, the efficiency of such shaft sealed pump motor is generally 10% ~ 20% higher than that of the canned pump motor.

The motor is provided with the anti-reversion device, which prevents the reverse rotation in case that one main pump is stopped and other pumps are operating. The motor is provided with flywheel at its rotor to increase the rotational inertia of the rotor and ensure adequate flow coast down in case of loss of power to the main pump set. In this way, the flywheel drives the main pump to provide coolant to the reactor core and ensures the reactor safety by means of the flow coast down.

The motor is provided with two water/air coolers, which are installed in a symmetric way. Air circulates from the motor to the coolers and then to the motor again. The capacity of the coolers shall meet the requirement to maintain optimal operation temperature of the motor. The residue heat from the motor to the containment will be removed to the environment via the ventilation system in the reactor hall.

② Inner bearing and main shaft

Either rigid shafts or flexible shafts can be used to connect the pump shaft and motor shaft. The rigid shafts are simple in structure, but with poor alignment as compared with the flexible shafts. The flexible shafts, though complex in structure, have better alignment, thus providing convenience for field installation and maintenance.

The RCP pump set is provided with upper and lower radial bearings for motors, double-direction thrust bearings and lower radial bearings for pumps. The hydrodynamic oil-lubricated sliding bearings are used as the upper and lower radial bearings for motors. The double-direction tilting-pad thrust bearings are adopted to bear the axial thrust from the hydrodynamic force brought by the impellers, the weight of rotary components and the system pressure. The hydrodynamic-water-lubricated sliding bearings are used as lower radial bearings for pumps.

③ Hydraulic components

The hydraulic components of the main pumps mainly include the pump casings, impellers and guide vanes. Reactor coolant is suctioned by the impellers via the inlet pipes along the axis of the impellers, with the pressure increased. It is boosted by the guide vanes, and then discharged from the horizontal discharge pipes along the radial direction. The pump casings can be SS castings, or low-alloy steel forgings with extra-low-carbon SS overlaid on their inner surface. The impellers and guide vanes can be SS castings, or SS forgings. In most cases, axial flow impellers and guide vanes are SS forgings.

④ Heat screen

Heat screens are varied in structure. One type of heat screens is composed of insulation jackets and the heat exchanger with multi-layer serpentine coiled tubes. Another type of heat screens consists of solid thermal insulations and high-pressure coolers. The heat screens are cooled by the component cooling water. The main pumps remove the heat from the hot coolant around the shafts above the impellers, providing a low temperature condition for normal operation of the pump shaft seals and bearings.

⑤ Shaft seal assembly

There are two kinds of proven shaft seal assembly solutions, namely the hydrostatic seal solution and the hydrodynamic seal solution. Both kinds of solutions have been applied in the shaft sealed pumps. However, the hydrodynamic seal solution represents the direction for future development, and has certain advantages over the hydrostatic seal solution.

The shaft seal system consists of the three-stage hydrodynamic seals and shutdown seal, which are arranged in series. The three-stage seals have the same structure. Each of the three seals withstand one-third of the system pressure during normal operation, and each one is designed to be able to withstand the full system pressure. The shutdown seal is provided above the third-stage seal, and it is closed for sealing purpose only upon the shutdown of the RCP, and is kept open during normal operation of the RCP.

3.5.3 *Reactor Coolant Piping*

(1) Characteristics of reactor coolant piping for marine application

The reactor coolant piping (also called the main pipes) is a key part of the marine nuclear power plants. As crucial parts of the reactor coolant pressure boundary, the main pipes are used to connect the main components such as the reactor, SGs and main pumps as well as several high-energy auxiliary systems, forming the flow paths for the reactor coolant circulation. Therefore, the main pipes are also called the "aorta of the RCS.

With extremely adverse operation conditions during the reactor operation, the main pipes are exposed not only to the mechanical damage due to the combined loads and low-cycle high-frequency fatigue, but also to the hazards from the high-temperature, high-pressure and high-flowrate medium or chloride corrosion in the marine environment. As a result, more attention shall be paid in particular to the design, materials and manufacturing technologies. In addition, more stringent requirements shall be put forward on their safety and reliability, in view to their increasing design lifetime. Their operation performance has significant impacts on the safe and reliable operation of the whole nuclear power plant.

Compared with the nuclear power plants on land, the pipes for marine nuclear power plants have the following concerns:

① The exposure of the pipe materials to the radioactivity or corrosion should be considered, as the primary loop system conveys the radioactive medium and works under high temperature and high pressure.
② Working in the limited space and harsh marine environment and required to withstand their dead loads, the thermal expansion and the external dynamic loads, the pipes shall have excellent mechanical performance and be provided with proper supports and expansion joints to keep them from failure.
③ The pipe materials shall be with high comprehensive mechanical properties, excellent corrosion resistance, radiation stability and good processing performance.

To ensure the safety and reliability of the main pipes for marine nuclear power plants, plenty of research and development work has been conducted constantly in the world, to finding alternate materials with better performance and advanced manufacturing technologies. So far, there are three kinds of materials available for the main pipes, including stabilized Austenitic SS, standard 304 or 316 Austenitic SS, and extra-low-carbon 304L or 316L SS. The stabilized Austenitic SS, such as ANSI 321 SS and ANSI 347 SS, has poor weldability as compared with the 304 or 316 SS, and is susceptible to many defects during subsequent treatment due to its contents of TiN and NbN, though its strength is acceptable. The standard 304 and 316 Austenitic SS are with satisfactory strength, but are susceptible to intergranular corrosion after large-scale welding due to its high carbon content. The extra-low-carbon 304L and 316L SS have excellent weldability and workability and are resistant to intergranular corrosion, but also have obvious weakness, i.e., insufficient strength.

Research on nuclear grade 316LN SS has been conducted internationally since 1970s. It is similar to the nuclear grade 316L SS mentioned above from technical perspective, except for slight difference in carbon and nitrogen contents. Moreover, novel manufacturing technology (integrated forging) has been adopted for the main pipes, instead of previous manufacturing process in which the straight sections, bends and nozzles are made separately and welded together. This new technology represents the technical trends for future R&D of the main pipes.

④ The stress analysis of the main pipeline, as Class I nuclear component, needs to consider the strength under various loading effects in the design to analyze the rationality of the structure and layout. For such stress analysis, each operation condition is correlated to a specific set of load effects (pressure, thermal loads, radiation and corrosion). For the main pipes for marine application, the analysis shall be focused on the dynamic loads applied inside or outside the pipes.

(2) Calculation of inner diameter and wall thickness of main pipes

According to the temperature of the coolant conveyed, the main pipes are divided into the following parts: the hot leg (the pipe from the outlet of reactor pressure vessel to the inlet of the SGs), the transition section (the pipe from the outlet of the SGs to the inlet of the RCPs) and the cold leg (the pipe from the outlet of the RCPs to the inlet of the reactor pressure vessel).

The flowrate of the reactor coolant shall be set properly for the main pipe to avoid too large inner diameter of the main pipe due to excessively low flowrate, and to prevent the corrosion or erosion of the pipes due to excessively high flowrate [1].

The inner diameter of a main pipe is defined as

$$d_i = 2 \times 10^3 \sqrt{\frac{G_c}{\pi \rho \omega}} \text{ mm} \tag{3.1}$$

where, ρ is the average density of the coolant in kg/m^3, and ω is the flowrate of the coolant in m/s.

The wall thickness of a main pipe is defined as

$$t_m = \frac{pD_o}{2(S_m + yp)} + A \text{ mm} \tag{3.2}$$

Or

$$t_m = \frac{pd_i + 2A(S_m + yp)}{2(S_m + yp - p)} \text{ mm} \tag{3.3}$$

where, t_{m} is the minimum required wall thickness calculated in mm (only applicable to Eq. (3.3) for $d_i = D_{\mathrm{o}} - 2t$); p is the inner design pressure of the main pipe in MPa; D_{o} is the outer diameter in mm; S_{m} is the basic permissible stress strength of the material at design temperature in MPa; d_i is the inner diameter in mm; A is the additional wall thickness in mm; and y is the coefficient.

Verification of the flowrate ω' and ω'' for the cold and hot legs shall be preceded by the determination of specifications of the main pipe.

$$\omega' = \frac{4G_c}{\rho' \pi d_i^2} \text{ m/s} \tag{3.4}$$

$$\omega'' = \frac{4G_c}{\rho'' \pi d_i^2} \text{ m/s} \tag{3.5}$$

where, ρ' and ρ'' are the density of the coolant in the hot and cold legs in kg/m^3.

3.6 Reactor Coolant Water Chemistry

The high-temperature reactor coolant is a kind of corrosive medium for the infiltrated structural materials. So, its impacts on the operational safety of the reactor mainly include:

① Corrosion of fuel cladding; ② corrosion of materials at reactor coolant pressure boundary; ③ deposits on the surfaces of fuel cladding, resulting in lower heat transfer efficiency and accelerated local corrosion; and ④ increase in corrosion products, resulting in higher ex-core dose rate.

Therefore, the water chemistry in a RCS is closely related to the safe and reliable operation of the nuclear power plant and the selection of structural materials for primary loop equipment, and the water quality shall be controlled. The elements of water quality control for the reactor coolant are described below.

(1) pH

The value of pH indicates the concentration of $[H^+]$ in water, and is equal to the negative logarithm of the concentration of $[H^+]$.

The pH of the reactor coolant is controlled mainly by forming a compact protective oxidation film on the metal surfaces of the primary structural materials to reduce the corrosion rate.

The extensive study and experiments show that the corrosion rate of iron-based and nickel-based alloy in high-temperature water is much lower than that in acid environment when reactor coolant pH is neutral and alkaline. For deaerated water, the main oxidizing agent to metal corrosion is H^+ . When the coolant is slightly alkaline, a large amount of OH^- can passivate the metal on the surface of the metal material and corrosion products deposit on the metal surface with the structure of R_3O_4(R=Fe, Ni or Cr), thus forming a closely-packed protective film and lowering the corrosion rates.

For a nuclear power plant during operation, LiOH is often used to regulate the pH of the coolant. The pH shall neither be kept too low, in which case chemical corrosion will be accelerated, nor kept too high, in which case caustic corrosion may occur to the structural materials. In the 360 °C deaerated water, it has no obvious impact on

Table 3.1 pH indicators for coolant of PWR

Name	$pH_{25\,°C}$	Name	$pH_{25\,°C}$
Indian Point Power Station	7.5–8.5	Lenin	6–8
Novovoronezh Power Station	10–10.5	Mutsu	6–9
Shippingport Atomic Power Station	9.5–10.5	Savannah	6–7

the corrosion rate of zirconium alloy at the LiOH concentration of 2×10^{-3}mol/L (equivalent to the pH of 11.3 at room temperature). But its impact on the corrosion rate becomes obvious when the LiOH concentration increases to 10^{-2}mol/L (equivalent to the pH of 12 at room temperature), and will become more significant with the increase of LiOH concentration and pH.

Therefore, the neutral or slightly alkaline reactor coolant is preferred for marine nuclear power plants. Table 3.1 shows the indicators, based on which the pH of the coolant of PWRs shall be regulated.

Reactor coolant of marine nuclear power plants is generally neutral or slightly alkaline, and in either case, there are advantages and disadvantages.

① Compared with alkaline water, neutral water has the following advantages:

i No alkalizer is required, avoiding related chemical concerns caused by chemical addition such as local concentration effect and induced radioactivity brought by the alkalizer;
ii The requirement on the water chemistry regulation is simple, lowering the requirement on the operators; and
iii The configuration for water quality control and monitoring process systems is simple, with the demand on installation space reduced.

② Compared with neutral water, alkaline water has the following advantages:
i The corrosion rate of the structural materials and the release rate of the corrosion products are lowered, and thus the concentration of corrosion products in the coolant is decreased;
ii Deposits of corrosion products on the surfaces of fuel elements are reduced;
iii Activation of corrosion products is mitigated, and thus the radioactivity of the reactor coolant is lowered.

It can be seen from the operation experience of the marine nuclear power plants outside China that the reactor coolant, whether neutral or alkaline, can meet the requirements on the safe operation of the plants.

(2) Conductivity

Conductivity reflects the contents of salt in the reactor coolant, and is taken as an important indicator of its demineralization. Conductivity increases with the increasing of the impurity ions in the reactor coolant.

Highly demineralized water with the conductivity less than 2 μS/cm is generally used as the reactor coolant to prevent the activation of the impurity matters in the reactor core. It shall be noted that this is the case for the reactor coolant without chemical additives. When chemical is added, the conductivity of the coolant will increase significantly. In this case, it cannot truly reflect the demineralization.

(3) Oxygen and hydrogen

Oxygen is a sensitive factor for uniform corrosion and stress corrosion cracking of the primary loop materials, and is a main factor affecting the redox potential of the coolant. Therefore, the control of the dissolved oxygen in the reactor coolant is important for water chemistry, and is necessary for reducing the corrosion of the primary loop structural materials and the release of the corrosion products.

Reducing the content of dissolved oxygen in the reactor coolant during operation is beneficial, as described below:

① In high-temperature water with low content level of dissolved oxygen, magnetic spinel oxides occurs on the SS surfaces to form a closely-packed film, preventing the contact of the metal base with the water and its oxidation. In such water, magnetic oxides featuring the crystal structure of spinel occurs on the nickel-based alloy surfaces to form a protective film.
② As dissolved oxygen is the accelerant of chloride stress corrosion cracking for austenitic stainless steel, reducing the dissolved oxygen content in reactor coolant can decrease the possibility of stress corrosion cracking of structural materials
③ The dissolved oxygen interacts with neutron flux and thus has large impact on the corrosion of zirconium alloy. The effects of neutron irradiation on corrosion rates of zirconium alloy can be decreased by controlling the dissolved oxygen content during the operation of the reactor.
④ Chloride, fluoride and sulfate ions

As described below, impurity ions, such as chloride, fluoride and sulfate ions, are key concerns for the reactor coolant, as they have impacts on the stress corrosion of stainless steel and corrosion of zirconium alloy.

① The stress corrosion of stainless steel is one of the key contributors to equipment damage, and originates from either the residual stress resulting from loads or heating applied, or the equipment and the particular corrosion medium in the coolant, such as chloride or fluoride ions. For Austenitic SS, sulfate ions are also sensitive factors for its stress corrosion cracking.
② Impurity ions have adverse impacts on the corrosion resistance of zirconium alloy. Even a trace of fluoride ions can result in significant increase of initial corrosion rate and hydrogen absorption of the zirconium alloy.

Fluoride ions in the reactor coolant may come from the concentrated nitric acid solution and hydrofluoric acid solution that are used for the cleansing of the zirconium alloy surfaces, or a small amount of fluorine ions that are dissolved in the coolant by some sealing filler materials, or from chemical additives. Sulfate ions mainly come from hydrolysate of ion exchange resin debris or sulfate foreign matters of other origin.

(4) Suspended matter

Concentration of suspended matters in the reactor coolant is restricted to minimize the deposits on fuel elements and thus reduce the accumulated radioactivity in the primary loop systems due to entrainment and re-distribution of the activated corrosion products. Excessive suspended matters also result in aggravated corrosion and abrasion of the materials, or even affect the operation of control rod drive mechanisms.

References

1. Minjun, Peng. 2009. *Marine Nuclear Power Unit*. Beijing: Atomic Energy Press.
2. Hanhong, Sun. 2010. *Third Generation Nuclear Power Technology AP1000*. Beijing: China Electric Power Press.
3. Dafa, Zhang. 2009. *Operation and Management of Marine Nuclear Reactor*. Harbin: Harbin Engineering University Press.

Chapter 4
Nuclear Auxiliary Systems

4.1 Overview

As an important part of the marine nuclear power plant, the nuclear auxiliary systems are the supporting systems and the related service systems of the reactor coolant system. Their main function is to ensure the normal operation of the reactor and the primary loop system. The nuclear auxiliary systems generally consist of pressure safety system, residual heat removal system, coolant purification system, waste treatment system, coolant-charging system, component cooling water system and sampling system. The main functions of above primary auxiliary systems are as follows:

① During the steady-state operation and normal power change of the nuclear power plant, compensate for the expansion or contraction of coolant, limit the pressure change of reactor coolant system within the allowable range to avoid the excessively high system pressure, which may damage the pressure-bearing components and equipment, and to avoid the excessively low system pressure, which may cause the boiling of core beyond the safety design criteria.

② Remove some of the soluble and insoluble impurities in the reactor coolant, so as to make the concentration of the impurities in the coolant lower than the allowable value, reduce the radioactivity dose level in the reactor compartment and maintain the water quality of the reactor coolant system in the specified range;

③ Collect, store or discharge the radioactive wastes generated during the operation and repair4 of the primary loop system, and handle such radioactive wastes to ensure the safety of the personnel on board and the environment;

④ During the normal operation of the reactor coolant system, compensate for the loss of reactor coolant such as leakage to maintain the water inventory in the primary loop system within the normal range;

⑤ Cool the primary loop system equipment to be cooled (including pumps and various types of heat exchangers) to ensure that the equipment to be cooled can normally perform the functions;

J. Yu, *Marine Nuclear Power Technology*,
https://doi.org/10.1007/978-981-15-2894-1_4

⑥ Provide shielding against the radioactivity generated by the reactor to reduce the radioactivity dose to the operation personnel and the public to the acceptable range; and
⑦ Monitor the water quality of primary loop system and provide the basis for judging the resin failure and the fuel element damage.

This chapter introduces some of these systems.

4.2 Pressure Safety System

During the operation of a nuclear power plant, the pressure of reactor coolant system will change with the plant load and external factors. If the pressure is too high and exceeds the design pressure of the system, it may lead to a LOCA; if the pressure is too low, it may lead to core flow instability and insufficient core cooling that may endanger the reactor safety. The purpose of pressure safety system is to control the pressure fluctuation of reactor coolant system [1].

The pressure safety system of the nuclear power plants mainly includes steam pressurization system or nitrogen pressurization system. The steam pressurization system utilizes the electric heating or spraying of coolant to establish the pressure of reactor coolant system and maintain it in the specified range, while the nitrogen pressurization system utilizes the compressibility of nitrogen to compensate for the fluctuation of reactor coolant volume with the temperature change. Compared with the steam pressurization system, the nitrogen pressurization system has the advantages of simple operating principle, no electric energy consumption during the system operation, no need for automatic control of pressure and water level, quick start-up of the unit and no generation of waste water, simplifying the control and operation modes of nuclear power plant and significantly improving the maneuverability. In addition, the space for nitrogen pressurization system can be divided into multiple small nitrogen cylinders distributed in the reactor compartment, realizing the flexible layout. Since the nitrogen pressurization system is a passive system, the reactor coolant pressure will automatically follow the change within a certain range during the reactor start-up, shutdown and power operation and there will be no need of any control, thus reducing the self-energy demand during the reactor start-up. The steam pressurization system requires that the medium in the pressurizer is frequently in a two-phase equilibrium state and that the saturation conditions of water and steam can easily trigger the phase transition, so as to flexibly regulate the pressure change caused by the volume change.

When the nitrogen pressurization system is adopted, the reactor coolant is generally the ammonia type alkaline water. This weakly alkaline (ammonia) water chemistry system has the stability and the self-regulation function, facilitating the maintenance of alkaline water quality of reactor coolant. However, the nitrogen dissolution in the reactor coolant is a problem, that is, as the temperature of the coolant decreases, the solubility of nitrogen will decrease, and some nitrogen in the coolant will be released and cause the local accumulation, affecting the flow stability and damaging some equipment in the loop. At the same time, it is necessary to consider the effect of nitrogen in the coolant on the physical characteristics of core and on the heat transfer of fuel elements. In addition, the nitrogen leakage will also bring about adverse consequences.

The following mainly introduces the steam pressurization system.

4.2.1 System Description

(1) System functions

The pressure safety system has the following main functions:

① Pressure regulation: during the steady-state operation, transient power change, cold start-up and cold shutdown of the nuclear power plant, utilize the steam compressibility and spray condensation or heating with electric heaters to keep the amplitude of pressure fluctuation in reactor coolant system within the allowable range;

② Pressure protection: under the accident conditions with excessively high reactor coolant system pressure, actuate the safety discharge function to realize the overpressure protection; under the accident conditions with excessively low reactor coolant system pressure, start the engineered safety system to perform the safety injection.

(2) System composition and operation flow process

The pressure safety system consists of one pressurizer, spray valves, safety valves and their connecting pipes and measuring instruments. Figure 4.1 illustrates the flow process of pressure safety system.

① Pressure regulation flow process: When the pressure in the reactor coolant system increases, the spray valve will open and the reactor coolant in the cold leg of the primary pipe will be sprayed by the spray pipe into the steam space of pressurizer to condense the steam, thus decreasing the pressure in the reactor coolant system to the allowable range. When the pressure in the reactor coolant system decreases, the electric heating elements will be put into operation to heat the saturated water and convert it into saturated steam, thus increasing the pressure in the reactor coolant system to the allowable range.

② Pressure protection flow process: Under the accident conditions, three safety valves are installed in the pressure safety system to prevent the overpressure of reactor coolant system. The safety valves utilize the staged discharge. During an overpressure transient, if the first safety valve fails to open at its opening pressure setpoint or it opens but the pressure keeps increasing, when the system pressure increases to the opening pressure setpoint of the second safety valve, the second

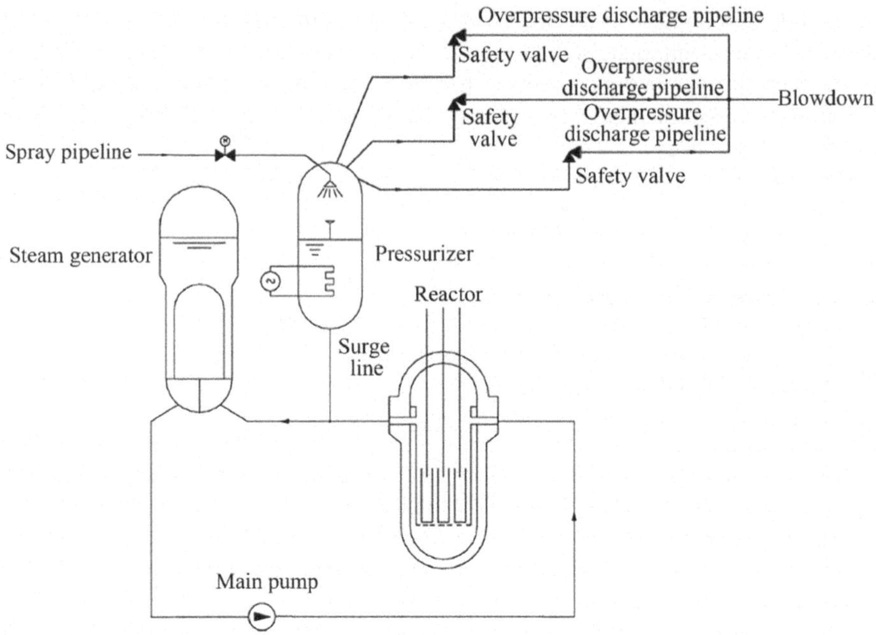

Fig. 4.1 Flow process of pressure safety system

safety valve will open. If the system pressure continues to increase, the third safety valve will open.

4.2.2 *Equipment Description*

(1) Pressurizer

The pressurizer is an important device for primary loop pressure control and overpressure protection. Its operating principle is as follows: at the rated power, the lower part inside the pressurizer is filled with saturated water and the upper part is filled with saturated steam, and the bottom of pressurizer (the liquid area) is connected to the hot leg of a loop of the reactor coolant system by a surge line. Since the reactor coolant system, except for the pressurizer, is full of water, the pressure in the pressurizer spreads over the entire system.

During the steady operation of pressurizer, the liquid phase and the vapor phase, which are in equilibrium, are respectively the saturated water and the saturated steam. The temperature of steam and water in the pressurizer is equal to the saturation temperature at this (absolute) pressure, thus the temperature in the reactor is lower than that in the pressurizer.

At the operating (absolute) pressure of the reactor coolant system, the density of water is several times that of steam. Therefore, when the electric heater of pressurizer heats the water to generate the steam, there will be a large volume change, so the steam pressure will inevitably increase, causing the pressure of the pressurizer to increase. On the contrary, when the steam is condensed by the spray water from the cold leg, the steam density will decrease, causing the pressure decrease.

① Design principles

The pressurizer volume must meet the following requirements: A. The pressurizer shall have enough water volume to prevent the exposure of heaters during the normal operation of pressurizer and adapt to the water level fluctuation caused by reactor coolant temperature change within the power range of 0 –00%. B. The pressurizer shall have enough steam volumecapacity to adapt to the water level fluctuation caused by shutdown at full load due to loss of the external power and prevent the opening of safety valve in case of the reactor scram caused by high pressure or high water level of the pressurizer.

② Structure of Pressurizer

The typical structure of a pressurizer is a vertical cylinder, the top and bottom of which are ellipsoidal heads, as illustrated in Fig. 4.2.

The lower part of pressurizer is the water space, which accommodates surge line nozzles, electric heaters, nuclear sampling ports and instrument nozzles. The upper part is the steam space, which accommodates spray nozzles, spray heads, instrument nozzles, pulse pipe nozzles and manholes.

On the bottom head, there are several electric heaters installed vertically, distributed on the concentric circles with the center line of bottom head as the center and inserted through the bottom head into the pressurizer water. In order to prevent the lateral vibration of electric heaters, two horizontal clapboards are arranged in the bottom headto support the electric heaters.

The surge line connects the bottom head of pressurizer to the hot leg of main pipe, allowing the exchange between the primary coolant and the pressurizer water. The baffle screen is arranged right above the inlet of surge line to uniformly mix the water from the surge line and the water inside the pressurizer, and prevent the impurities in the pressurizer from entering the reactor coolant system. Both ends of the surge line are arranged with thermal sleeves to withstand the thermal stress generated by the exchange of hot water or cold water.

The top head of pressurizer is provided with a manhole that is sealed with a flat bolted cover, and the bottom head is arranged on a cylindrical skirt. The upper circumference of the supporting skirt is arranged with ventilation holes.

(2) Safety valves

The safety valves of pressurizer are the overpressure protection devices of reactor coolant system and the key devices to ensure the safety of pipes and equipment in the reactor coolant system and other systems connected with reactor coolant system. They mainly consist of main valve, pilot valve and relief valve (see Fig. 4.3). The

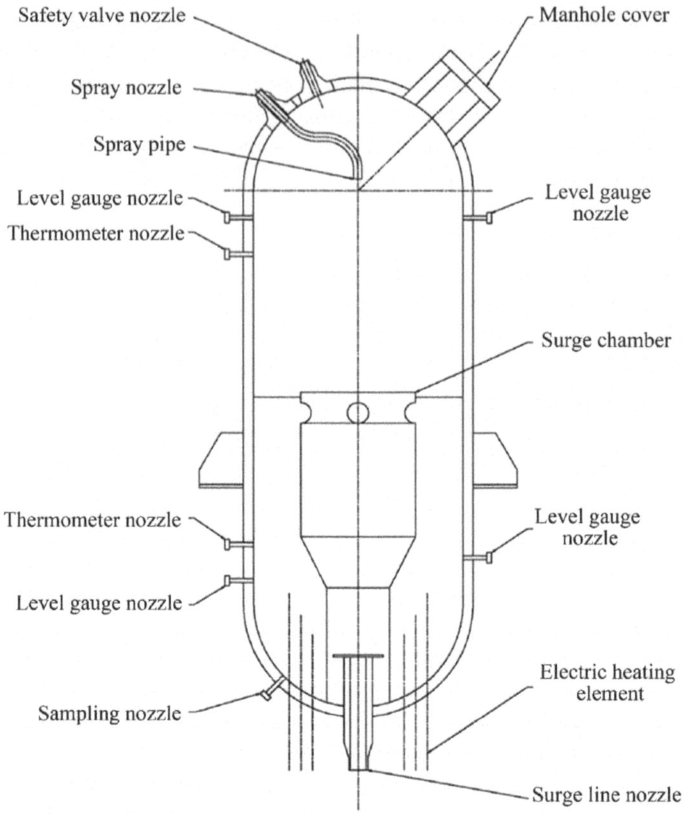

Fig. 4.2 Structure of electric heater type pressurizer

main valve consists of upper and lower valve bodies, valve seat, valve flap, upper and lower piston assemblies, main valve rod, main valve stem, preloaded spring, inlet and outlet flanges and nozzle. The pilot valve consists of valve body, valve seat, valve flap, bellow assembly, preloaded disc spring and setting assembly.

The inlet of pressurizer safety valve is connected by a pipe to the upper steam space of the pressurizer. After the pressurizer establishes the steam space, the inlet cavity and inlet pipe of the safety valve are filled with condensate to form the water seal section of safety valve. When the safety valve is closed, the safety valve will not be affected by the transient change of temperature in the reactor coolant system because of the water seal. In addition, the condensate can prevent the leakage of hydrogen from the sealing surface and damage to sealing surface under normal operating conditions. The pulse pipe from the inlet line of the main valve inlet line of safety valve is connected with to the pulse pipe nozzle of pilot valve. A filter is in series connection with the pulse pipe to prevent the impurities from entering the pilot valve and causing the pipe blockage.

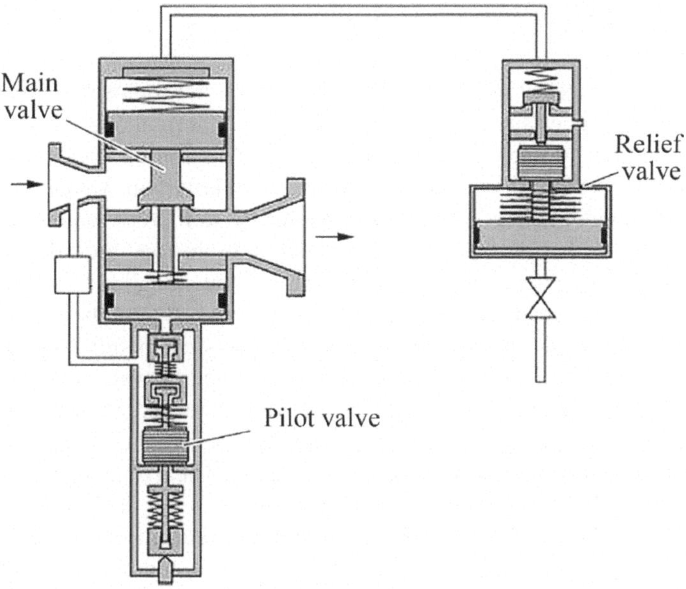

Fig. 4.3 Pressurizer safety valve

The pressurizer safety valve performs the automatic overpressure protection through the pilot valve. When the system pressure reaches the setpoint pressure of the pilot valve, the pilot valve will open to pressurize the lower cavity of lower piston of main valve, the main valve flap will be driven to open by the lower piston, and the system will relieve the pressure. When the system pressure decreases to the re-seating pressure of the pilot valve, the pilot valve will be closed, the lower cavity of the lower piston of the main valve will be depressurized, the pressure balance between upper and lower cavities of lower piston of main valve will be restored, the main valve flap will be reset by spring force and medium hydrodynamic force, and the main valve will be closed. The pressurizer safety valve performs the remote forced relief function through the relief valve. When the compressed air solenoid valve is open, the relief valve piston will, through the push rod, push and open the relief valve flap to relieve the pressure of upper cavity of upper piston of main valve, the main valve flap will be pulled to open by the upper piston, and the system will relieve the pressure. When solenoid valve is closed and the relief valve is reseated, the upper cavity of upper piston of main valve and the discharge pipe will be disconnected, the pressure balance between upper and lower cavities of upper piston of main valve will be restored, the main valve flap will be reset by spring force and medium flow force, and the main valve will be closed.

4.3 Residual Heat Removal System

One of the main issues in nuclear safety is to ensure the removal of heat released from nuclear fuel under any circumstances. Under the normal operating conditions, the heat generated by nuclear fission and decay of fission products is transferred from primary loop through steam generator to secondary loop. When the reactor is shut down, although the heat generated by nuclear fission drops to zero quickly, the heat released by the fission fragments generated by fission and their decay products during the radioactive decay will still exist. This is the so-called "residual power". Figure 4.4 illustrates the principle of residual heat removal system.

The system composition is relatively simple. The system mainly consists of residual heat removal heat exchanger, residual heat removal pump, valves and corresponding piping systems. When the residual heat removal system is put into operation, the main pump will operate at a low speed to maintain the circulation of coolant in the system at a certain flow, the coolant entering the residual heat removal heat exchanger will be cooled by the component cooling water ore seawater at the opposite side of the heat-transfer surface, and the cooled coolant will return to the pipe at the main pump inlet side. The cooling rate of reactor coolant system can be controlled by adjusting the flow of coolant entering the residual heat removal heat exchanger. In order to reduce thermal shock, before the residual heat removal heat exchanger shall be preheated by a small coolant flow before it is put into operation, a small amount of flow shall be preheated.

During the process of cold shutdown of the reactor, the pressure and temperature of reactor coolant system will decrease with the removal of residual heat, and the makeup water must be timely provided to the reactor coolant system to ensure that the system pressure will not become lower than the allowable operating pressure of the main pump.

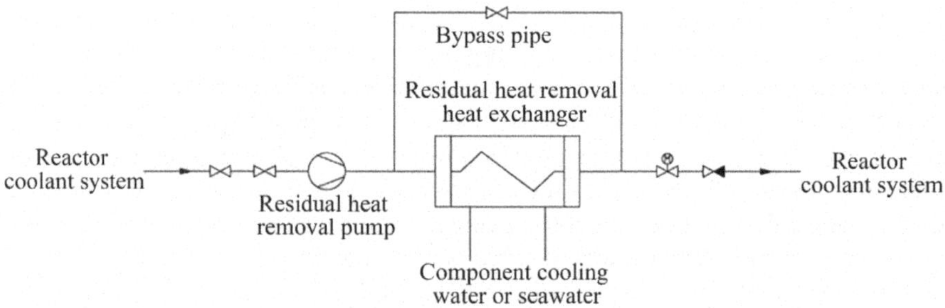

Fig. 4.4 Principle of residual heat removal system

4.3.1 System Description

(1) System functions

The main function of the residual heat removal system is to, when the primary loop temperature drops below 180 °C and the absolute pressure drops below 3.0 MPa after the normal shutdown of reactor, remove the residual heat of core, the sensible heat of primary coolant and equipment and the heat generated in the primary loop by the main pump, and bring the reactor into a cold shutdown state.

(2) System composition and operation flow process

The residual heat removal system consists of residual heat removal pumps, residual heat removal heat exchangers and related valves and pipes connected to the equipment.

At the first stage of the cold shutdown, the reactor coolant system utilizes the steam generators to transfer the heat to the main steam system. At the second stage of cold shutdown, the residual heat removal system reduces the temperature of reactor coolant system by removing decay heat and sensible heat from the core and the reactor coolant system.

From cold shutdown to restart, the nuclear power plant relies on the residual heat removal system to remove the heat from the core and the reactor coolant system.

4.3.2 Equipment Description

(1) Residual heat removal pump

During the normal shutdown of reactor, the residual heat removal pump provides a head for the residual heat removal system to facilitate the circulation of reactor coolant, and the residual heat removal cooler removes the residual heat of core and the latent heat of equipment, thus guaranteeing the safety of reactor.

The residual heat removal pump can be of a single-stage single-suction vertical centrifugal pump design, and its main parts and components are made of stainless steel.

(2) Residual heat removal heat exchanger

The function of residual heat removal heat exchanger is to remove the residual heat of core and the latent heat of equipment during the normal cold shutdown of reactor.

The typical residual heat removal heat exchanger is the U-tube heat exchanger, the tubes and tube sheets of which are welded together. During the operation of heat exchanger, the reactor coolant flows through the tube side and the seawater flows through the shell side.

4.4 Coolant-Charging System

4.4.1 System Description

(1) System functions

The main functions of coolant-charging system are as follows:

① During the cold start-up of reactor, perform the initial water filling and initial boosting of reactor coolant system and nuclear auxiliary system;
② During the normal operation of reactor coolant system, compensate for the loss of reactor coolant due to leakage, sampling and other losses to maintain the pressurizer water level within the normal range;
③ During the cold shutdown or accidental shutdown of reactor, provide makeup water to the reactor coolant system to compensate for the volume contraction of coolant; and
④ Supply water for other purposes.

(2) System composition and operation flow process

The coolant-charging system generally consists of makeup water pump, makeup water cooler, makeup water ion exchanger and makeup water pipeline. The system flow process is illustrated in Fig. 4.5.

The water source for the coolant-charging system is generally of dual configuration. After the makeup water is cooled by the makeup water cooler and the dissolved oxygen and soluble and insoluble impurities in it are removed by the ion exchanger, the makeup water is delivered to the reactor coolant system by the makeup water pump via makeup water pipeline.

During the normal operation, according to the pressurizer water level signals, when the level drops to a certain value, the makeup water pump will be started automatically for the makeup of reactor coolant system. When the level rises to a certain value, the makeup will be stopped automatically. When other consumers

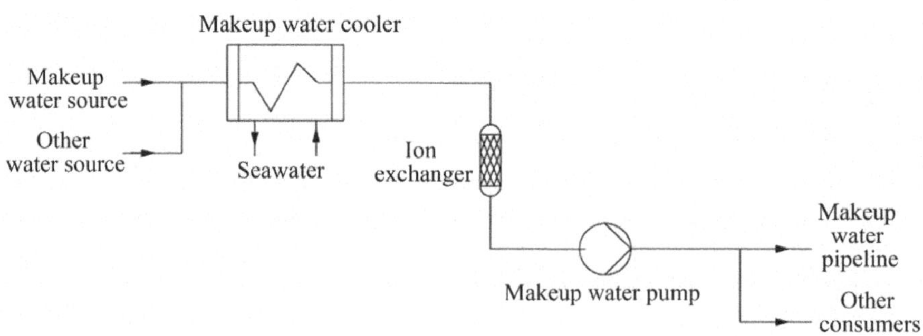

Fig. 4.5 Flow process of coolant-charging system

require the makeup, the relevant valve will be opened and the makeup water pump will be started for the makeup. After the makeup is completed, the system state will be restored.

4.4.2 Equipment Description

(1) Makeup water pump

During the normal operation, the makeup water pump performs the makeup for the reactor coolant system to maintain the water inventory of the reactor coolant system. During the start-up of nuclear power plant, the makeup water pump performs the initial water filling and boosting of reactor coolant system and primary loop system, and injects the chemical additives into the reactor coolant system.

The makeup water pump may adopt a multi-stage centrifugal pump design, and its main parts and components are made of stainless steel.

(2) Ion exchanger

The ion exchanger is of low temperature and low pressure design and installed upstream of the makeup water pump. It achieves different functions by different types of resin contained. This equipment is a typical vertical container. The medium enters it from the top head, flows through the filter into resin bed, flows through the outlet filter, and flows out from the bottom head. The ion exchanger is mainly made of stainless steel.

4.5 Component Cooling Water System

4.5.1 System Description

(1) System functions

The function of component cooling water system is to provide the cooling water for the equipment to be cooled in the primary loop. In order to prevent the seawater from directly contacting the primary equipment and prevent the radioactive substances from releasing to the seawater and polluting the environment, the component cooling water system is designed as an intermediate closed-loop circulation system, which isolates the primary loop system from the seawater system.

(2) System composition and process flow

The component cooling water system is mainly composed of component cooling water heat exchanger, component cooling water pump and component cooling water surge tank. The component cooling water pump drives the circulation of component

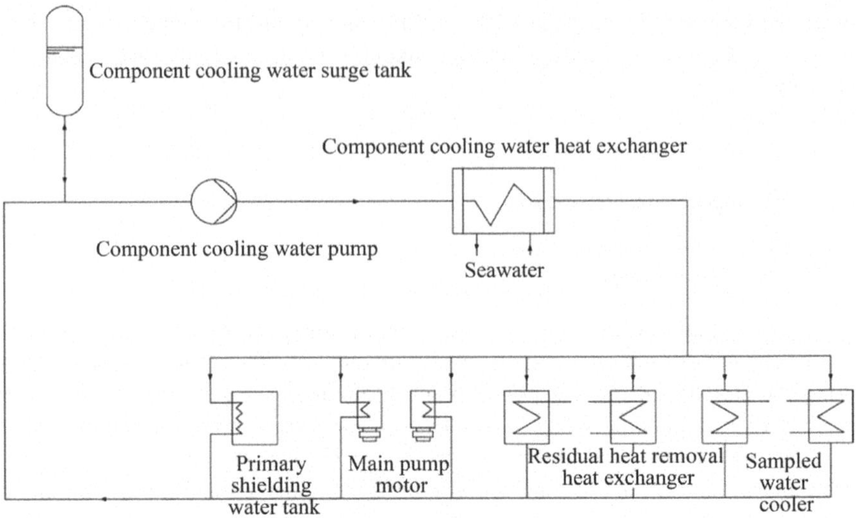

Fig. 4.6 Component cooling water system

cooling water in the closed loop of the system. The component cooling water heat exchanger is used to transfer the heat removed from the equipment cooled by the component cooling water to the seawater. The component cooling water surge tank is used to maintain the system operating pressure and regulate the volume fluctuation of component cooling water.

Before the reactor and the primary loop system are started up, the component cooling water system shall have been started up and kept in the normal operational state.

After the reactor and the primary loop system are shut down and the reactor reaches the cold state, the component cooling water system will be shut down.

The cooling water in the system is the demineralized water. The component cooling water flows into the equipment to be cooled, removes the heat and transfers the heat to the seawater on the shell side of component cooling water heat exchanger. The circulation of seawater is driven by the seawater pump. The typical flow process of component cooling water system is illustrated in Fig. 4.6.

4.5.2 Equipment Description

(1) Component cooling water pump

During the normal operation, shutdown and various accident conditions of reactor, the component cooling water pump delivers the component cooling water to cool the equipment to be cooled in the primary loop system.

The component cooling water pump is generally of vertical centrifugal pump design, and its main parts and components are made of stainless steel.

(2) Component cooling water heat exchanger

The function of component cooling water heat exchanger is to transfer the heat removed from the equipment to be cooled by the component cooling water to the seawater, and provide the "consumers" of component cooling water system with cooling water meeting the flow and temperature requirements.

A typical component cooling water heat exchanger is the fixed tube-sheet heat exchanger, with the tubes and tube sheets welded together. During the operation of heat exchanger, the seawater flows through the tube side, and the component cooling water flows through the shell side.

4.6 Coolant Purification System

The marine PWRs use high-purity demineralized water as the moderator and coolant. The demineralized water works under irradiation conditions with high temperature, high pressure, high flowrate, high heat flux and high neutron flux. If the water quality fails to meet the specified requirements, the corrosion of structural materials will increase, thus resulting in the fouling of the heat transfer surface and affecting the heat transfer efficiency and operation reliability. At the same time, the corrosion products shall be activated by irradiation, increasing the radioactivity level of the plant.

Therefore, the primary loop system shall be equipped with coolant purification system to purify the reactor coolant to ensure that the water quality of reactor coolant meets the indicator requirements.

4.6.1 System Description

(1) System functions

The main function of coolant purification system is to continuously filter out some soluble and insoluble impurities in the reactor coolant by means such as filtration and ion exchange, so as to keep the concentration of impurities in the reactor coolant below the allowable value and reduce the radioactivity level of reactor coolant [1].

(2) System composition and operation flow process

Figure 4.7 illustrates the principle of coolant purification system.

The coolant purification system consists of reactor coolant pump (or purification pump), filter, regenerative heat exchanger, non-regenerative heat exchanger, ion exchanger and corresponding piping.

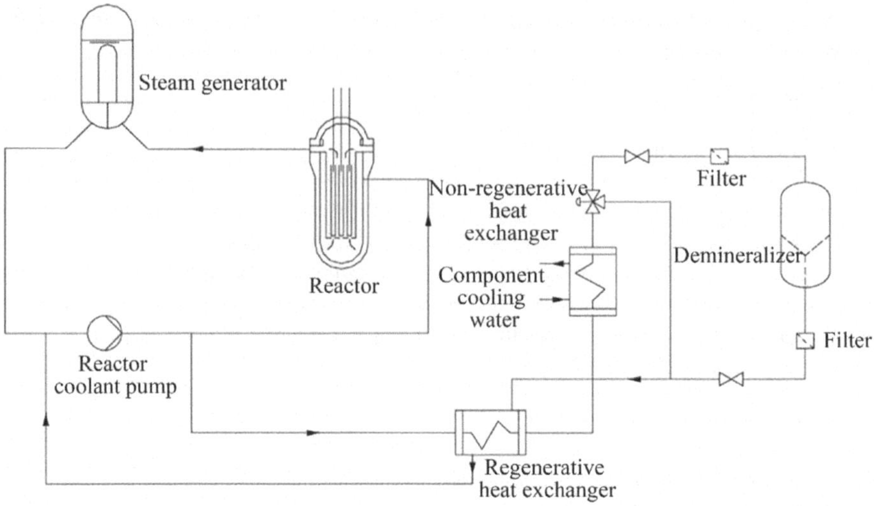

Fig. 4.7 Coolant purification system

During the operation of coolant purification system, the reactor coolant to be purified is diverted from the reactor coolant system to the tube side of regenerative heat exchanger. The heat exchange takes place between the high-temperature coolant at the tube side of regenerative heat exchanger and the purified low-temperature coolant flowing at the shell side, thus decreasing the temperature of the coolant. The coolant flows at the tube side of non-regenerative heat exchanger, is further cooled by component cooling water flowing in the shell side to the operating temperature of ion exchanger resin, and flows into the purification ion exchanger. The purification ion exchanger removes some soluble ionic impurities in the reactor coolant, and filters some insoluble impurities. The purified coolant enters the shell side of regenerative heat exchanger, and is heated at the shell side and then returned to the reactor coolant system.

The coolant purification system is characterized by simple flow process, little equipment and compact layout. However, the equipment shall meet the requirement of withstanding high pressure, so the manufacturing costs are rather high. Therefore, this system is usually used in cases with rather strict equipment space requirements.

4.6.2 Equipment Description

(1) Purification pump

The purification pump delivers the reactor coolant into the coolant purification system at the specified flow to make the water quality of reactor coolant meet the requirements. The purification pump can be of a vertical single-stage single-suction centrifugal pump design, and its main parts and components are made of stainless steel.

(2) Regenerative and non-regenerative heat exchangers

The function of regenerative heat exchanger is to cool the reactor coolant before it is purified, to heat the purified reactor coolant, and to preheat the makeup water from the coolant-charging system to reduce the thermal stress caused when the makeup water enters the reactor coolant system.

A typical regenerative heat exchanger is the shell-and-tube heat exchanger, the tubes and tube sheets of which are welded together. During the operation of heat exchanger, the high-temperature reactor coolant to be purified flows through the tube side, and the purified low-temperature reactor coolant flows through the shell side.

The function of non-regenerative heat exchanger is to, under normal operating conditions, use the component cooling water as the cooling source to further cool the reactor coolant to be purified, which has already been cooled by the regenerative heat exchanger, to the temperature required by the process, so as to provide the purification ion exchanger with working medium meeting the temperature and flow requirements.

A typical non-regenerative heat exchanger is the tubular heat exchanger. Its tubes and tube sheets are welded together. During the operation of the heat exchanger, the reactor coolant to be purified flows through the tube side and the component cooling water flows through the shell side.

(3) Purification ion exchanger

The function of purification ion exchanger is to remove the soluble and insoluble impurities in the reactor coolant by using the nuclear class ion exchange resin in the purification ion exchanger, so as to make the concentration of impurities in the coolant lower than the allowable value and reduce the radioactivity dose level in the reactor compartment.

The purification ion exchanger is a vertical container. During the normal operation of system, some reactor coolant enters the equipment from the medium inlet pipe on the top head, flows through the inlet filter, and enters the purification resin bed. The soluble and insoluble impurities in the reactor coolant are retained in the resin bed by the absorption action of the resin bed, and the purified coolant flows out from the outlet filter and finally flows out from the medium outlet pipe.

The filters of purification ion exchanger are metal filters made of laminated metal sheets. The thickness of the metal sheet is selected according to the filter diameter requirements.

Since the resin in the purification ion exchanger absorbs the impurities with radioactivity dose in the reactor coolant, the radioactivity dose in purification ion exchanger is relatively high. It is necessary to provide the shielding outside the purification ion exchanger to reduce the radioactivity dose in the reactor compartment.

4.7 Valves

4.7.1 Overview

The reactor coolant system, the nuclear auxiliary systems and the steam power conversion system introduced in this book later have, besides the equipment, pumps and pipes, a large number of valves. These valves perform such functions as isolation, return prevention, pressure relief protection, throttling and regulation. Their operation reliability is directly related to the stability, safe and reliable operation of nuclear power plant.

Because of the special requirements for weight, arrangement space and so on, the valves of nuclear power plant are mainly the non-standard products specially designed according to the system requirements. Because of the characteristics of nuclear power plant, these valves have the following features when compared with the valves used in other industries.

① The connections to valves in the primary loop system are mostly performed by welding. On the one hand, under the high temperature and high pressure conditions, the flanges are made of a large amount of metal and are heavy, so they cannot properly meet the requirements of the nuclear power plant on limiting the size and weight in the primary loop system. On the other hand, the flange connections may face certain risk of leakage because of the effects of temperature fluctuations, alternate cooling and heating, large operating stress and relaxation process.

② Valve seat seals are mostly the metal seals. In order to ensure high corrosion resistance and erosion resistance, the buildup welding is performed by hard alloys on the sealing surface between the valve seat and the valve flap to extend the service life of valves.

③ The reliability requirement is high. Under marine environmental conditions and during design basis accidents, the system start-up, switching and shutdown require relevant valves to act reliably (to open or close).

④ The design analysis requirements are strict. The valves of nuclear power plants shall maintain the structural integrity and reliable operation under severe ambient conditions such as vibration, operating condition transition, transient conditions and external dynamic loads. Therefore, the impact of ambient conditions on valves shall be fully evaluated in the design process.

According to the function classification, the valves of nuclear power plants mainly include shut-off valves, check valves, safety and protection valves, regulating/throttle valves.

4.7.2 *Shut-off Valves*

Shut-off valves have two states in operation, i.e. fully open or fully closed. Currently, the shut-off valves of nuclear power plants mainly include gate valves, stop valves, solenoid valves and ball valves.

(1) Gate valves

① Structure of gate valves: Gate valves are one of the most common types of shut-off valves. The gate, the sealing component of a gate valve, is installed on the lower end of valve stem. It is driven by the valve rod to insert into the bottom of valve body for closing the flow channel. According to the sealing types, the gate valves can be divided into two types: parallel gate valves and wedge gate valves. Each type can be further divided into single-plate type and double-plate type [2].

② Features of gate valves: Compared with shut-off valves of other structures, the main features of the gate valves [3] are as follows: A. The flow resistance is small. The medium channel in the valve body is of the straight-through type, so the medium does not change its flow direction when flowing through the gate valve. B. The medium flow direction is not limited. When the medium flows through the gate valve from either side, the opening or closing can be realized, hence it is easy for installation. C. At the same time, because of the rather large stroke, the gate valves are also characterized by long opening/closing time and large torque requirement at the moment of opening.

(2) Stop valves

Stop valves are the most widely used shut-off valves in nuclear power plants. The valves of this type are mostly of the straight-through type, and the valve stems often adopt packing seals or combined seals of packing and bellow.

① Structure of stop valves: A stop valve generally consists of actuator (electric device or hand wheel), valve stem, valve stem nut, valve body, valve flap and packing seals. The actuator is connected with the valve stem and drives the valve flap to move along the axis direction of valve seat, so as to realize the opening or closing of valve.

The valve seat/valve flap seal of stop valve is a forced seal. When the valve is to close, sufficient force must be applied to the valve flap to realize the sealing. The sealing pairs often adopt planar sealing, conic sealing or spherical sealing.

i Planar sealing

The valve seat/valve flap seal structure of stop valves in primary loop system of nuclear power plant is mainly the planar sealing. The advantages of this structure are as follows: the valve flap can automatically align itself and match the valve seat sealing surface during the assembly, so the guiding requirement for valve flap is not high. In addition, this structure has low requirement for anti-abrasion property of sealing material, and the change of circular degree of the inner hole on the valve seat does not influence the sealing. The disadvantage is that the solid particles and precipitates in the medium will easily damage the sealing surface [4].

ii Conic sealing
The sealing surface of conic sealing is a cone, the contact surface becomes narrower and narrower. Compared with the planar sealing, the conic sealing has a significantly increased specific seal pressure under the same sealing force, proving it with better sealing performance. However, because of the narrow sealing surface, it is not easy to correctly guide the valve flap to the valve seat surface. In order to improve the sealing performance, the valve flap shall be guided. In addition, when the circular degree of valve seat hole changes, the sealing performance will be significantly degraded [4].

iii Spherical sealing
The valve flap of spherical sealing is a sphere and the valve seat is conic. The valve flap sphere can rotate freely and has a certain self-regulation function. The sealing pair is linear sealing, so the sealing stress is very high and it is easier to realize the sealing. However, the linear seal structure is easily damaged by erosion, so the requirement for erosion resistance of valve seat material is high [4].

② Features of stop valves: Compared with other shut-off valves such as gate valves and ball valves, the biggest disadvantage of stop valves is the large flow resistance caused by twists and turns in the flow channel. Therefore, the stop valves are not suitable for applications with rather high requirement for resistance coefficient. However, since they have the following advantages, they are still widely used in the nuclear power plants.

Compared with the gate valves, the stop valves are characterized by rather simple structure and rather convenience for maintenance; there is no relative sliding between the valve flap and the valve seat sealing surface during the opening/closing, so the sealing surface wear and abrasion are minor; the stroke of valve flap is short, so the height is rather small.

(3) Ball valves

The ball valves evolve from the plug valves. The opening/closing component is a ball. The function of opening/closing is achieved by rotating the ball around the axis of valve rod by 90° [4]. A ball valve mainly consists of actuator, valve body, ball, valve rod and other parts and components.

The ball valve seat is usually a soft seal, that is, the rubber, nylon or Teflon are used as the material for seal ring of valve seat. Therefore, the applications are limited by the temperature. In addition, there are some other disadvantages such as rather many internal parts and components. The main applications of ball valves are as follows:

i. Where the system temperature is not high; ii. where the requirement for internal sealing is not particularly strict; iii. where the flow resistance coefficient is rather small; and iv. where the opening/closing time required is short.

(4) Solenoid valves

Solenoid valves use the electromagnetic force as the power source, that is, they use the electromagnetic force generated by energized electromagnet to drive the movement of relevant parts and components of solenoid valves to realize the opening, closing or switching [4]. According to the actuation principle of valve flap, the solenoid valves can be divided into three types: direct-acting solenoid valves, pilot operated solenoid valves and recoil solenoid valves.

The solenoid valve seat is similar to stop valve seat, so the flow resistance is also rather large. However, compared with stop valves and other types of shut-off valves, the solenoid valves have the following advantages: A. the opening/closing time is short (generally from a fraction of a second to 3 s); B. the size is small and the weight is light; and C. there is no external leakage. Therefore, the solenoid valves are often used in applications where quick opening/closing and no low flow resistance are required by the nuclear power plants.

4.7.3 Safety Valves

When the pressure in the equipment (or device, container, pipe and so on) exceeds the limit value, the safety valves will automatically open to release the excessive medium in the equipment, thus preventing the pressure from keeping increasing. When the pressure decreases to the specified value, the safety valves will automatically close, thus guaranteeing the normal operation of equipment.

Safety valves used in nuclear power plants are mainly pilot operated safety valves (such as pressurizer safety valves) and spring direct-acting safety valves. Compared with the spring direct-acting safety valves, the pilot operated safety valves use small-diameter pilot valves with rather high sensitivity to control the opening and closing of the main valves, so the accuracy in opening pressure and opening/closing differential pressure is higher. In addition, the main valves can realize the large discharge more easily, so pilot operated safety valves are especially suitable for the medium environment with high temperature and high pressure [5].

(1) Pilot operated safety valves

A pilot operated safety valve consists of a main valve and a pilot valve. The function of the pilot valve is to sense the pressure of the pressure-bearing system and to open/close the main valve.

For the characteristics and description of pilot operated safety valves in the pressurizer, refer to Sect. 4.2.2 of this chapter.

(2) Spring direct-loaded safety valves

A spring direct-loaded safety valve directly relies on the acting force generated by medium pressure to overcome the spring force applied on the valve flap to open the safety valve.

According to the opening height of the opening/closing component, the direct-loaded safety valves can be divided into the following types:

① Full lift safety valve: during each actuation, this type of valve can reach the fully-open position of the valve flap (there is usually a special mechanism to help to increase the valve flap opening height); in this case, the circular seam cross section is rather larger than or equals to the minimum cross section of valve seat hole.

② Low lift safety valve: during each actuation, the circular seam cross section is smaller than the minimum cross section of valve seat hole (the opening height is 1/40 to 1/20 of the flow channel diameter, and it changes with the pressure in the system; there is no special mechanism to increase the valve flap opening height).

4.7.4 Check Valves

Check valves belong to protection valves. Most outlets of pump type equipment in nuclear power plants are equipped with check valves to prevent the backflow of fluid in the pipes. The check valves commonly used in nuclear power plants are swing check valves and lift check valves. The swing check valves are generally used in piping systems with a nominal diameter larger than 50 mm, and the lift check valves are generally used in piping systems with a rather small diameter.

(1) Swing check valves

A swing check valve is a check valve in which the valve flap rotates around the rocker shaft inside the body cavity. The closing component is a valve flap or disc with a diameter equivalent to that of the pipe and it is suspended in the valve cavity. During the forward flow of fluid, the valve flap is opened under the action of fluid pressure. When the pressure decreases, the valve flap will be closed under the action of self-weight and the backward flow pressure.

(2) Lift check valves

A lift check valve is a check valve in which the valve flap is lifted along the axis of valve seat sealing surface.

When the valve is at the open position, the medium will flow upwards from the area below the valve flap. In order to make the check valves more sensitive to the change in flow direction and to accelerate its closing action, a spring or additional weight is installed on the valve flap. However, this will increase the head loss and energy loss of the flow of medium in the pipe.

The lift check valves can operate reliably, but the flow resistance is rather large. They are suitable for applications with small diameters.

4.7.5 Regulating Valves

The regulating valves are the devices controlling the fluid flow and pressure by changing the cross section of the flow channel inside the valve body. When the valve flap sealing surface and the valve seat sealing surface closely contact each other, the closing function is realized. When the valve rod drives the valve flap to lift or rotate along the axis of valve rod, the valve flap, as the regulating component, blocks or opens the flow channel or changes the cross section of flow channel to regulate the flow area. The control of regulating valves can be operated by the operating personnel or the relays triggered by various kinds of sensors in the system. The capacitor or sliding rheostat for valve position feedback is arranged inside the electric device to send the valve position signals to the control room [5].

4.7.6 Valve Reliability

The valves of nuclear power plants have their specific operating environment and working medium. Compared with valve products used in conventional industries, the reliability of such valves has an especially outstanding impact on the safety performance and maintenance cost of the system. In general, the initial purchase price of a valve is equivalent to 20% of the total expenses for its entire service life. About 3/4 of the total expenses are for the maintenance, major overhaul and repair and the indirect costs incurred by system operation suspension caused by the valve failure. Therefore, the valve reliability is the focus of nuclear power plants.

(1) Common failure modes or problems of valves

The common failures and problems of various types of valves of nuclear power plants are as follows:

i. Failure of external seal; ii. failure of the internal seal; and iii. actuator performance abnormality.

(2) Valve reliability improvement measures

Under most circumstances, the poor quality or reliability of a valve is not caused by a single factor such as design, manufacture, test, operation or maintenance. A valve failure is the result of the comprehensive effect of design, manufacture, test, operation, maintenance, management and human performance. In order to fundamentally improve the reliability of valves, all stages in the valve lifetime shall be comprehensively considered, the possible causes of failures shall gain full attention and analysis, and corresponding measures shall be taken.

① Improvement and enhancement of design methods and means

The design of valves of nuclear power plants shall mainly consider the actual operating environment such as contamination of impurities in the medium, ambient temperature, operating temperature, ambient humidity, radioactivity, power supply and voltage fluctuations and vibration environment.

During the design, advanced design methods and means shall be used. Besides the conventional strength design and structural design, the stress limit calculation and fatigue performance analysis related to system operation and operating condition transition shall also be performed.

② Improvement of materials and key manufacturing process performance

The new research on materials of valve actuators and the research on surface treatment processes shall be performed to improve the corrosion resistance and wear resistance. For the phenomena of valve sealing surface such as the susceptibility to flushing and wearing, the new research or performance improvement research on valve sealing materials (including surfacing materials) shall be performed to improve the wear resistance and abrasion resistance of valve sealing surface.

③ Improvement of valve function and environmental qualification tests

According to valve functions and environmental qualification requirements, the valve function improvement and environmental qualification tests shall be performed. In addition, the implementation of the operation and performance tests of some key valves under the simulated actual operating environment and actual operating conditions shall be considered in order to grasp the reliability under actual operation environment conditions.

④ Improvement of valve use and maintenance requirements

On-line valve failure monitoring refers to the measurement of related characteristic parameters during the valve operation (such as valve rod displacement, opening/closing time, driving torque and vibration frequency) without disassembling or dismantling the valve and the direct judgment of valve operation performance according to such characteristic curves. In the structural design of valves, the consideration shall be given to the provision of "intelligent" components of valves to realize the functions of self-monitoring, evaluation and failure prediction of the valves, thus improving the reliability of valves.

The requirements for storage after delivery, storage before installation, cleanliness maintenance during installation, cleanliness maintenance after installation, maintenance before system operation, daily maintenance and maintenance during system outage shall be defined for the valves to reduce the failure rate caused by improper installation and maintenance after the delivery of valves.

⑤ Research on data management system for valve whole lifetime reliability

Tracking of valve data during the use and maintenance shall be strengthened. It is essential to collect the performance data of valves during the design, manufacture, test and use, and the failure phenomena and maintenance records during the valve operation and maintenance, establish a valve reliability database, perform an in-depth research on the causes and natures of valve failures, and gradually turns the valve failure research from qualitative research to quantitative research to guide the reliability improvement for future valve products.

References

1. Fengge, Pang, and Peng Minjun. 2000. *Marine Nuclear Power Unit*. Harbin: Harbin Engineering University Press.
2. Peiwen, Lu. 2011. *Valves of Nuclear Power Plants*. Beijing: China Machine Press.
3. Peiwen, Lu. 2001. *Valve Selection Manual*. Beijing: China Machine Press.
4. Qingshuang, Zhang. 2012. *Valve Manual—Type Selection*. Beijing: Chemical Industry Press.
5. Guoyuan, Jiang. 2009. *WWER-1000 Nuclear Power Plants—Mechanical and Electrical*. Beijing: Atomic Energy Press.

Chapter 5
Engineered Safety System

5.1 Overview

A nuclear power plant may fail or be damaged for various reasons, such as human errors in its design, manufacture and operation, and equipment aging. Without timely countermeasures or appropriate safety facility, the deterioration of the heat transfer of the core, the failure of the reactor pressure boundary and even the most severe accidents jeopardizing the public and environment with serious radioactive contamination may occur. Therefore, the principle of defense in depth is widely applied in the design of land and marine nuclear power plants, and the necessary engineered safety system is provided. One of the defensive measures is the provision of the engineered safety system to guard against various hypothetical accidents that are almost impossible but must be considered from the perspective of safety. The engineered safety system shall be able to mitigate the consequences of hypothetical accidents to an acceptable level. This is an important symbol of the safety of a reactor [1].

The engineered safety system mainly includes residual heat removal system, safety injection system, and reactor compartment heat removal system, which are used for cooling the core after normal or accident shutdown. In case of an accident, the engineered safety system is used for flooding and cooling the core to prevent the damage of the integrity of the third safety barrier and to limit the development of the accident and mitigate the consequences of the accident.

Improving the inherent safety of nuclear power plants and adopting passive technique in the engineered safety system are the current development trend of advanced nuclear power technology in the world, which can not only simplify the system and reduce the misoperation caused by human intervention, but also greatly improve the operational reliability of the safety system to enhance the safety and economy of the whole plant.

J. Yu, *Marine Nuclear Power Technology*,
https://doi.org/10.1007/978-981-15-2894-1_5

5.1.1 *Design Principles for Engineered Safety System*

The operation safety of the marine nuclear power plants shall be guaranteed by the basic principle of safety design and specific preventive measures from the perspective of the whole ship and taking full consideration of the characteristics of the MNPP as a mobile nuclear facility such as narrow space, severe environment, remote from mainland, limited maintenance capacity and lack of emergency support, while insisting on the design principle of defense in depth to set up multiple barriers.

The design criteria or application principles of the engineered safety system mainly include the following points.

(1) Single failure criterion

The single failure criterion is one of the important principles that must be followed in the safety design of marine nuclear power plants. This criterion is of great significance to the comprehensive evaluation of the system reliability and has been widely adopted in the design of important safety systems at present.

(2) Diversity principle

Multiple systems with different mechanisms are provided to realize the same safety function, to reduce common cause failures or common mode failures and improve the reliability of certain systems.

(3) Independence criterion

In order to improve the reliability of marine nuclear power plants and avoid common cause failures or common mode failures, functional isolation and physical isolation shall be adopted in the design to realize the independence in the system layout and design.

Obviously, a complete physical isolation is impossible based on the characteristics of marine nuclear power plants. Whether the physical isolation is required for any specific system and equipment and what extent to which the physical isolation is required shall be further defined through analysis and research.

(4) Inherent safety principle

The inherent safety generally refers to the natural safety and passive safety of the reactor that can control the reactivity or remove core heat in case of abnormal operating conditions of nuclear power reactors, to realize the normal operation and safe shutdown of the reactor without the mandatory intervention from manual operation or external equipment.

5.1.2 *Basis for Determining Engineered Safety System*

For the marine nuclear power plants, the safety system used for mitigating the design basis accidents shall be determined according to the analysis results of the design basis accidents. These systems used for mitigating the design basis accident shall be designed according to the requirements of the engineered safety system.

The design of the engineered safety system and its equipment (including its parameters, design capacity and some setting valves) shall be evaluated and checked through accident analysis [1].

The selection of design basis accidents shall be generally based on the necessary nuclear safety requirements, the technical specifications of marine nuclear power plants and the available nuclear technology, while taking comprehensive consideration of various initial events and accident spectrum.

5.1.3 Design Characteristics of Engineered Safety System of Marine Nuclear Power Plants

(1) Dual-purpose functions

From the perspective of the overall design of ships, the nuclear power plant shall possess small volume, light weight, simple system configuration and simplified equipment. However, what makes a safety system an "engineered" one shall be discussed. If the function is "specific" and can be "specific" to its application when required, the safety system qualifies an "engineered" one. There are many engineered safety systems that are compatible with operational and functional system at land NPPs now. For example, the chemical and volume control system is also with the function for high-pressure safety injection; and the residual heat removal system is also with the function for low-pressure safety injection. Therefore, from the perspective of the characteristics of marine nuclear power plants, the dual function principle for engineered safety features shall be given more emphasis, without the effect on the safety function. Besides the dual-purpose function of the system, the dual function of equipment can also be considered. For example, the main pump can also be used as the driving pump of the residual heat removal system when the lift of the main pump is large enough.

(2) Combination of active and passive systems

No matter how its type changes, the safety related system of PWRs in the world is classified into as active and passive systems. The active safety system refers to the system that realizes its safety function relying on the external triggering and power source (electricity, compressed air, etc.). One advantage of the active safety system is that it has a relatively high effectiveness in the aspects of prompt termination of accidents and prevention of the spread of the accident and can provide operational flexibility under different accident conditions. Its disadvantage is that it may deteriorate the accidents due to misoperation or operator's incomprehension of the nature and process of accidents.

The passive safety system refers to the system that realizes function by relying on the inherent characteristics of natural substances such as density difference and gravity-driven injection, rather than on the external power source. The passive mode has improved the reliability of system operation. In recent years, the passive safety system is widely used in the design of nuclear power plants, such as AP1000 NPP of Westinghouse Corporation and MRX marine nuclear power plant of Japan [2].

The passive safety system has been used more and more extensively in nuclear power plants and has played an active role in improving the inherent safety of reactors and reducing the probability of nuclear accidents. The adoption of the passive safety system is an important trend of technological development of marine nuclear power plants, and an important and feasible measure for improving its safety.

However, a relatively long period time is required for the termination of an accident, and there are uncertainties during the operation of the passive system, which arise from the deviations from expected behaviors, i.e. the relevant physical phenomenon violates the system performance or the initial conditions and boundary conditions change, and the passive system fails [3]. Moreover, the marine environmental conditions such as swing pose higher requirements of the application of the passive safety system in the marine nuclear power plants. Therefore, the sufficient test verification and assessment shall be implemented before the installation and service of the passive safety system on board, to improve its environmental adaptability and reliability.

For the conditions above, the active and passive combined safety system designs have been considered and implemented in AP1000 NPP.

5.2 Emergency Core Cooling System

5.2.1 Safety Injection System

The safety injection system is one of the engineered safety systems of nuclear power plants. It can rapidly inject the cooling water into the reactor core when the loss of coolant accident (LOCA) occurs at a reactor coolant system or the steam pipe rupture occurs at a secondary loop, to ensure the removal of heat from the core. The safety injection system is also called the emergency core injection system.

The requirements of the safety injection vary with the types or severity of the accidents in nuclear power plants. The safety injection system is usually composed of high pressure safety injection subsystem, medium pressure safety injection subsystem and low pressure safety injection system, which are respectively put into operation under different pressures according to the pressure drop in the reactor coolant system caused by accidents.

(1) Direct injection

In the event of LOCA or the steam pipe rupture at secondary loop, the safety injection system is started to provide the cooling water for the reactor core quickly.

(2) Recirculation cooling

The liquid gathered at the bottom of reactor compartment after LOCA is pumped, cooled and re-injected into the core, for recirculation to continuously cool the reactor core.

The safety injection system is usually composed of two series designed as the redundant one to the other. Each series is composed of a high pressure safety injection pump, one safety injection tank, a low pressure safety injection pump, a recirculation cooler and the connected water tanks, pipes, valves, etc. The system schematic diagram is shown in Fig. 5.1.

When the reactor coolant pressure boundary is damaged due to small breaks and LOCA occurs, the pressure and water level of the reactor coolant system would decline slowly. Once the pressure and water level are lower than the specified values,

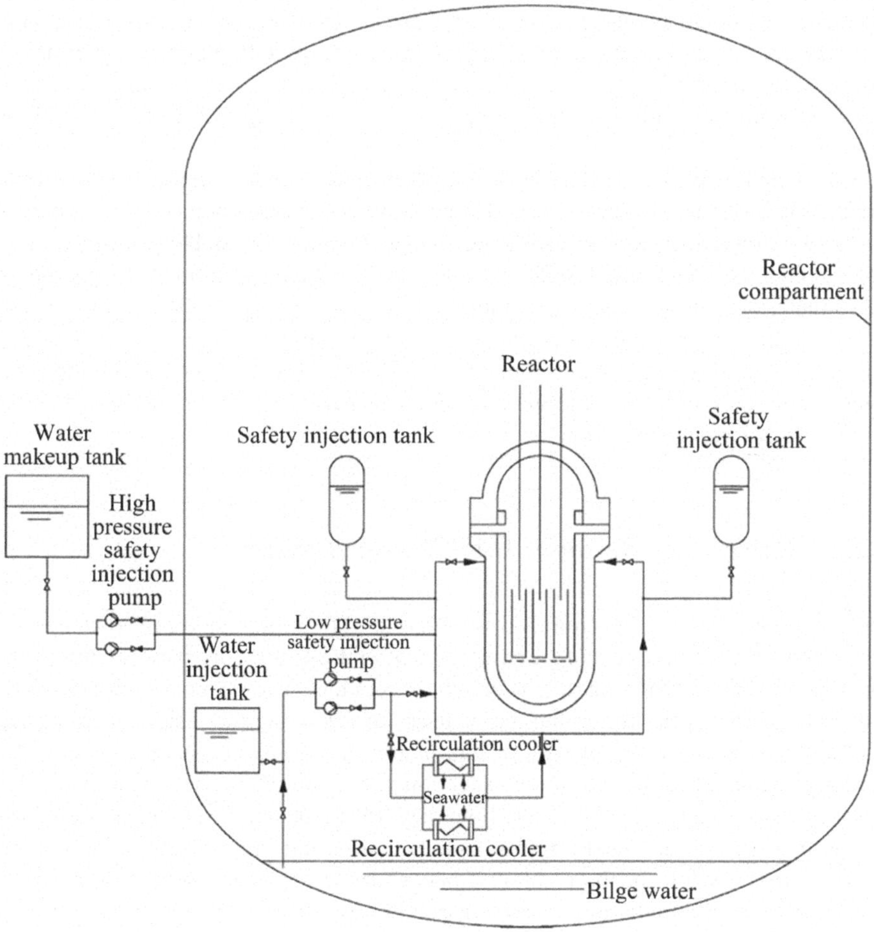

Fig. 5.1 Schematic diagram of safety injection system

the high pressure safety injection subsystem is put into operation. The high pressure safety injection pump is automatically started to inject the water into the core at a certain flow from the cold leg of the main pipe. At this time, the operating personnel should judge that a rupture occurs based on the signals for declining of the pressure and water level, the dose alarm at reactor compartment as well as the increasing of temperature at the reactor compartment.

Under LOCA conditions due to a relatively large break, the pressure of the reactor coolant system and the water level of the pressurizer decline relatively rapidly. When the pressure of the reactor coolant system declines below the pressure of the safety injection tank, the water stored in the safety injection tank is injected into the cold leg of the reactor coolant system by the pressurized nitrogen. In the event of LOCA due to relatively large break, the medium pressure safety injection subsystem can be used for flooding the core within a short time to avoid the melting of the fuel rods.

When the pressure declines continuously to a certain value, the low pressure safety injection subsystem is put into operation. The low pressure safety injection pump and relevant valves in the reactor compartment start up automatically when being triggered by the starting signals of low pressure safety injection, to draw the water from the water injection tank and inject it into the cold leg of the reactor coolant system.

When the low pressure safety injection subsystem has been put into operation for a relatively long time, the pressure and temperature of the reactor coolant system and the reactor compartment would have declined and a relatively stable working condition would have been formed. When the water injection tank is emptied, the operating personnel would initiate the recirculation cooling depending on the circumstances, to switch the low pressure safety injection pump to pump the liquid gathered at the bottom of reactor compartment after LOCA. The liquid is re-injected into the core after being cooled by the recirculation cooler to participate in the recirculation to continuously cool reactor core.

5.2.2 Emergency Residual Heat Removal System

The emergency residual heat removal system is used for removing the residual heat of the core and the sensible heat of the equipment after the scram caused by the ship-wide blackout and the loss of reliable power supply accident. The system is automatically put into operation after the scram and is used for removing the residual heat of the core and the sensible heat of various equipment of the reactor coolant system at a certain rate under the premise of not exceeding the specified fuel design limit and the design conditions of coolant pressure boundary, to put the reactor coolant system into a state that the residual heat removal system can be put into operation and ensure the operational safety of the reactor. If necessary, the system can also be put into operation manually under other accident conditions.

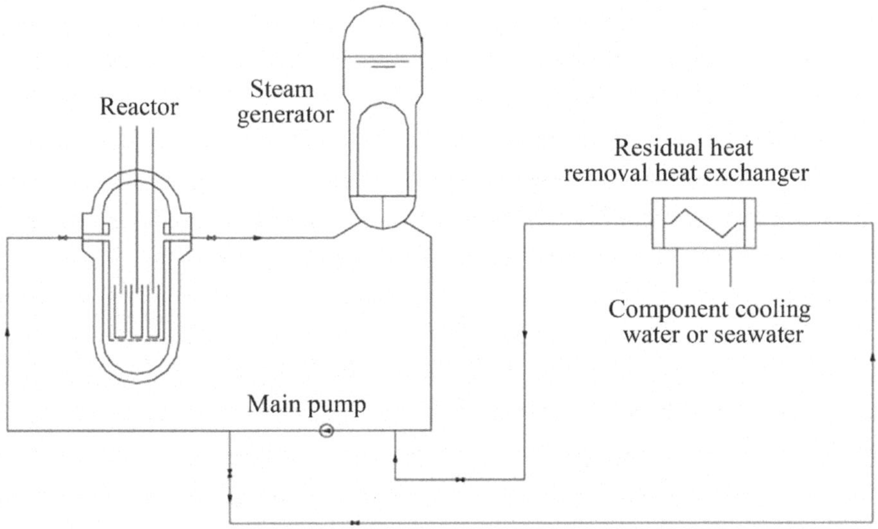

Fig. 5.2 Principle of (Active) emergency residual heat removal system

The emergency residual heat removal system can be set as active or passive depending on the situation. The active emergency residual heat removal system usually functions through the main pump (low speed) or the emergency residual heat removal pump. The design pressure of the system is the same as that of the reactor coolant system. The system can be put into operation immediately at the shutdown of the reactor. By maintaining the circulation of a certain flow of coolant in the system, the heat of the coolant is transferred to the component cooling water or seawater through the emergency residual heat removal cooler.

The schematic diagram of an (active) emergency residual heat removal system is shown in Fig. 5.2.

The passive emergency residual heat removal system can be classified into the direct cooling (removing the residual heat directly from the core) and the indirect cooling (removing the residual heat from the steam generator) passive residual heat removal systems according to the arrangement of loop.

The direct cooling passive residual heat removal system can be located near the reactor to transfer the residual heat of the core to the ultimate heat sink (seawater) directly through the intermediate loop instead of removing it from the steam generator. The system can be used for easier establishment of a natural circulation within a limited arrangement space of marine nuclear power plants and has a simple configuration, a faster transient response and a stronger ability of removing the heat. However, leakage may occur on the heat exchange equipment of the system due to seawater corrosion. When seawater enters the reactor and primary loop system or the reactor coolant enters the sea, the reactor and primary loop system and external

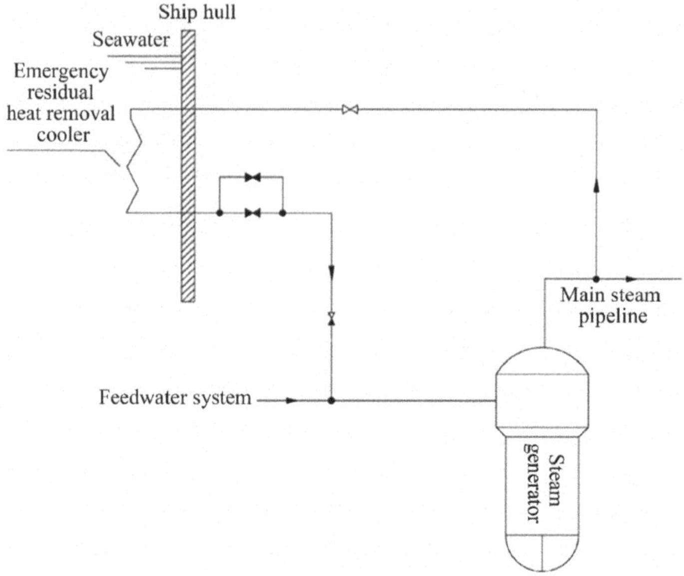

Fig. 5.3 Schematic diagram of secondary side passive residual heat removal system

environment would be affected, resulting in severe consequence. Therefore, the seawater corrosion resistant materials should be adopted, and the relevant pipe system and heat exchanger shall be provided.

The indirect cooling is equivalent to the provision of an intermediate loop between the reactor coolant and the seawater. The cooling medium in the passive residual heat removal system does not contact with the reactor coolant and the leakage on the heat exchange equipment would not result in the entering of the seawater into the reactor and the primary loop system, or the entering of the coolant in the sea directly. This guarantees the safety of the system, equipment and environment, however leads to a complex system with more factors influencing the safety, and the slow starting up of natural circulation.

The following mainly introduces a kind of indirect cooling passive residual heat removal system, also called the secondary side passive emergency residual heat removal system.

The configuration of the secondary side passive emergency residual heat removal system shall meet the single failure criterion. It mainly includes emergency residual heat removal cooler, secondary side of steam generator, main steam isolation valve, pipes, valves, and instruments.

The emergency residual heat removal cooler is directly immersed in seawater and installed above the auxiliary feedwater pipe with an elevation higher than the main feedwater pipe. When the outlet valves of the system open after receiving the signals of protective action, the system is automatically put into operation. The condensate water filled in the emergency residual heat removal cooler is injected into the secondary side of the steam generator under the effect of gravity and is turned into steam after absorbing the residual heat of the core. The steam then enters the emergency residual heat removal cooler, exchanges the heat with the seawater and is condensed into water after transferring heat to the seawater. The condensate water returns to the secondary side of the steam generator under the effect of gravity to complete the natural circulation of steam and condensate water loop. The system flowchart is shown in Fig. 5.3.

5.3 Reactor Compartment Heat Removal System

In the event of LOCA or the main steam pipeline break in the reactor compartment, high temperature and high pressure coolant or steam would enter the reactor compartment, resulting in the rapid increasing of the temperature and pressure of the reactor compartment. If the temperature and pressure cannot be decreased in time, the sensors in the reactor compartment and the leakproofness of the reactor compartment may be damaged, causing the spreading of the radioactive substance to other compartments. Therefore, the reactor compartment heat removal system shall be provided, as shown in Fig. 5.4.

The system is mainly composed of spray pump and spray pipeline. The spray pipeline is arranged at the top of the reactor compartment and installed with a number of nozzles to ensure a uniform distribution of the spray water on the cross section of the reactor compartment.

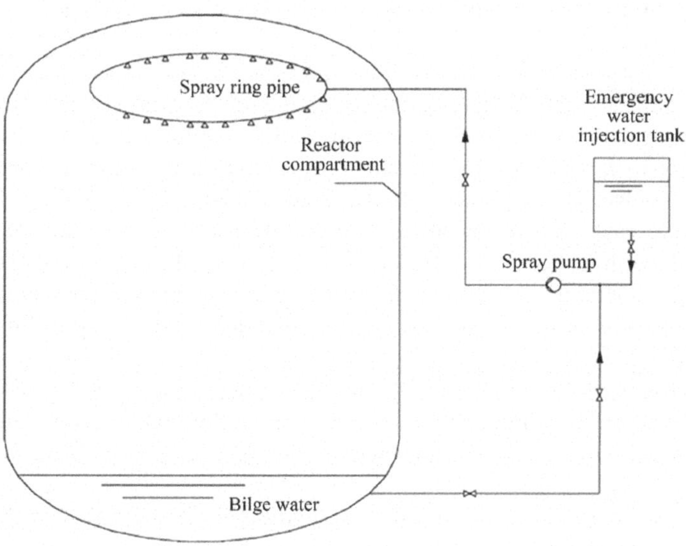

Fig. 5.4 Reactor compartment heat removal system

When the pressure in the reactor compartment is increased to a certain threshold value and the signals of safety injection and high pressure in the reactor compartment are sent out simultaneously, the reactor compartment heat removal system is automatically put into operation under the control of the reactor protection system to start the spray pump, which can be used for spraying the water in emergency water injection tank into the reactor compartment to decrease its pressure and temperature.

When the system has been put into operation for a period, the pressure and temperature of reactor coolant system are decreased and then maintain stable, or when the water sources required by spraying are unavailable, the discontinuous recirculation cooling can be initiated.

5.4 Dehydrogenation System

The control of hydrogen usually includes "recombination" and "inerting". "Recombination" refers to the removal of hydrogen through the active or passive hydrogen and oxygen recombination devices, while "inerting" refers to the introduction of nitrogen to reduce the oxygen concentration to a certain degree.

The hydrogen removal system of AP1000 is composed of passive hydrogen recombiner and igniter, and can be used to prevent hydrogen detonation or explosion and provide defense in depth in the event of design basis accidents and severe accidents. For BWR NPPs, the primary containment is filled with nitrogen to control the oxygen concentration and the secondary containment is provided with the hydrogen and oxygen recombiner. In the event of accidents, the exhaust fan is started to feed the gas in the primary containment into the reaction chamber of hydrogen and oxygen recombiner.

In order to provide a friendly working and living environment for the crew, the radioactive substance shall be strictly sealed in the reactor compartment. The hydrogen removal system is used for the monitoring and controlling of the overall hydrogen concentration in the reactor compartment in case of normal operation and design basis accident of nuclear power plants; it is also used for reducing the hydrogen concentration in the reactor compartment in case of severe accidents to keep the overall concentration of hydrogen below the combustible limit. The reactor compartment is similar to the containment of NPP and is the third pressure barrier. The hydrogen removal system is an important measure for maintaining the integrity of the third safety barrier.

The hydrogen removal system of the marine nuclear power plant is mainly composed of the active dehydrogenation unit, the passive hydrogen recombiner and the hydrogen measurement unit. The flowchart of the hydrogen removal system is shown in Fig. 5.5.

During the operation of the reactor, the hydrogen concentration in the reactor compartment is continuously monitored by the hydrogen measurement unit. When it is increased to the specified value during the normal operation, the alarm signals would be sent out from the hydrogen measurement unit and the operating personnel

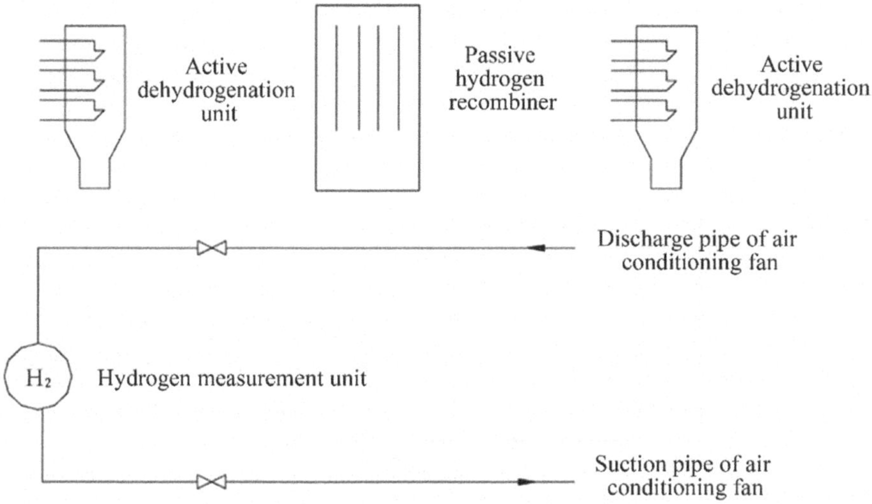

Fig. 5.5 Hydrogen removal system

would start up the active hydrogen removal unit. When the hydrogen concentration is reduced below the specified value, the hydrogen removal unit is shut down. In case of accidents, the hydrogen concentration in the reactor compartment would increase rapidly. When the hydrogen concentration in the reactor compartment exceeds the specified values, the alarm signals would be sent out from the hydrogen measurement unit and the operating personnel would start up the active hydrogen removal unit. When the hydrogen concentration in the reactor compartment is increased to the starting threshold of the passive hydrogen recombiner, the hydrogen removal unit is automatically put into operation.

5.5 Backup Reactor Shutdown System

The backup reactor shutdown system is an important system of the reactor and the primary loop system, and is used for providing multiple shutdown means for the reactor. The backup reactor shutdown system is used for injecting enough boric acid into the reactor to realize chemical shutdown when the reactor control rod is stuck outside the reactor core due to accidents and the reactor cannot be shut down. The function of the backup reactor shutdown system is safety-related.

The backup reactor shutdown system is mainly composed of the boric acid dosing tank and the relevant pipes, valves, and instruments. When the operation of the backup reactor shutdown system is required, the boric acid solution in the boric acid dosing tanks would be transported to the reactor by using the injection pump to realize the

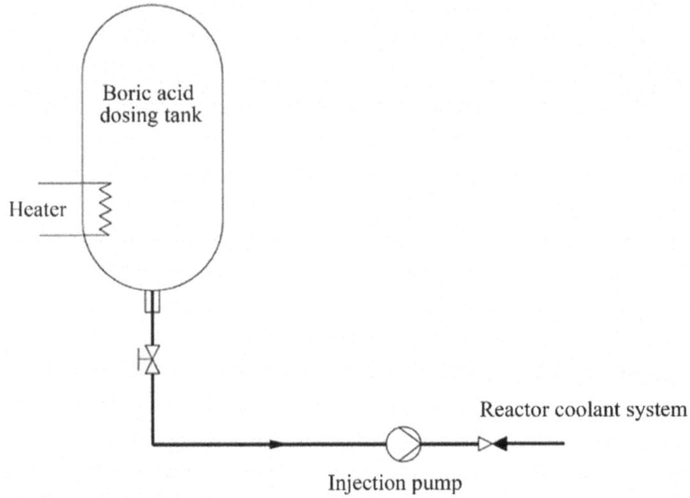

Fig. 5.6 Backup reactor shutdown system

safe shutdown of the reactor. Figure 5.6 is the schematic diagram of the backup reactor shutdown system.

The main equipment of the backup reactor shutdown system is the boric acid dosing tank, which is a vertical cylindrical container with heating devices.

During the normal operation, the system is on standby. When the reactor control rod gets stuck due to the deformation of the control rod or the failure of driving mechanism, the operating personnel shall start the backup reactor shutdown system and inject the boric acid into the reactor by using the injection pump.

5.6 Case Analysis of Design Flow of Safety Injection System

The safety injection system is taken as an example in the following paragraphs to introduce the design flow.

(1) Determination of system functions

The first step is to make clear definition of functions. The requirements for the endowed functional missions shall be made clear firstly before implementing the system analysis demonstration or design. The names and the corresponding requirements of different systems shall vary with their functions, which is the basis for the design of relevant systems of all types of reactors. For the safety injection system, its function is to rapidly inject the cooling water into the reactor core when the loss of coolant accident (LOCA) occurs at the reactor coolant system or the steam pipe rupture occurs at the secondary loop, to ensure the removal of heat from the core.

Therefore, the safety injection system is also called the emergency core injection system.

(2) Analysis and demonstration of system capacity

Generally, the safety injection system is classified as high pressure, medium pressure and low pressure safety injection systems, which are respectively put into operation at each stage of the accident. Therefore, the analysis of the break spectrum of the specific reactor shall be implemented at first. The design basis for large, medium and small break spectrum shall be determined through the analysis of break spectrum and by combining with the allowable volume of the reactor compartment. The capacity design of the safety injection system is determined based on the analysis of the maximum break accident in each break spectrum. The system design shall be provided with enough equipment and pipe redundancy to realize the safety function and satisfy the acceptance criteria of LOCA even at the single failure of components.

(3) Argument of protection signals

In the event of abnormal operation or accident conditions, the reactor control and protection system shall be triggered in time to actuate the safety system to protect the reactor. High pressure, medium pressure and low pressure safety injection systems, respectively, shall be put into operation automatically after LOCA. The cooling water injection shall be maintained for a long time during the declining of the reactor coolant pressure. During the design of the reactor, a series of protection signals (mainly pressure signals) shall be designed respectively for high pressure, medium pressure and low pressure safety injection systems to distribute the water injection tasks at different pressure stages of LOCA mitigation process to the systems above. The selection of protection signal parameters to trigger the above systems, and the rationality and timeliness of parameters shall be demonstrated through the safety analysis of accidents.

(4) Analysis and demonstration of process system schemes

Based on the system capacity and protection signals determined through analysis and demonstration and considering the characteristics of the nuclear power plant designed, multiple design schemes shall be proposed according to the relevant standard specifications, design principles and criterion requirements of the engineered safety system. Based on the demonstration principles, the analysis and demonstration of schemes are implemented from different prospective such as system configuration, process flow, system operation mode, layout conditions, and reliability to determine which scheme is feasible and reasonable to form the technical scheme for the system demonstration.

(5) Determination of main parameters of system and equipment

The analysis and calculation of process system parameters and equipment parameters are implemented according to the system design schemes determined through the analysis and demonstration implemented to provide basis for determining of the main parameters of system and main equipment. For the determination of system and equipment parameters, the steady operation conditions shall be studied and the transient operation conditions shall be analyzed.

The professional fluid system simulation software shall be adopted based on the working medium and the characteristics of the system to implement the system analysis. The operating characteristics of the system are obtained through the simulation analysis and calculation of the system at the rated and transient working conditions.

The results of the analysis and demonstration of the system capacity as well as the system parameters and equipment parameters obtained through simulation analysis can be used as the design inputs for the selection or non-standard design of equipment, pipes, valves, etc.

(6) Reliability design and analysis

Based on the initial technical schemes for the system formulated through the analysis and demonstration, the process design of the safety injection system shall be implemented in accordance with the relevant standard specifications and design criteria of the engineered safety system as well as the relevant design specifications of the process system, including the measurement and control requirements of the system as well as the requirements for the internal and external interfaces of the system and power demands (such as water, electricity, gas and oil).

In the design of the process system, the qualitative design analysis and the quantitative reliability design and analyses can be adopted for the important safety system and equipment such as the safety injection system. The reliability design and analysis refers to a comprehensive design and analysis of the occurrence and development of various possible accidents of the complex system using the system reliability evaluation techniques, usually including the pre-stipulation of reliability design criteria, failure modes, influence effect and hazard criticality analysis, and fault tree analysis. The relevant reliability requirements are converted into the specific requirements that are practically operable in the design of hardware and software through the formulation of the reliability design criteria of the system and equipment of nuclear power plants to satisfy the qualitative reliability design requirements. The designers shall follow the reliability design criteria, apply criteria to the design one by one by combining with the specific characteristics of the nuclear power plants designed and provide the corresponding design guarantee measures for each design criterion. The potential failure modes of constituent unit (functional unit or structural unit) of the system and their effects on the system function shall be analyzed through the analysis of failure mode, effect and criticality. Each potential failure mode shall be classified according to its severity and shall be provided with the adoptable prevention and improvement measures to improve the reliability of the system. The fault tree analysis can be used for determining the potential system failure modes and catastrophic hazards and finding the weakness in reliability and safety to improve the design.

The works above can be conducted to confirm that the relevant design and analysis meet the reliability and other design requirement, and finally the process design scheme for the safety injection system is formed.

References

1. Fengge, P., and P. Minjun. 2000. *Marine Nuclear Power Unit [M]*. Harbin: Harbin Engineering University Press.
2. Jizhou, Z. 2000. *Safety Analysis of Nuclear Reactor [M]*. Xi'an: Xi'an Jiaotong University Press.
3. Juan, C. 2013. Comparison of Reliability Analysis Methods for Passive System [J]. *Huadian Technology* 35 (2): 14–21.

Chapter 6
Instrumentation and Control System

6.1 Overview

6.1.1 Functions of I&C System

The functions of instrumentation and control system ("I&C system" for short) of the nuclear power plant mainly include information functions, control functions and auxiliary function, so as to provide the constituent systems and process equipment with various control & protection and monitoring information, thus guaranteeing the safe and reliable operation of the nuclear power plant.

(1) Information functions
The information functions are as follows:

① Monitoring operational state and diagnosing equipment;
② Monitoring important safety parameters of the nuclear power plant;
③ Monitoring technical facilities for ensuring the implementation of the nuclear reactor shutdown, safety, human behaviors and post-accident measures;
④ Monitoring relevant information for supporting the operators.

(2) Control and protection functions
Control and protection functions are as follows:

① Control of the Safety class safety class systems, mainly the accident protection of the reactor;
② Control of the non-Safety class systems, including reactor power control, pressurizer pressure and water and level control, steam generator feedwater control and steam dump control;
③ Protection of the process equipment;
④ Sequential control and interlock protection of the systems and equipment;
⑤ Manual control in the control room or on site.

J. Yu, *Marine Nuclear Power Technology*,
https://doi.org/10.1007/978-981-15-2894-1_6

(3) Auxiliary functions
Auxiliary functions are as follows:

① Diagnosis and testing of control system and equipment state from the field instruments to actuators;
② Diagnosis and testing of the system hardware and software;
③ Diagnosis of the state of the measurement channel;
④ Diagnosis of the process equipment and electrical equipment and their operational states;
⑤ Supervision of the states, safety limits and conditions of the operational safety indicating instruments for nuclear power plant.

6.1.2 Design Principles

The I&C system of nuclear power plant shall meet the following design principles:

① Combination of centralized operation and field operation. For the entire nuclear power plant, the monitoring and control equipment installed in the main control room can be used to realize the start-up, shutdown, normal operation and treatment of the abnormal conditions or accident conditions of the nuclear power plant. For the system and equipment that do not have a close relationship with the overall operation of the primary and secondary loops of the reactor and that often require the operators to be present, the decentralized and grouped monitoring and control is adopted. For the systems and equipment that do not have a close relationship with the overall operation of the primary and secondary loops of the reactor and that do not require the frequent intervention of operators or that will be used only during the reactor shutdown or occasionally used in an external emergency, the local monitoring and control is adopted.
② The operational degrees of nuclear power plant can be intervened. The I&C system shall be capable of providing the normal operation control means and the backup control means in case of the failure of normal operation control means. When the main control room of nuclear power plant does not have the operational conditions, the emergency control means shall be provided to shut down the reactor and monitor the core cooling process.
③ Safety classification of I&C system. The safety classification of I&C system and equipment is based on the safety importance of such system and equipment. The design of I&C system shall be based on the safety classification requirements to meet the requirements for different degrees of safety.
④ The I&C system shall have the capabilities of controlling the nuclear power plant under normal operation, deviation from normal operation and the processes under the design basis accidents, generally meeting the overall maneuverability specifications and ensuring the nuclear safety and radiation safety.

⑤ The design of I&C system important for safety shall meet with the single failure criterion.
⑥ The design of I&C system shall meet the six requirements for safety, reliability, maintainability, testability, supportability and environmental adaptability of I&C system and equipment.
⑦ I&C system shall have the capabilities of system automatic diagnosis and automatic testing necessary to support the maintenance and testing.

6.1.3 Overall Structure and Characteristics

(1) Safety classification
The I&C system of the nuclear power plant can be classified into safety class systems and non-safety class systems according to the requirements of the safety classification.

① Safety class I&C system. Safety class I&C system include reactor protection system and other safety systems. Generally, safety class systems shall consist of redundant channels, and physical isolation and electrical isolation shall be provided between the channels.
② Non-safety class I&C system. Non-safety class I&C system include various control systems, information display systems and some special instrument and control system of the nuclear power plant.

The data communication interface between safety class systems and non-safety class systems shall be designed in such a way that the data communication of a system at a lower safety level will not jeopardize the data communication and operation of a system at a higher safety level.

(2) Function classification
The structure of the nuclear power plant I&C system can be classified into three levels according to the functions.

① Field level. The field level of I&C system has interfaces with the process equipment. It mainly includes sensors, actuators, power supplies, power amplifier components and other field equipment.
② Control and protection level. The control protection level of I&C system has interfaces with both the field level and the operation and display level. It mainly consists of secondary instruments, electrical drive equipment (including electrical measuring devices), control protection equipment and related communication and network equipment.
③ Operation and display level. The operation and display level I&C system mainly provides the man-machine operation and display interfaces. It mainly consists of communication and information processing devices, integrated display consoles, emergency control panels and compartment TV monitoring devices.

(3) Main constituent subsystems

The I&C system of the nuclear power plant mainly consist of nuclear measurement system, process measurement and control system, reactor power control system, reactor protection system, control rod control and rod position measuring system, electrical control systems for pumps and valves, man-machine manipulation system and field actuators and instruments of secondary coolant circuit, etc.

The typical overall structure of the nuclear power plant I&C system is illustrated in Fig. 6.1.

(4) Main features

The main features of the marine nuclear power plant I&C system are as follows:

① The system design is closely related to the mission requirements of the marine nuclear power plant. The design of the marine nuclear power plant shall satisfy the requirements for safety, reliability and, especially, maneuverability. On the one hand, compared with the I&C system of common nuclear power plant, the automatic control for more operations shall be considered for the control systems of marine nuclear power plant. On the other hand, the specifications to be met under normal transient operation shall be higher.

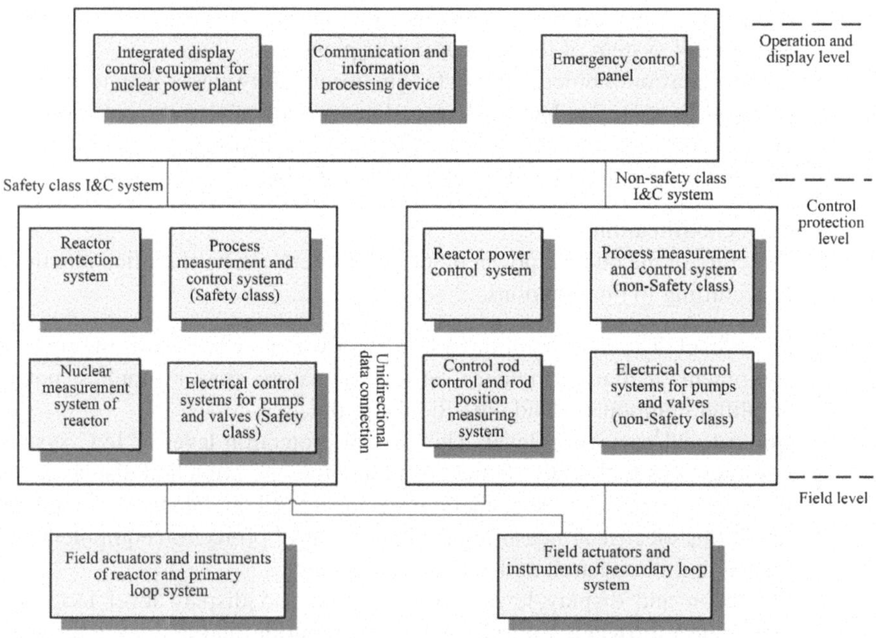

Fig. 6.1 Typical overall structure of I&C system in marine nuclear power plants

② The characteristics of the marine environment must be considered in the system design. The ship tilt, swing, vibration and shock resistance conditions as well as the influence of marine atmospheric conditions such as salt mist and mildew shall be considered in the design of the I&C system.
③ The rather high reliability must be considered in the design of I&C system. Besides the redundant channels considered in the design of safety class I&C system, the reliability requirements for I&C system of marine nuclear power plant also reflect in several aspects such as the redundant configuration of equipment or devices in the systems implementing rather important functions and the derated use of components. For the overall structure of the I&C system of marine nuclear power plants, see Fig. 6.1.
④ The equipment arrangement is compact. The arrangement space of marine nuclear power plant is limited, so the I&C system and equipment shall meet the requirements of the maintainability with compact arrangement and effective utilization of the compartment space.

6.2 Nuclear Measurement System

6.2.1 System Functions

The ex-core nuclear measurement system is Safety class system. Its system functions include the operation functions and safety functions.

(1) Operation functions

① A series of detectors arranged outside the reactor pressure vessel are used to continuously monitor the neutron fluence rate (i.e. reactor power) and reactor period signals of the neutron source level from shutdown conditions to a certain level higher than the full power of reactor (150% of full power ~200% of full power), and provide the operators with information with status of the reactor during shutdown, start-up and power operation.
② Provide the redundant reactor power signals in the power range to the reactor power control system for reactor power control.
③ Provide the relevant signals to the data acquisition and alarm equipment of the man-machine operation system of the nuclear power plant.

(2) Safety functions
The safety functions of ex-core nuclear measurement system are to provide the nuclear power and reactor period signals to the reactor protection system, and trigger the reactor shutdown when the reactor power or reactor period is higher than the setpoints.

6.2.2 Basic Principle of Ex-core Nuclear Measurement Detectors

The nuclear measurement technology is based on the detection of neutrons or γ-rays, or both. Currently, the most widely used detectors in nuclear power plant are counter tubes, fission chambers and ionization chambers, etc. The operating principles of all these detectors are based on the gas ionization. The neutron-sensitive medium is used to trigger nuclear reactions and further cause the gas ionization, and the ionized ions are collected in different electric fields to measure the electrical signals.

(1) Boron counter tubes

Boron counter tubes are usually used in the source range of ex-core nuclear measurement. Boron counter tubes mainly consist of boron-contained (BF_3) proportional counter tubes and boron-coated proportional counter tubes. The structure of the boron-coated proportional counter tube is illustrated in Fig. 6.2 [1]. The counter tube is a metal cylinder, at the center of an anode metal filament stretched by insulators. The inner surface of the cathode is coated with 10B, and the electrodes are insulated from each other. The working gas filled in the counter tube is a mixture of argon (Ar) and carbon dioxide (CO_2). During the operation of counter tube, a high voltage is applied to the anode to collect the ionization pulses. BF_3 proportional counter tubes are filled with the gas of boron trifluoride, and their operational principles are similar to those of boron-coated counter tubes.

The detector makes use of neutron nuclear reactions for measurement. The nuclear reactions between incident neutrons and boron are as follows [1]:

$$
\begin{aligned}
&{}_0^1n + {}_5^{10}\mathrm{B} \rightarrow {}_3^7\mathrm{Li} + {}_2^4\mathrm{He} + 2.793\,\mathrm{Mev} \rightarrow {}_3^7\mathrm{Li}^* + {}_2^4\mathrm{He} + 2.316\,\mathrm{Mev} \\
&{}_3^7\mathrm{Li}^* \rightarrow {}_3^7\mathrm{Li} + \gamma + 0.48\,\mathrm{Mev}
\end{aligned} \tag{6.1}
$$

The above reactions are ^{10}B(n, a)^{7}Li reactions, and the product nucleus is ^{7}Li in the excited state. The energy during the transition is released in the form of γ-rays. Lithium ions and a particles generated by nuclear reactions ionize the argon and generate electrons and positive ions. Under the effect of external electric field, the electrons move towards the anode and the positive ions move towards the cathode,

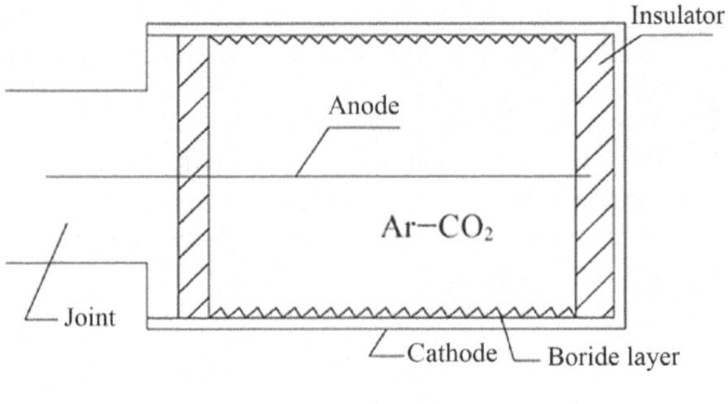

Fig. 6.2 Structure of boron-coated proportional counter tube

thus generating the detectable electric pulses. Compared with boron-coated proportional counter tubes, BF_3 proportional counter tubes have better sensitivity, pulse amplitude distribution and plateau length though lower radiation damage resistance and shorter life. Therefore, the boron-coated proportional counter tubes are used more widely in the PWR nuclear power plant.

(2) Fission ionization chambers

A fission ionization chamber has a sensitive layer formed by coating one electrode of the detector with fissionable materials such as ^{235}U so as to detect the neutrons. The slow neutrons are transformed by fission reaction to ionized products that can be detected. That is, the detection of neutrons is realized by the measurement of ionization effect of fission fragments. The fission ionization chambers are generally used under conditions with rather large γ radiation background, because the pulse signal amplitude generated by the fission is much larger than that generated by α, β and γ radioactivity and thus easy to be identified. When the nuclear power plant is at a relatively low power level, the neutron flux density is larger than the source range and the γ radiation background is large, the fission ionization chambers are usually used as detectors for the intermediate range or used as measuring instruments for core power distribution.

(3) Ionization chambers

Generally, the ionization chambers are divided into uncompensated ionization chambers and γ-compensated ionization chambers. The output signals are direct current directly proportional to neutron flux density. The operation principles of ionization chambers are based on the fact that the incident particles (such as neutrons) entering the ionization chamber will directly or indirectly interact with the material filled in the chamber, causing ionization of the material. By detecting the positive and negative ion pairs generated during the ionization, the radioactive intensity or the energy of the incident particles can be measured. The non-compensated ionization chambers are usually used in situations where the neutron flux density is rather large and the γ radiation background can be neglected, such as the detectors in the power range. A γ-compensated ionization chamber is actually composed of two ionization chambers. One ionization chamber is sensitive to both neutrons and γ-rays and generates neutron current and γ current under the effect of high voltage. The other ionization chamber is sensitive only to γ-rays and only generates γ current under the effect of offset voltage. The measured current finally obtained by the detector is the current difference between these two ionization chambers. If these two ionization chambers have the same sensitivity to the γ-rays, then full compensation is realized. The γ-compensated ionization chambers are generally used in situations where the γ radiation background is rather large and its effect cannot be neglected, such as the detectors in the intermediate range. In addition, when the detectors are arranged slightly far away from the reactor but the decrease of the γ radiation background is not large, the γ-compensated ionization chambers can also be used as the detectors in the power range.

6.2.3 *Description of the System and Equipment*

(1) System composition

The ex-core nuclear measuring system consists of source range measuring channel, intermediate range measuring channel and power range measuring channel. Each channel consists of corresponding detector, signal processing electronic equipment, cables and so on. In order to meet the single failure criterion of safety I&C system, each of these three channels is designed with redundancy.

① Source range measuring channel.

The neutron flux density range monitored by the source range measuring channel is from the neutron source level to the level with 5– 6 orders of magnitudes higher. Estimated as per the counting rate, the range is about $1\text{–}10^5$ counts per second (c_{PS}). The neutron level monitored by the source range measuring channel is low, and the detector signals are counted in the form of pulses.

Signals from source range detectors (e.g. boron counter tubes) are transmitted through cables to the counter, and are amplified, screened and shaped by the preamplifier of the counting device. The pulses caused by γ-rays and other interference pulses can be screened out in the screening process. Then the signals can be sent to the operation and display system for displaying counting rate, and the reactor period signals will be outputted and used as the protection parameters and displaying parameters after the calculation and processing.

② Intermediate range measuring channel.

The intermediate range measuring channel is used to measure the neutron fluence rate and reactor period of reactor state from subcriticality to full power operation. Generally, there is an overlap of 1–2 orders of magnitude between intermediate range measuring channel and the source zone measuring channel. In the range of the intermediate range measuring channel, the neutron level change is rather large, and the γ background effect is rather significant. Therefore, the fission chambers or γ-compensated ionization chambers are usually used as the detectors.

The signals from intermediate range detectors are transmitted through cables to the preamplifier for amplifying, screening, calculation and processing, after which the period signals are outputted to the reactor protection system and the man-machine operation and display system, and the counting rate signals are transmitted to the man-machine operation and display system.

③ Power range measuring channel.

There is an overlap of 1–2 orders of magnitude between the power range measuring channel and the intermediate range measuring channel. Generally, the measurement range of the power range measuring channel covers 3 orders of magnitude. The neutron current signals from the detectors (γ-compensated ionization chambers or uncompensated ionization chambers) in the power zone measuring channel are transmitted through cables to the micro-current amplifier for amplification and transformation to obtain the nuclear power signals of the reactor. The nuclear power signals, as protection signals and control signals, are transmitted to the reactor protection system and the reactor power control system.

The typical composition of ex-core nuclear measuring system is illustrated in Fig. 6.3.

(2) Main equipment

① Source range measuring channel.
The source range measuring channel equipment mainly includes source range detectors, detector supports, cables connecting detectors and electronic equipment, and chassis for signal conditioning and processing. The typical source range detector is the boron counter tube assembly, which consists of three parts, i.e. boron proportional counter tube, integrated radiation resistant inorganic cables and outer sheath. The coaxial cables connecting the detectors and the electronic equipment transmit high voltage power from the electronic equipment to the connectors outside the detectors, and transmit the pulse signals from the connectors to the electronic equipment. The chassis for signal conditioning and processing mainly accommodates the pulse amplifier module, high-voltage power supply and pulse conditioning module, pulse counting and interface module, signal conditioning power supply module and microprocessor CPU module.

② Intermediate range measuring channel.

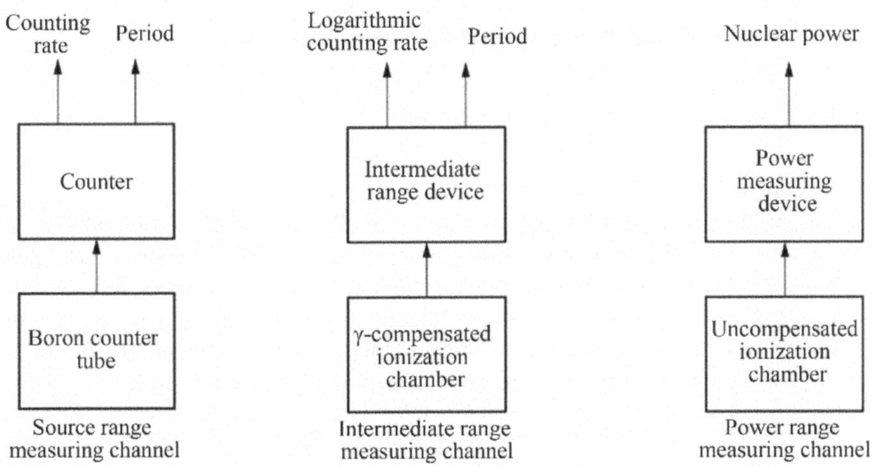

Fig. 6.3 Composition of ex-core nuclear measuring system

The intermediate zone measuring channel equipment mainly consists of intermediate range detectors, detector supports, cables connecting detectors and electronic equipment, and instrument chassis for signal conditioning and processing. The typical intermediate range detector is the boron-lined compensated ionization chamber assembly, which consists of three parts, i.e. ionization chamber, integrated radiation resistant inorganic cables and outer sheath. The coaxial cables are also used to connect the detectors and the electronic equipment, and to transmit high voltage, offset voltage and detector current. The instrument chassis mainly accommodates signal conditioning power supply module, high-voltage power module, current amplification module and microprocessor CPU module.

③ Power zone measuring channel.
The power zone measuring channel equipment mainly consists of power range detectors, detector supports, cables connecting detectors and electronic equipment, and instrument chassis for signal conditioning and processing. The typical power range detector is the boron-lined uncompensated ionization chamber consisting of several neutron-sensitive sections. The coaxial cables connecting the detectors and the electronic equipment transmit the current signals and high voltage signals of the detectors from the connectors to the power range instrument chassis. The instrument chassis mainly accommodates signal conditioning power supply module, high-voltage power module, current amplifying and neutron noise amplifying module and microprocessor CPU module.

6.3 Process Measurement and Control System

6.3.1 Process Measurement System

(1) System functions
The process measurement referred in the nuclear power plant generally is the measurement of temperature, pressure, water level, flow and so on. The systems involved mainly include reactor coolant system, pressure safety system, residual heat removal system, safety injection system, coolant charging system, component cooling water system and coolant purification system. The functions of process measurement system of nuclear power plant are as follows:

① Perform the whole-process monitoring of variables and system states of various systems and equipment under normal operation, anticipated operational occurrences and accident conditions of the nuclear power plant, and provide the operators with real-time numerical display of various process parameters to analyze and judge the operational states of the primary loop system and equipment.

② Give automatically the alarm signals when the process parameters are beyond the specified variation range, so as to remind the operators to pay attention to the abnormal change of the thermal parameters, analyze the causes and take the treatment measures in time.

③ Provide necessary process parameter signals to process control system, reactor power control system and reactor protection system for automatic control and protection.

④ Send process parameters important to system safety to relevant systems for screen display, alarm and record, and provide the basis for safety operation and accident analysis of the nuclear power plant.

(2) Description of the system and equipment

① Principles of system design.

i The design of process measurement system shall be based on the safety requirements of the nuclear power plant. The necessary measuring instruments and the requirements for their measurement range, accuracy, response time and output signals shall be determined according to various operational states;

ii Redundant measurement channels shall be provided for variables related to reactor protection system, and the channel independence shall be maintained. The output signals from the channels can be directly used in the protection system, and the measurement signals shared by other systems with the protection system shall be isolated before output;

iii The sensors, transmitters and instruments in the process measurement system must be able to work normally under the specified ambient conditions. The measuring instruments used in the reactor protection system shall be able to work under the accident conditions without any failure;

iv In the design of instrument and instrumentation system, the possibility of field inspection and maintenance of marine nuclear power plant must be taken into account, so that the failures of instrument can be easily identified and repaired, and the periodic tests and calibration can be performed according to functions, anticipated drift rate and reliability requirements of the instruments.

② System composition.

i Primary detecting elements that convert the characteristics of measured parameters to the easily detected signal, such as temperature sensors, pressure transmitters and throttling devices;

ii Carriers used for signal extension or signal transmission, such as instrument tubes and cables;

iii Equipment that distributes, isolates or converts electrical signals, such as isolators, distributors and bus adapters;

iv Equipment that distributes or isolates pressure signals, such as tee joints and instrument valves;

v Other auxiliary equipment, such as mounting brackets and hangers used for equipment support.

③ Main equipment.

i Temperature measuring equipment.

The temperature measurement of the nuclear power plant mainly includes the temperature measurement of reactor coolant, motor stators of important pumps, steam and component cooling water. The main temperature measuring equipment includes the electrical resistance thermometers and thermocouple thermometers. The electrical resistance thermometers may have one element, two elements or three elements. The thermocouple thermometers are used to measure the temperature at the reactor core outlet.

ii Pressure measuring equipment.

The pressure measurement of the nuclear power plant mainly includes the pressure measurement of the pressurizer, equipment inlets and outlets and steam. The measuring equipment is capacitance-type pressure and differential pressure transmitters, such as the 1151 series of capacitance-type pressure transmitters.

iii Flow measuring equipment.

The flow measurement of the nuclear power plant mainly includes the flow measurement of reactor coolant system, safety injection system, component cooling water, steam generator feedwater, steam generator steam and purification water. The flow measurement is mainly the instantaneous flow measurement. The generally used flow measuring equipment is differential pressure flowmeter which is a combination of throttling devices and differential pressure transmitters. Proper throttling devices are selected according to the pressure loss requirements of the system loops and the features such as pipe diameter and flowrate. Throttling devices can be standard ones such as orifices, nozzles and Venturi tubes, or non-standard ones such as bends, V-cones and uniform velocity tubes.

iv Level measuring equipment.

The level measurement of the nuclear power plant mainly includes the liquid level measurement of pressurizers, steam generators and the containers of the nuclear auxiliary systems. The level measurement can be divided into continuous measurement and discrete measurement. For continuous measurement, the pressure or differential pressure type level measurement is adopted. The differential pressure type level measurement usually requires the setting of reference level. If the measured liquid density varies greatly, the pressure, temperature of the liquid shall be measured to calculate the density for the compensation of level measurement, such as the measurement of the liquid level of the steam generators. The discrete level measurement is generally performed by electric contact level probes, level switches and other sensors, such as for the bilge water level measurement of the reactor compartment.

v Process measurement signal processing devices.
The process measurement signal processing devices mainly consist of microprocessors, memories, A/D converters and D/A converters, etc. They have self-diagnosis, compensation, calculation and other functions, and are used to further process the field measurement signals and send them to relevant systems.

6.3.2 Process Control System

The process control system of the nuclear power plant can be divided into manual control system and automatic control system according to the degree of automatic control. Manual control is mainly the control of temperature, pressure and level of relevant pumps, valves and containers of nuclear auxiliary systems and secondary-loop auxiliary systems, etc. This type of control often does not require too much control logic operation, and it is manually completed by the operators. Automatic control is mainly the control of pressurizer pressure and level, steam generator level and so on. This type of control often requires multiple control inputs for the logic operation, to produce output control quantity to control the controlled objects, and to realize the control operation procedure with the controller. This section mainly describes the process control system of automatic control in the nuclear steam supply system.

(1) Pressurizer pressure control system

① System functions.
The pressurizer pressure control system ensures that the pressure of reactor coolant system is maintained within the normal pressure fluctuation range under the steady-state operation. The pressurizer pressure is maintained within the specified limits under normal transient operation to exclude the accidental shutdown of reactor and the actuation of safety valves, while the alarm signals will be sent out when the pressurizer pressure is beyond the specified limits.

② The system and equipment description.

i System control principle.
Various transient processes in the operation of nuclear power plant will result in an imbalance between the power generated by the reactor and the power used by the steam generators. Therefore, the temperature of reactor coolant changes along time, causing the thermal expansion or cold-contraction of water in primary loop. As a result, the coolant will flow through the surge lines from the reactor coolant system to the pressurizer or in the reverse direction, causing the volume change of water in the pressurizer and further leading to the pressure change in pressurizer

steam space and reactor coolant system. The pressurizer pressure control system uses the pressurizer pressure measuring signals for automatic control of pressure during the reactor operation. The control principle can be simply summarized as follows: when the pressure in the pressurizer steam space is lower than the setpoint, the electric heater will be turned on to boil the water to make up the steam and increase the pressure of the pressurizer steam space; when the pressure is higher than the setpoint, the mist spray will be used to condensate the steam to suppress the pressure increase. The principle of pressurizer pressure control system is illustrated in Fig. 6.4.

ii Main equipment.
The pressurizer pressure controller is the main equipment of pressurizer pressure control system. This controller collects the pressure signals from the process measurement system, performs the calculation and processing, and sends the control signals and alarm signals to corresponding actuators to realize the control.

(2) Pressurizer level control system

① System functions.
The functions of pressurizer level control system are to maintain the pressurizer level within the specified limits, maintain certain coolant inventory in the coolant system, ensure the function of the pressurizer to maintain the specified reactor coolant pressure, and send out alarm signals when the pressurizer level is beyond the specified limits.

② Description of the system and equipment.

i System control principle.
The water inventory of the reactor coolant system shall be stable to minimize the volume of fluid entering or leaving the reactor coolant system

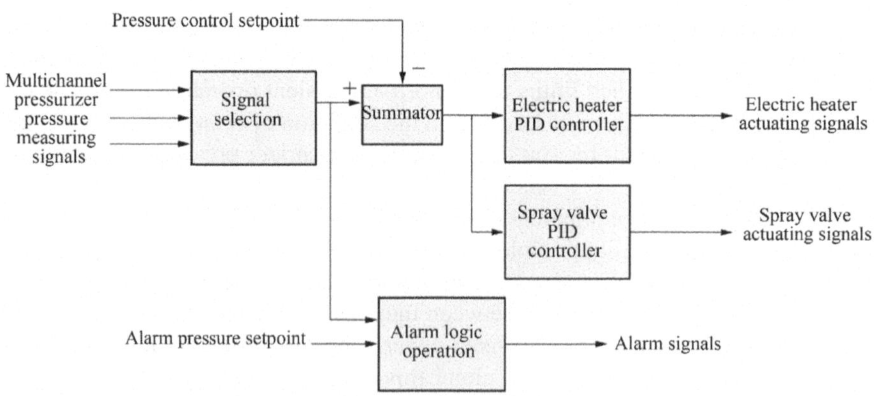

Fig. 6.4 Principle of Pressurizer pressure control system

during the power change. Therefore, the setting of the pressurizer level setpoint must take account of temperature change in the reactor coolant system. This temperature change at constant mass will cause the water to enter or leave the pressurizer. The level setpoint shall be calculated in its variation range as a function of reactor coolant temperature. The principle of pressurizer level control system is illustrated in Fig. 6.5.

ii Main equipment.

The main equipment of pressurizer level control system includes the pressurizer level controller. This controller collects the level signals from the process measurement system, performs the calculation and processing, and sends the actuating signals and alarm signals to corresponding equipment to generate actuating signals of corresponding actuators and level alarm signals.

(3) Steam generator feedwater control system

① System functions.

The functions of steam generator feedwater control system are as follows: under the steady-state operation, it ensures the feedwater flow required by the steam generator and maintains the level of steam generator within the specified range; under the transient operating conditions, it enables the feedwater flow to follow the change of the steam flow from steam generator in time, so that the level of steam generator can be maintained within the specified range.

The steam generator feedwater control system consists of the steam generator level control system and the feedwater pump speed control system.

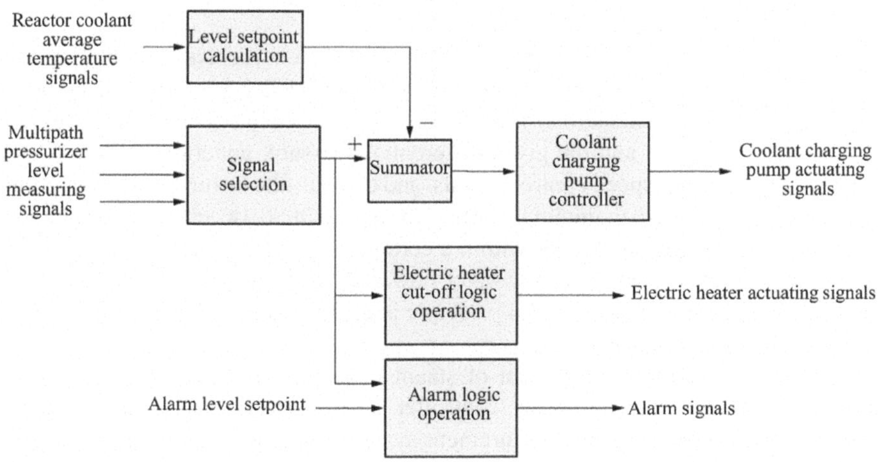

Fig. 6.5 Principle of Pressurizer level control system

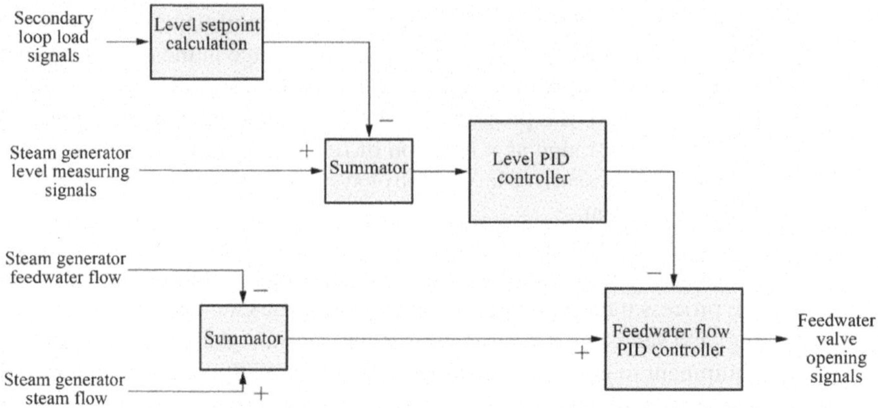

Fig. 6.6 Steam generator level control principle

② Description of the system and equipment.

i System control principle.

a. Steam generator level control.

The level controller receives the signals of steam generator level, steam flow and feedwater flow (called "three elements"), outputs control signals according to input signal changes and given control law, and changes the opening of feedwater regulating valve and thus changes the feedwater flow so that the steam generator level can be maintained within the specified range. The steam generator level control principle is illustrated in Fig. 6.6.

b. Feedwater pump speed control.

The feedwater pump speed control system receives pressure signals from the steam header and feedwater header, performs the comparison calculation of the difference value between these two pressure signals and the given differential pressure, generates the feedwater pump speed control signals, and controls the turbine feedwater pump speed to maintain the differential pressure between steam header and feedwater header within a certain range so that the requirements of steam generator for water flow and head can be met. The feedwater pump speed control principle is illustrated in Fig. 6.7.

ii Main equipment.

The main equipment of steam generator feedwater control system includes the feedwater controller. This controller collects the signals from the process measurement system, such as signals for steam generator feedwater flow, steam flow, steam header pressure and feedwater

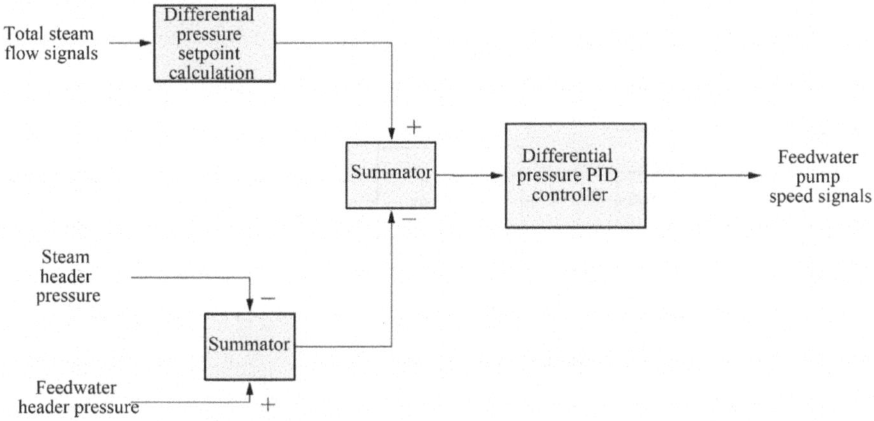

Fig. 6.7 Feedwater pump speed control principle

header pressure, and performs the calculation and processing and sends the actuating signals to drive the corresponding equipment.

(4) Steam dump control system

① System functions.
The functions of steam dump system are to directly dump the excess steam in main steam system when there is a rather large unbalance of the energy between primary and secondary loops during the sharp drop of secondary loop loads so that the amplitudes of temperature and pressure changes in nuclear steam supply system caused by the sharp drop of loads can be decreased.

② Description of the system and equipment.

i System control principle.
The automatic control process of steam dump for nuclear power plant can be classified into three forms according to the control signals. The difference value is obtained by comparing the reactor power signals and the secondary loop steam loads. When the difference is larger than the setpoint, the steam dump system will start up automatically to dump the steam. When the reactor is operating above the predetermined high power and the load rejecting signals are sent out, the steam dump system will start up automatically to dump the steam. When the main steam pressure is higher than the preset pressure value for starting steam dump, the steam dump system will start up automatically to dump the steam. The control principle of steam dump control system is illustrated in Fig. 6.8.

ii Main equipment.
The main equipment of steam dump control system is the steam dump controller. This controller collects the signals such as reactor power, main

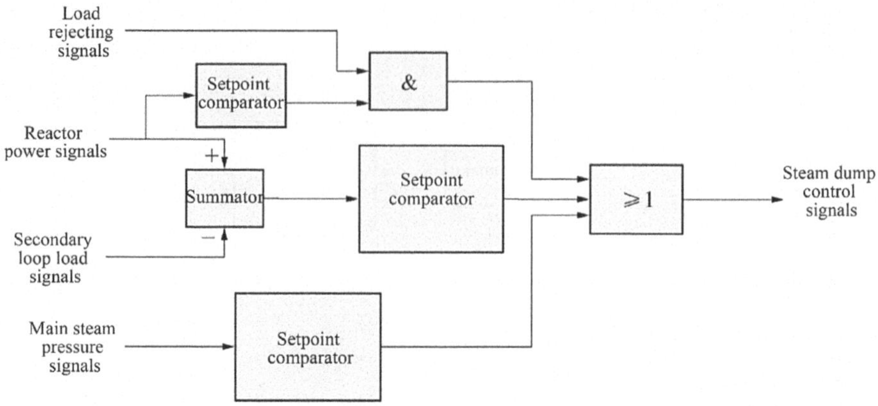

Fig. 6.8 Control principle of steam dump control system

steam pressure and quick shutdown of turbine, performs the calculation and processing, and sends the actuating signals to drive the corresponding equipment.

6.4 Reactor Power Control System

6.4.1 System Functions

The energy of reactor is generated by the sustained fission chain reactions of uranium atomic nuclei in the core fuel elements. The reactor power is in direct proportion to the core neutron density. In order to change the reactor power to meet the change of energy demand, it is necessary to introduce the required reactivity to cause the corresponding change in the core neutron multiplication during the chain reactions, so as to change the core neutron density and finally change the reactor power. It is necessary to introduce the positive reactivity to raise the reactor power and introduce the negative reactivity to reduce the power.

On the other hand, the sustained nuclear reactions in the core always change the operational states of the reactor. Such changes are mainly caused by the following effects inside the PWR [2]:

(1) Temperature effect

The temperature effect is the effect that reactivity changes due to the change of reactor temperature. The PWRs are generally designed to have a negative temperature effect, that is, the reactivity decreases as the reactor temperature increases. The fuel reactivity temperature effect and moderator reactivity temperature effect play a leading role. The fuel reactivity temperature effect lasts for a rather short time, generally a fraction of a second, while the moderator water reactivity temperature effect lasts for a rather long time, generally about several seconds. In addition, as the reactivity temperature coefficient is used to measure the reactivity temperature effect, the temperature coefficient of moderator water is generally one order of magnitude higher than that of fuel. Therefore, the temperature effect of moderator in the PWRs plays a decisive role.

(2) Pressure effect of moderator

During the power operation of reactor, the moderator has a positive pressure effect, that is, the reactivity increases as the pressure increases. However, the reactivity change caused by the pressure change is very small when compared with the temperature effect of moderator, so the influence of the pressure effect is negligible.

(3) Void effect of moderator

If the excessive local energy of core causes the local boiling of moderator and generates bubbles, the moderation effect will be influenced and the reactivity will change. During the normal operation, the reactor power change is limited and the boiling is forbidden inside the reactor, so void effect actually does not work.

(4) Poisoning effect

The fission products of the core, i.e. xenon and samarium, absorb a large number of thermal neutrons and thus cause the reactivity change. This process is rather complex and may cause the xenon oscillation within a certain frequency range. However, the oscillation amplitude is rather small and the oscillation frequency is low, about 0.2–2 period/day.

(5) Burnup

With the consumption of nuclear fuel uranium in the reactor, both the fission number and the effective multiplication factor decreases, so it is necessary to compensate for the reactivity decrease caused by burnup. However, the reactivity change caused by burnup change is much slower than that caused by the temperature effect.

When the energy demand of the nuclear power plant changes, it is necessary to change the reactivity so that the corresponding reactor power level is achieved. Among the several effects inside the reactor during the operation, the temperature effect, poisoning effect and burnup are rather significant for the reactivity change. However, such effects occur at different time scales, so the reactivity shall be changed to overcome the adverse impacts of such effects. For the frequent power change of nuclear power plant, the temperature effect is the most significant, while the temperature effect of moderator is the most obvious. In addition, the PWRs have a certain excess reactivity as operation margin for overcoming the poisoning, compensating for the reactivity decrease caused by the increase in reactor temperature to the operating temperature, burnup and absorption of other fission product poisons. However, certain excess reactivity shall be compensated by certain means.

In order to solve the above problems, the most common method for PWRs is to change the number of absorbed neutrons to change the reactivity. One way is to insert the control rods made from material with higher thermal neutron absorption cross section into the active region of reactor core, and move such control rods to change their positions in the active area of core, so as to change the absorption of neutrons, thus changing the reactivity. Another way is to compensate the reactivity by adjusting the concentration or total volume of boric acid solution in the moderator. On the one hand, this compensates for the rather slow reactivity decrease caused by the temperature change from the cold state to the hot state, the burnup and the poisoning. On the other hand, the negative reactivity is introduced when the reactivity increases because of reactor shutdown, reload and xenon decay.

A number of control rods can be arranged for one a reactor. They can operate individually or in group and move in different speeds. Most control rods are firstly used to start up and shut down the reactor. Withdrawal or insertion of several groups of control rods with rather large worth can change the reactivity to realize the start-up or shutdown of reactor. If such control rod groups with rather large worth are moved rapidly, the reactivity will change too rapidly, which will affect the reactor safety. Therefore, such control rods are generally moved slowly. In addition, such control rods only play a rough control role in the reactor operation, that is, they are used to compensate for rather slow reactivity changes caused by the burnup and the poisoning. Since there are no precise control requirements, both the manual control of operators and the sequential automatic control realized by established computer programs can be adopted.

In order to adapt to the frequent power change of the nuclear power plant, it is necessary to use one control rod or a group of control rods with rather small worth as the regulating rod(s). A regulating rod can automatically and rapidly change its position to quickly change the reactivity, make the reactor power consistent with the load change of nuclear power plant, compensate for the reactivity change caused by the temperature change of moderator, maintain the reactor power at the constant level required for the load after the position of regulating rod is finally fixed, and keep the reactor temperature at the setpoint to eliminate any additional temperature reactivity. This regulation process is very complex, involving many factors of primary and secondary loops of reactor. Meanwhile, it is closely relevant to the dynamic characteristics and temperature effect of reactor and there are control requirements with rather high precision for reactor power and temperature, so the process is mainly realized by the automatic control of regulating rod.

According to above factors and requirements, the main functions of reactor power control system are as follows: Firstly, match the output power of reactor with the load demand, and keep the energy balance between primary and secondary loops; secondly, eliminate the effect of internal and external disturbances in the reactor and the primary loop system,, make the reactor operate according to the given steady-state operation characteristics, and keep the main parameters of nuclear power plant within the specified ranges; and thirdly., during the transient operation, overcome the reactivity disturbances within the scope of design, follow the load changes within the range of design, make the reactor power to quickly and automatically follow the large-scale and high-speed transient changes of loads, and ensure that neither accidental shutdown nor pressurizer safety valve actuation will be triggered.

6.4.2 Principles of Reactor Power Regulation

(1) Dynamic characteristics of reactor

In the PWRs, the main factor affecting the dynamic characteristics of reactor is the reactivity change caused by temperature effect of fuel and moderator. In case of reactor power change, the fuel temperature will change first. However, the fuel temperature change has a very small delay in response time, and there is no obvious time delay between the reactivity change and the fuel temperature change. Therefore, when the fuel temperature changes, the reactivity change caused is relatively instantaneous. However, the moderator temperature change is rather slow because the thermal capacity of moderator is generally large. Therefore, the reactivity change of moderator has a rather large time delay after the fuel temperature change is transferred to the moderator. The reactivity temperature coefficient of PWRs is always designed to be a negative value, thus establishing the negative feedback of temperature effect in the reactor dynamics, as illustrated in Fig. 6.9 [3].

In Fig. 6.9, $K_R G_R(s)$ means the kinetic transfer function of reactor, $K_T G_T(s)$ means the transfer function for temperature feedback, $\Delta\rho_{ex}$ means the externally introduced reactivity change, such as the reactivity change caused by control rod movement, $\Delta\rho_T$ means the reactivity change caused by the changes of fuel temperature and moderator temperature, $\Delta\rho$ means the difference between $\Delta\rho_{ex}$ and $\Delta\rho_T$, and ΔP_n means the change in reactor power (or neutron density). When the control rod moves in the core and causes $\Delta\rho_{ex}$, $\Delta\rho$ will change accordingly. Therefore, if the reactivity change of the reactor is actually introduced, the reactor power will have a corresponding change ΔP_n, and the power change will further cause the change in fuel temperature and moderator temperature successively. Because of the existence of a negative reactivity temperature coefficient, the total reactivity change $\Delta\rho$ will gradually approach zero, thus stabilizing the reactor power at another power level. The reactor closed-loop transfer function with temperature reactivity feedback is as follows:

$$\frac{\Delta P_n(s)}{\Delta P_{ex}(s)} = \frac{K_R G_R(s)}{1 + K_R K_T G_R(s) G_T(s)} \tag{6.2}$$

(2) Self-stabilization and self-regulation of reactor

The negative feedback mechanism inside the PWRs caused by negative temperature reactivity feedback effect of reactor is the basis for the stable operation of reactor

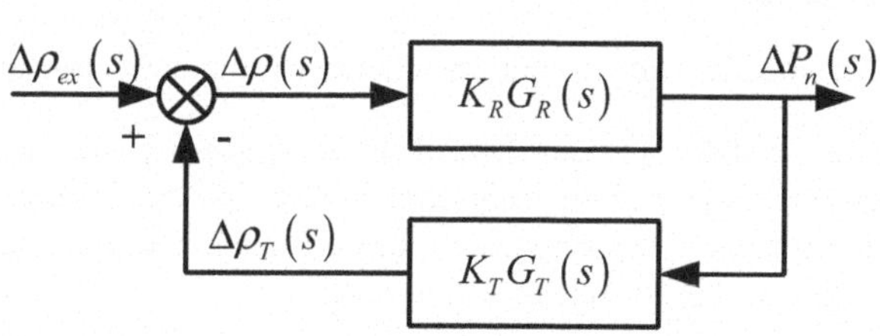

Fig. 6.9 Reactor transfer function with temperature feedback

and the inherent safety of nuclear power plant, and provides the reactor with the characteristics of self-stabilization and self-regulation, which is beneficial to the design of reactor power regulation system.

"Self-stabilization" means that the reactor can maintain the original power level when the reactor has any internal or external disturbances. For example, when a positive reactivity disturbance ($\Delta\rho_{ex}$) is introduced into the reactor, the neutron density will suddenly increase by (Δn), reactor power will suddenly increase by (ΔP_n), the fuel temperature will increase by (ΔT_f), and the average temperature of moderator will increase by (ΔT_{av}). Since a negative reactivity is generated by the temperature effect to offset the positive reactivity disturbance, finally, the neutron density will basically restore to the initial value, and the reactor power will keep unchanged.

"Self-regulation" means that when the load changes, the reactor itself can quickly reach the thermal balance and the reactor power can follow the loads without the intervention of any external regulation system. For example, when the turbine load suddenly increases (ΔP_H), the turbine speed will decrease (Δf), the regulator will increase the opening of turbine valve (ΔK), and the steam flow will increase (ΔF_s). As a result, the steam pressure (ΔP_s) and the steam temperature will decrease (ΔT_s), thus the average primary coolant temperature will decrease (ΔT_{av}). Since a negative reactivity is generated by negative temperature coefficient to increase the neutron density (Δn), the reactor power (ΔP_n) and the fuel temperature (ΔT_f), a negative reactivity will be generated to offset the positive reactivity generated by the decrease in average coolant temperature. Finally, the reactor will reach a new balance state and the reactor power will be consistent with the load demand [2].

For a marine nuclear power plant with frequent load change, the self-stabilization and self-regulation of reactor can help the reactor to follow the load of nuclear power plant, but will make the operational state of the reactor deviate from the designed initial state, such as the change in average temperature of the reactor. Therefore, on the one hand, the introduction of external reactor power regulation system for the regulation of control rods can, based on the self-stabilization and self-regulation of reactor, regulate the reactor power more effectively, and make the reactor reach the target load demand more quickly and stably, thus making the nuclear power plant meet higher overall performance indicators; on the other hand, the reactor power regulation system will, by the regulation of control rods, recover the reactor to the designed steady state operation after a transient process, such as maintaining the average temperature of reactor at a setpoint, thus meeting the design and operation requirements of the reactor and finally facilitating the stable and safe operation of the reactor.

(3) Steady-state operation scheme of nuclear power plant

The steady-state operation of nuclear power plant means an interrelationship to be followed by operating parameters such as temperature, pressure and flow under the steady-state operating conditions of reactor and nuclear power plant, with the load power or reactor power as the core.

The output power P_{H} of nuclear power plant and the temperature difference between the primary side and secondary sides of the steam generator has the following relation [4]:

$$\mathrm{PH} = (\mathrm{UA})\mathrm{s}\ (\mathrm{T}_{\mathrm{av}} - T_{\mathrm{s}}) \tag{6.3}$$

$$T_{\mathrm{av}} = \frac{T_{\mathrm{h}} + T_{\mathrm{c}}}{2} \tag{6.4}$$

where, $(UA)_{\mathrm{s}}$ means the equivalent heat transfer coefficient (W/°C) from primary side of steam generator to its secondary side, T_{av} means the average primary coolant temperature (°C), T_{h} means the reactor inlet temperature (°C), T_{c} means the reactor outlet temperature (°C), and T_{s} means the saturated steam temperature (°C) at the secondary side of steam generator.

The reactor output power P_{n} can be expressed as follows:

$$P_{\mathrm{n}} = FC_p(T_{\mathrm{h}} - T_{\mathrm{c}}) \tag{6.5}$$

where, F means the primary coolant flow (kg/s), and Cp means the specific heat capacity of coolant [J/(kg · °C)].

The operational objective of nuclear power plant is $P_{\mathrm{n}} = P_{\mathrm{H}}$. To perform this regulation, the variable that can reflect the difference between the reactor power P_{n} and the output power P_{H} of nuclear power plant shall be selected as the main regulating variable. As can be seen from Eqs. (6.3) and (6.5), the change of anyone of the parameters will influence the output power P_{n} of reactor and the output power P_{H} of nuclear power plant. However, in fact, during the design and operation of PWR nuclear power plant, it is required that all parameters other than the average coolant temperature T_{av} and saturated steam temperature T_{s} at the secondary side of steam generator remain unchanged. If one of these two parameters is kept unchanged and a certain relation is created between the change of the other one and the output power, the basic steady-state operation mode is established. The PWR nuclear power plant generally adopts the operation mode with a constant average coolant temperature T_{av}, as illustrated in Fig. 6.10 [2]. In this operation mode, when the primary coolant flow remains unchanged, the average coolant temperature will not change with the load. This is most advantageous for the primary loop system, especially for a reactor with a negative temperature coefficient, which can fully use its self-stabilization and self-regulation under this operation mode. Moreover, the average coolant temperature is constant, the coolant volume change is rather small, and the pressurizer level is almost unchanged, so the pressurizer volume can be reduced.

(4) Principle of reactor power regulation

The ultimate purpose of a nuclear power plant reactor is to match the reactor power in the form of heat transfer with the load, so the variable that can reflect the difference between the thermal power and the load shall be selected as the main regulation variable. Generally, the reactor power regulation system of a PWR uses the average

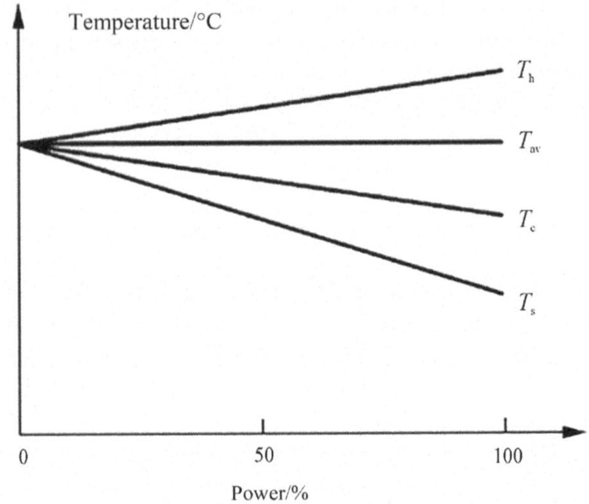

Fig. 6.10 Steady-state operation mode with constant average coolant temperature

temperature as the main regulation variable: the deviation between the measured value and the setpoint of average coolant temperature is used to control the movement of control rods and regulate the introduced reactivity.

It is necessary to use the reactor power regulation system to regulate the position of the special regulating rod banks to realize the change in operational state of reactor with the change in reactor power, and the precise load following of reactor power. In view of the inherent stability, self-stabilization and self-regulation of reactor, the balance between reactor power and output power of nuclear power plant can be automatically realized. Therefore, in the most common steady-state operation scheme with constant average coolant temperature, the most basic function of the moving the regulating rods is to change the average coolant temperature to return it to the setpoint. However, it is a long process from the change in output power of nuclear power plant to the change in average coolant temperature: change in steam flow, change in steam pressure, change in saturated water state at secondary side of steam generator and change in heat transfer at from primary side and to secondary side; change in coolant temperature at inlet and outlet at primary side of steam generator after coupling to primary loop; change in temperature at reactor inlet and outlet due to coolant flow; and finally change in average coolant temperature. This process has many intermediate steps and contains rather large thermal inertia and hysteresis, and the coupling factors and nonlinearity are also very complex. If the reactor power is regulated only according to the change of the average coolant temperature, the regulation quality will be limited by the lag of regulation action and various disturbances and difficult to improve. Therefore, based on the regulation of the average coolant temperature, the secondary loop load is introduced. It is expected that the response of the reactor power to the output power of nuclear power plant will be accelerated by direct following the secondary loop load. The control principle of this control scheme [5] is illustrated in Fig. 6.11.

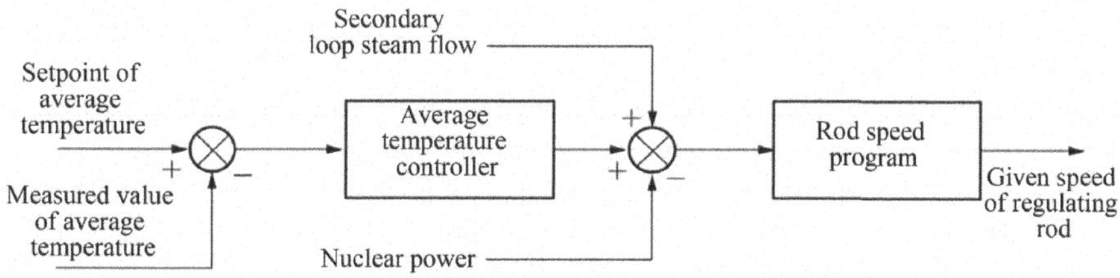

Fig. 6.11 Control principle of regulating rod

In Fig. 6.11, the setpoint of average coolant temperature is predetermined according to the steady-state operation scheme, and its measured value is calculated according to the measured values of temperature at reactor inlet and outlet. The deviation between the measured value and setpoint of average temperature is regulated by the average temperature controller, and added to the deviation between the current load and the reactor output power to obtain the final power deviation. Determined by the rod speed program, the relation curve between the power deviation and the given rod speed is used to obtain the given speed of the current regulating rod. The average temperature controller can be the most common proportional-integral (PI) controller. The proportional action of PI controller reflects that the system instantaneously regulates the amplitude of control action in direct proportion to the deviation, while the integral action can generate a continuous and non-transient regulation action based on the deviation integral to gradually reduce the deviation to zero. The typical rod speed program specifies the linear change range for the change in rod speed with power deviation. Meanwhile, the positive sign or negative sign of the rod speed determines the withdrawal or insertion of regulating rod (positive sign for rod withdrawal and negative sign for rod insertion).

6.4.3 Design Constraints

Although the reactor power control system does not perform the safety functions, it still plays an important role in the operation of nuclear power plant. It shall, under the premise of adapting to the object characteristics and ambient conditions, meet various operation control requirements and match with the interfaces of the related systems. During the design, the above constraints must be comprehensively considered. The main design constraints of reactor power regulation system are as follows.

① Basic functional requirements: under normal conditions, realize the control functions required by nuclear power plant;

② Performance requirements: static specification, deviation requirements and dynamic requirements of the controlled variables;

③ Signal range: the amplitude range and change rate range of the measured variables in the input under various conditions;

④ Operation requirement: the operation limits and control action setpoint of the controlled variables;

⑤ Object characteristics: the physical, thermal and hydraulic characteristics of reactor and relevant parameters required for the design;

⑥ Environmental conditions: the design of reactor power control system must meet the environmental conditions, mainly including impact, vibration, tilt, swing, mildew, salt mist, high temperature and low temperature;

⑦ Overall structure of I&C system: during the design of reactor power control system, the signal standard and number shall match the overall structure of I&C system of nuclear power plant, and the adopted communication protocols shall be compatible;

⑧ Electromagnetic compatibility: the controller in reactor power control system are electronic devices and placed in the same space with many kinds of other electronic devices, so their electromagnetic emission and sensitivity must be within the range required by electromagnetic compatibility.

The reactor power control system shall keep the reactor operating within the allowable range and meet the load demand, reserve a certain margin between the static and dynamic performance specifications (such as steady-state deviation and overshoot) and the safety limits, ensure that the input and output signals are within the specified normal range and maximum range, limit the rod speed range and prevent the malfunction.

6.4.4 Description of the System and Equipment

(1) System overview

A complete set of reactor power control system is a closed-loop control system that includes setpoint module, parameter measuring devices, controllers and actuators and controlled process, as illustrated in Fig. 6.12.

The reactor power measured by nuclear measuring devices and the temperature of reactor hot leg and cold leg, the secondary steam flow or the turbine impulse stage pressure measured by process measurement devices are sent to the reactor power controller, and processed by the internal setting device to generate the setpoints. Comparisons are then made between the setpoints and the measured values, in which, the related logic processing and calculation are completed according to the design of control scheme before generating the automatic rod speed signal that is sent to the manual/automatic selectors. The manual/automatic selector and the manual rod speed switch are controlled by the operators. The signals for automatic output rod speed or manual rod speed setting are selected and sent to the control rod drive mechanism controller for processing and generating the controlled pulse current. This current drives the magnetic coils in the control rod drive mechanism to generate the lifting force.

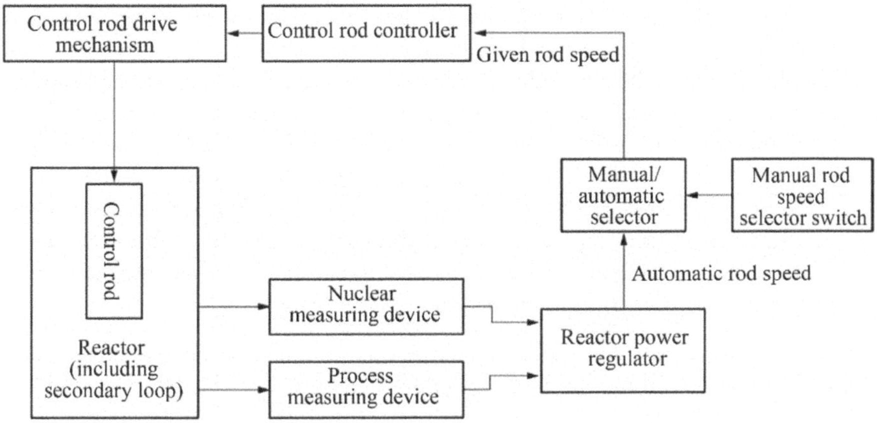

Fig. 6.12 Basic composition of reactor power control system

This lifting force drives control rods upwards or downwards and finally changes the core reactivity.

(2) Equipment description

In the composition of reactor power control system, the control rod drive mechanism, the control rod and the reactor are independent objects; the manual/automatic selector and the manual rod speed selector switch are constituent parts of operation and display interface of main control room; the rod position measuring device, the nuclear measuring device, the process measurement device and the control rod controller are the independent subsystems in the I&C system of nuclear power plant, and have, besides realizing the specific functions of reactor power control system, their own system functions. Relevant description has been given in specific chapters of this book. Therefore, the equipment description for reactor power control system in this section only covers the reactor power controller.

① Reactor power controller.

Analog and digital input/output modules, processor modules, communication modules and data buses are used to realize its functions and send relevant control rod movement signals and rod speed signals to the control rod controller.

② Control rod assemblies.

The reactor has many groups of control rod assemblies. Relevant description has been given in "Fuel assemblies and associated assemblies".

(3) System operation

During the power operation of reactor, when the load of the secondary loop decreases, the steam flow will decrease at the initial stage of the transient process. Since the reactor still maintains original power, the thermal balance in primary and secondary loops will be broken, the average coolant temperature will increase, and the reactor power will be higher than the secondary loop load. In this case, on the one hand, the reactor power starts to decrease due to the load reactivity feedback effect of coolant temperature. On the other hand, the deviation in average coolant temperature are corrected by temperature deviation channel and power deviation channel to generate the corresponding rod speed signals and rod insertion signals, after which the regulating rod banks are inserted downwards and the reactor power decreases. When the reactor power and the load are the same and the average coolant temperature returns to the set-point range, the control rod will be fixed at the proper position, and the nuclear power plant will operate stably at a new power level.

When the secondary loop load increases, the action of reactor power control system is opposite to above controlling action.

6.5 Reactor Protection System

6.5.1 System Functions

The reactor protection system refers to the system that generates the necessary output signals for triggering the actions of the safety actuators and safety system's auxiliary facilities, prevents the reactor state from exceeding the specified safety limits or mitigates the consequences of exceeding safety limits. It includes all the equipment and channels from the sensitive elements to the input terminals of safety actuators (or to the input terminals of safety system's auxiliary facilities) [6]. The reactor protection system of marine nuclear power plant has functions such as scram, actuation of engineered safety features, safety interlock and safety alarm, which are detailed as follows:

① Automatically trigger the protection actions of relevant systems (including the scram system) to ensure that the operation states will not exceed the specified operational limits and conditions in the event of anticipated operational occurrence to prevent from exceeding the safety limits.
② Monitor abnormal occurrences or accident conditions and trigger the relevant system actions to prevent the abnormal events from developing into accidents or mitigate the consequences of accident conditions, including prevention of core meltdown, removal of residual heat during the anticipated operational occurrence or accident conditions, etc., wherein the most important function is to maintain the integrity of primary loop pressure boundary under all the operation.
③ Trigger any required safety actions via the auxiliary facilities of safety system.
④ Implement the safety interlocking and restrain the unsafe actions of the control system.

⑤ When the operating parameters approach the setpoints of warning or the setpoints of protection actions, the reactor protection system will send out alarm signals to remind the operators to perform appropriate operations.

6.5.2 System Design Principles

The general design principle for the reactor protection system is to consider various kinds of factors during the system design, ensure that the system is safe and reliable, and meet the requirements of the lower rejection rate and reduce the malfunction rate as much as possible. The main design principles and design requirements to be satisfied are as follows:

(1) Single failure criterion

In the protection system, the single failure and the consequential failure caused by the single failure or the multiple failures caused by single event shall not obstruct system from completing protection actions [6].

(2) Redundancy

In order to meet the single failure criterion, the redundant design must be considered for protection system. It includes the redundancy of monitoring parameters and monitoring equipment and the redundancy of safety logic processing units, etc.

(3) Coincidence

The required protection actions shall be triggered after adopting certain logical coincidence for the redundant channels. This can ensure the reliability requirements of the triggered protection actions and properly reduce mal-operation rate to ensure the safe operation of the nuclear power plant. The compliance is further classified into the overall logical coincidence, the local logical coincidence or the combination of the two.

(4) Principle of independence for protection channels

Physical separation and functional isolation shall be ensured during the design and installation of the systems to avoid common cause failures.

(5) Diversity principle

The diversity design (including functional diversity and equipment diversity) shall be adopted to effectively reduce the common cause failures. The functional diversity includes the monitoring of a certain accident condition by using multiple protection parameters according to the results of safety analysis (multiple settings of protection parameters), as well as the acquisition of the roughly similar results through the application of different types of protection actions. The equipment diversity includes the application of different types of equipment for the measurement of the same protection parameter, the conduction of the same logical judgment, or the implementation of the same protection action.

(6) Fail safe principle

The possible failures of equipment shall be taken into full consideration during the design. In the event of failure, the probability of unsafe actions shall not be increased and the safety actions shall not be rejected. For example, a selection between the power-on triggering action and the power-off triggering action shall be made during the system design.

(7) Non-interference between protection system and control system

The interconnection between the protection system and the control system shall be avoided. The signals delivered from the protection system to the control system shall pass the isolation equipment. The isolation equipment shall belong to a part of the protection system.

(8) Measurement of variables

The measurement of variables of the nuclear power plant must be precise and must satisfy the performance requirements specified in the design basis. The variables used for the protection actions shall be continuously measured, obtained from the direct measurement and used separately.

(9) Bypass

The bypass of the protection system includes operational bypass and maintenance bypass. For the operational bypass, the bypass shall be started when the allowable conditions are met and shall be automatically cut off when the allowable conditions are not met. For the maintenance bypass, the reactor shall still be protected after the introduction of bypass. Both kinds of bypass shall be provided with striking instructions. Any requirements (such as redundancy and independence) of the protection system shall not be compromised due to the bypass method.

(10) Test and calibration

The protection system shall be able to implement periodic test. The test shall cover all the parts from sensitive elements to input terminal (may be extended to output terminal) of safety actuator. The protection system can be used for calibrating its equipment. During the online test and calibration, the system shall not be obstructed from implementing the required protection functions and the malfunction shall not be caused.

(11) Completion of protection actions

The protection system shall be designed as follows: once triggered, the system-level protection action shall be completed. The resumption to the state before action shall be realized by the operators [6].

(12) Manual triggering

Each system level protection action can be initiated manually. The number of equipment used in the manual initiation must be reduced as much as possible and the single failure criterion shall be met.

6.5.3 System and Equipment Description

(1) Design basis for protection system

The design basis for the reactor protection system includes the function requirements for the safety analysis of the nuclear power plant, the selection of design basis accidents, the determination of protection actions, the selection of protection variables and sensitive elements, the protection system response time, the environmental conditions, etc.

① Safety analysis

The safety analysis is used for determining the function requirements of protection system. The main steps are as follows [7]:

① Determine the possible (or the design basis) accidents; ② check the consequences; ③ specify the safety margin of reactor characteristics in the event of accidents; ④ select the variables and sensitive elements used for triggering protection actions; ⑤ determine the type of protection actions as well as the necessary performance characteristics of the protection system.

② Selection of design basis accidents.

According to the operation conditions of nuclear power plant, the typical accidents are selected as the design basis accidents of reactor protection system. During the determination of design basis accidents, usually the typical malfunction or failure shall be analyzed.

③ Determination of protection actions.

The determination of the protection actions shall be based on the selected typical accidents. The influence of the hypothetical accidents on marine nuclear power plant and marine personnel without the action of reactor protection system shall be analyzed. The determined protection actions shall be able to improve the safety features of marine nuclear power plant, reduce the accident damages, restrict the expansion of accidents and interlock the actions of other systems.

④ Selection of protection variables and sensitive elements.

The protective variables used by the protection system shall be selected and the setting values thereof shall be determined based on the safety analysis. During the determination of these variables, sufficient protection of the initial events shall be considered, mainly including core protection, protection for departure from nucleate boiling in PWR core, protection for the pressure boundary of primary coolant, restriction for dangerous extension, reactivity protection, etc.

The number of the sensitive elements used for protecting parameter measurement shall be determined according to the requirements for redundancy and the spatial effect related to measurement. Meanwhile, the sensitive elements shall be placed at the position that minimizes the measuring errors caused by time lag, flow transient, pressure transient effect, backflow effect, etc.

⑤ Protection system response time.

The actions of reactor protection system must be implemented within a certain time interval after the variables exceed the setting values of protection actions to prevent the nuclear power plant from reaching the safety limit. The accident progress speed is closely related to the response speed required by the reactor protection system. The response time of the protection system depends on the maximum progressing speed of the design basis accidents.

⑥ Environmental conditions.

The protection system can be used for implementing its protection functions under the normal environmental conditions or even the extreme environmental conditions. For example, the maximum tilting and swing conditions of ship shall be considered during the design of marine reactor protection devices.

(2) System composition

According to the implemented protection functions, the reactor protection system can be classified into 3 subsystems (including the reactor scram system, the actuating system of engineered safety features and the safety interlocking control system). The reactor scram system is used to realize the scram. When the shutdown signals are generated, the power supplies of the control rod drive mechanism will be cut off to make the control rods fall into the core by the action of their gravity and acceleration springs. The actuating system of engineered safety features is used to generate the actuating signals of engineered safety features for triggering the engineered safety features (such as safety injection system) during the shutdown triggered by reactor LOCAs or intermediate steam pipe breaks, such as safety injection system. The safety interlocking control system mainly has two functions: one is to generate the safety interlocking control signals and allow some equipment or safety system states to ensure the operation safety during the normal shutdown, closing or raising power of reactor; the other is to generate the safety interlocking actions to limit reactor power and avoid triggering shutdown protection when some abnormal circumstances occur and the reactor trip shall be avoided.

The reactor protection system can also be classified into three parts, from the sensors of obtaining protection parameters to the actuators of implementing protection actions, including the instrument part of protection system, the logical processing part of protection system and the actuator part of protection system. The composition of the reactor protection system by these three parts is shown in Fig. 6.13.

(3) Main equipment

The equipment of protection system includes a complete set of equipment from the protection parameters measurement sensitive elements to the implement of the protection action.

① Measuring channel equipment.

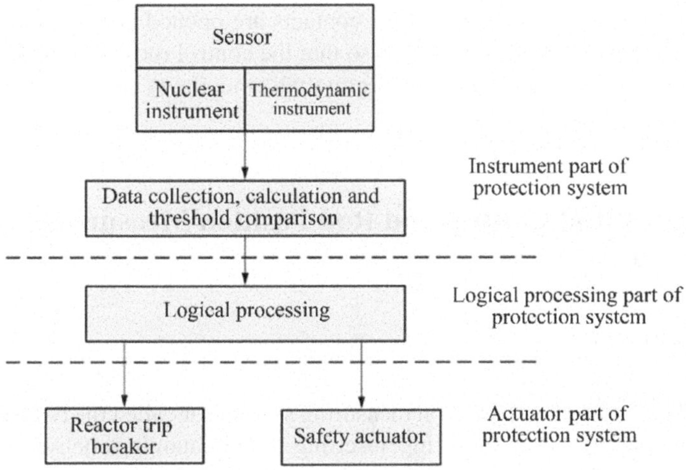

Fig. 6.13 Composition of reactor protection system

The equipment of measuring channel mainly includes detectors, measuring instruments, etc. They are used for monitoring the variables of nuclear power plant and conveying measurement signals to reactor protection devices. According to the redundant design requirements for the protection system, each type of variables shall be provided with two or more channels. Their coverage, measurement accuracy, response time and other technical indexes shall be in accordance with the requirements for protection system.

② Reactor protection devices.

The reactor protection devices belong to electrical equipment of nuclear safety class, including power modules, main processor modules, communication modules, isolation modules and input/output modules. According to the redundant design requirements for the protection system, the reactor protection devices are usually configured with redundant settings. From the perspective of protection functions, the contained components of reactor protection devices can be classified into the components with out-of-limit judgment function, the components with logical judgment function, etc. The components with out-of-limit judgment function are used for determining whether the monitored signals exceed the allowable ranges and reach the setpoints and sending out the safety alarm signals. The logical judgment components (i.e. the safety logic devices) shall be used to realize the logic algorithm functions of protection parameters and protection channels, which usually include coincidence logic, "or" logic, "and" logic, etc.

③ Reactor trip breaker.

The reactor trip breaker receives the shutdown actuating signals sent out by the safety logic devices in reactor protection devices to realize the shutdown function. When the system sends out the shutdown actuating signals, the reactor trip breaker contacts are opened and the power supply of rod control system is cut off so that the control rods are inserted into the reactor core under the effect of gravity to shut down the reactor.

6.6 Control Rod Control and Rod Position Measuring System

6.6.1 Functions

The control rod control and position measuring system (or called the rod control and rod position system) is used for lifting, inserting and maintaining control rod clusters as well as monitoring the position of each control rod cluster. It is the automatic executive control system and rod position monitoring system for the normal start-up, operation, manual shutdown, automatic power operation or reactor scram. Although the system belongs to a non-safety system of the reactor control, its functions are very important.

(1) Control rod control

Under the normal operation, it first keeps the shutdown rod clusters in the fully withdrawn position, moves the relevant control rod banks according to the predetermined rod withdrawal sequence and then adjusts the automatic rod speed instructions sent out by the reactor power control system which simultaneously controls the power compensation rod banks and the regulating rod banks.

When the reactor scram is required, all the rod banks (no matter settled or moving) must rapidly fall into the core. When a shutdown is triggered by the reactor protection system, the system shall be used for cutting off the power supply of control rod drive mechanism to make all the shutdown rods and control rods fall into the core under the effect of gravity so that the reactor turns into subcritical and is shut down.

(2) Control rod position monitoring

This function is used for displaying the measured position of each rod bank in the main control room to make the operators notice the real position of control rod in core so that the operators can validate the position of a certain control rod bank according to the measured position and given position of rod banks. This function also includes the provision of various indications for identifying rod stuck and rod drop of control rod in the main control room.

6.6.2 System and Equipment Description

The control rod control and position measuring system can be classified into the control rod control system and the rod position monitoring system according to its functions.

(1) Control rod control system
The control rod control system mainly includes control rod drive mechanism and control equipment, wherein the control equipment includes logical units and power supply units. The control rod drive mechanism has been described in the dedicated sections, therefore, only the control equipment is described here.

① Logical unit.
The logical unit is used for completing the overall management and coordination, selecting the control rod clusters and providing the relationship with the rod clusters drive equipment. The logical unit is composed of controllers, printed circuit board plug-in boxes with electronic components. The main modules and functions are as follows:

①The digital input module, used for collecting the logical data from the external part of control logic equipment; ② the digital output module, used for transferring the logical data to other units; ③ the analog input module, used for collecting the changing analog signals and converting these signals into digital format; ④ the analog output module, used for converting the digital quantity signals into the analog voltage or current and outputting them to other equipment; ⑤ the interface circuits, used for isolating the relevant signals and then inputting or outputting them to other relevant equipment.

The logical unit receives the operation mode, operation direction (manual selection or automatic command) and interlocking signals selected by the operator. After processing, the moving rod clusters and power supply units shall be selected to make the corresponding control rod groups move.

② Power supply unit.
The power supply unit is used for completing the lifting and inserting loop program according to the operation instructions of logical unit. According to the rod cluster drive function s, each power supply unit shall be used for controlling a control rod sub-bank. Each power supply unit is further classified into two parts: one part is the loop monitoring logic unit and the other part is the power supply drive unit.

The loop monitoring logic unit receives the lifting or inserting instructions from logical unit and generating the excitation sequence for each kind of coil. The power drive part of each rod cluster is composed of three parts, which are used for providing regulating current to three coils of drive mechanism. Each of these three parts is composed of the electronic plug-ins with electronic components and the current regulating circuit.

Electronic plug-ins: the electronic plug-ins is the interfaces between logical part and power supply drive part. They convert the logic signals into the analog setpoints so that the current regulating circuit can be used for providing the required current to the coil. The failures of the power supply part are also transferred to the loop monitoring unit.

The current regulating circuit delivers the required current to the relevant coils of drive mechanism, and the circuit is the three-phase half-wave rectification (three silicone controlled rectifiers) circuit based on the large power silicon controlled rectifier circuit. The conduction angle of silicon controlled rectifier is controlled through the current regulating circuit. The corresponding reference current is generated by the current reference signal generator under the control of command signals output by electronic plug-ins.

(2) Control rod position monitoring system

The control rod position monitoring system is composed of rod position detectors and rod position measuring devices.

① Rod position detectors.

The rod position detectors are installed outside of the drive mechanism stroke casing and are easy to install, remove and maintain. The induction coils are adopted and are classified into upper and lower sections. The first coil at the upper end of upper section and the first coil at the lower end of lower section are all used as the signal generators of control rod stroke limit protection switch. Other coils located between these two coils are used to indicate the position of control rods in reactor core.

A rod position detector is mainly composed of outer protective shell, indicator coil, screw, support ring, sealing washer, etc. When the drive mechanism is in operation, the primary coil of rod position detector is provided with AC excitation power supply by the rod position measuring device. When the control rod rises and falls, the induced voltage variation at secondary coil and the induced current variation at primary coil can be caused by using induced voltage or induced current. The electric signals for these changes can be processed and then used for detecting the actual position of control rod in core, which is the working mechanism for the induced voltage type rod position detector or the induced current type rod position detector.

The impedance of coil winding of the rod position detectors will change with temperature. When the excitation voltage is constant, the winding resistance increases with the rising of temperature, and then the coil current will decrease. Therefore, the relevant excitation parameters shall be adjusted when the reactor is at cold state and hot state.

② Rod position measuring device.

The rod position measuring device is usually composed of multiple rod position measuring chassis and a signal output chassis, of which the rod position measuring chassis correspond to the control rod banks and are used for measuring, calculating and processing the measured rod positions and the limiting positions, while the signal output processing chassis is used for dealing with the centralized processing, status display and signal outputs of the information of the whole device. The digital processing of the measured rod position signals has been realized. The rod position signals, temperature signals and induced voltage or current fluctuation signals of control rod can be set and isolated, and these signals can be sent to the signal processing chips for integrated computation and processing after being collected by A/D. After processing, the measured rod position signals indicated as standard analog signals and the limiting position signals indicated as Boolean values may directly pass through bus network and be delivered to the relevant equipment.

(3) System operation

① Control rod control system.
Under the condition of the reactor normal start-up, shutdown and various operating conditions, the action of rod bank can be operated manually according to the specified operation operating procedures. During the normal power operation, an automatic operation shall be implemented by using the regulating rod banks to follow the load change and maintain the reactor power at the required level.

In order to check whether the drive mechanism is in normal operation, or when a single rod or multiple rods step-out occurs at normal operation which shall be rectified, the manual operation of a single rod can be adopted and the rising and falling of the rod shall be operated in the main control room after selecting a single control rod.

When the reactor is in operation, the control power supply of all the drive mechanisms in static or in operation can be cut off immediately through the manual operation of shutdown button or the acceptance of the scram signals sent out by the reactor protection system. Thus all the control rods fall into the bottom of core rapidly under the effect of gravity and acceleration spring to realize the reactor scram.

② Control rod position monitoring system.
In order to check whether the position indication of each control rod is normal before starting up the reactor, the single-rod test run can be conducted through manipulating the lifting and inserting of control rod drive mechanism at a relatively low rod speed, and the upper and lower limiting positions of each control rod as well as the measured rod position can be adjusted.

During the start-up, operation and shutdown of reactor, this system can be used to automatically collect and process the signals, such as measured rod position, upper and lower limiting positions and temperature, to provide the control rod position information for the operators. In addition, this system can also be used for calibrating position indications in the event of single rod step-out.

6.7 Electrical Control System for Pumps and Valves

6.7.1 System Functions

The electrical control system for pumps and valves of the nuclear power plant (the electrical control system for short) is one of the important systems for the safe and reliable operation of the nuclear power plant. The system is mainly used for the power distribution, control, protection, information of the electrical equipment such as motor driven pumps, electrical valves at primary and secondary process systems of the nuclear power plant.

① Power distribution: performing the power supply distribution of the corresponding electrical equipment according to the requirements of power supply for electrical equipment (such as pumps and valves) of various process systems.
② Control: performing the operation of electrical equipment according to the control requirements for various electrical equipment (such as pumps and valves) of various process systems, including the realization of field control corresponding to the upper-layer operation instructions and the corresponding interlocking control.
③ Protection: providing the abnormal state alarm and fault protection for electrical equipment (such as pumps and valves) of various process systems, including the electrical protection such as overload, open-phase, phase unbalance, locked rotor, and overtime.
④ Information: measuring and collecting the system equipment and electrical equipment state, including the electrical parameters, such as current, voltage and operation state, displaying the necessary electrical information for operator and providing the relevant electrical information for the data collecting and storing equipment.

This system involves the primary loop process systems (including reactor coolant system, pressure safety system, safety injection system, coolant purification system, water charging system, component cooling water system, etc.) and the secondary loop process systems (including main feed feed-water system, main steam system, steam dump system, condensate system, etc.). It has many control objects and complex interlocking relationship and assumes the responsibility of actuation and control of main engineered safety related systems.

The electrical control system can be divided into the safety class part and the non-safety class part according to the requirements for safety classification. The safety part mainly includes pumps, valves, etc. of the relevant engineered safety systems as well as the corresponding power distribution and control channels of secondary side main feed-water isolating valves, main feed feed-water regulating valves, steam dump valves, etc. in secondary side. The safety part is an important constituent for ensuring the participation of the relevant engineered safety systems of the nuclear power plant in the disposal of relevant accidents. Except for the safety part above, the rest parts of the electrical control system belong to the non-safety, mainly including the corresponding power distribution and control channels of the electrical equipment such as pumps and valves of the auxiliary process systems except for the engineered safety systems.

6.7.2 System and Equipment Description

(1) System composition
The electrical control system is mainly composed of safety interlocking control devices, engineered safety electrical drive devices, electrical drive control devices of nuclear auxiliary process system, secondary side pump and valve electrical control devices, etc.

① Safety-related electrical drive part.
This part is mainly composed of safety interlocking and field drive devices.

The safety interlocking part is mainly used to process the interlocking logic of the process systems related to the engineered safety. Under the accident conditions, based on the corresponding interlocking signals (including the relevant interlocking signals of reactor protection system, main coolant pump control system, etc.) that have been processed by the safety interlocking logic devices, the corresponding actuation signals are sent to the relevant engineered safety electrical drive parts to realize the drive functions of the corresponding engineered safety systems.

The field drive part is mainly used to realize the control and drive of the electrical equipment of engineered safety related process systems. Under the accident conditions, the drive of the corresponding electrical equipment of engineered safety systems shall be realized according to the actuation signals sent out by the safety interlocking control part and the reactor protection system to realize the drive functions of the corresponding engineered safety systems. Under normal conditions, the control and drive of the corresponding pumps and valves (including spray valves, release valves, water charging devices, water discharge devices, etc.) shall be automatically realized according to the pressurizer pressure and water level control signals of pressurizer in coolant system etc. to realize the pressure and water level control of pressurizer. At the secondary side, the control and drive such as adjustment of the feed-water regulation valve opening, the opening, closing and regulating of steam dump valves shall be realized according to the control signals of the steam generator level control system and the steam dump control system. Meanwhile, the manual control of the corresponding pumps, valves, etc. can be realized according to the upper-layer operating instructions.

② Non-safety related electrical drive part.
This part is mainly composed of the electrical drive devices of the nuclear auxiliary process systems, and is mainly used to realize the electrical control of the related electrical equipment of non-safety nuclear auxiliary process systems, including waste water pumps, waste treatment system related valves, etc. The field control of the corresponding electrical equipment shall be realized through receiving the upper-layer operation signals.

(2) System operation

① Manual operation.
Accept the operation instructions for operating equipment of process system from the upper-layer, initiate the manual operation of the relevant electrical equipment such as pumps and valves, accept the backup, emergency and local operation instructions and initiate the corresponding manual operation.

② Automatic operation.
Accept the interlocking control signals from the relevant equipment, complete the required functions such as the required engineered safety drive and the non-safety related operation control and realize the automatic interlocking control of the system.

③ Fault protection.
Diagnose the operation state according to the collected information about the operating state of the relevant electrical equipment and the established fault diagnosis requirements as well as implement the protection actions in time and send out the corresponding protection action alarms when the corresponding fault protection conditions are met.

④ Abnormality warning.
Diagnose the operation state according to the collected information about the operating state of the relevant electrical equipment and the established abnormality diagnosis requirements as well as timely send out the corresponding abnormality warning information to remind the operators to pay attention to the operating state when the corresponding abnormality warning conditions are met.

6.8 Man-Machine Information Display and Operation System

6.8.1 Functions

The information display and operation system of nuclear power plant is used for providing the necessary man-machine interface equipment and the relevant resources to realize the operation targets of nuclear power plant for the operators.

The information display and operation system is a non-safety class system, wherein the main functions are as follows:

(1) Display functions

The information display and operation system shall be used for providing the required, real-time, effective display information and the corresponding data management necessary for the whole-process monitoring of the variables and system equipment state of nuclear power plant under normal operation, anticipated operation events and design basis events. They mainly include the equipment state and process parameter display under various operation conditions, the image display of the flowchart of each system, the safety state diagram display, and the radiation dose information display of nuclear power plant.

(2) Control functions

The operation of nuclear power plant is mainly subjected to the automatic control which is supplemented by the manual control. Under the normal conditions, the operation of system is automatically maintained. When the control parameters exceed the limit or the local control equipment are becomes abnormal, the operators can maintain the nuclear power plant within the allowable operating state through the adjustment of operation parameters or through the manual operation. The control room is also provided with two kinds of necessary operating buttons: one kind is for a few operations relating related to protection system, and the other kind is for the auxiliary operation for keeping the reactor safe, implementing safe shutdown, etc. when the computerized system is failed.

(3) Alarm functions

These functions are used for providing all the information required for monitoring the deviation of nuclear power plant from the normal operation conditions. The alarm information can make the operators with acquaint the accident state in time and will not bring a heavy load to the operators. Multiple kinds of dynamic methods such as those of sound, light, color and symbol are adopted to dynamically display the alarm information.

(4) Management and storage functions

The main functions of the data collection and processing system are to: receive and process the information from the control system and the protection system; collect, convert, calculate and process other measured information about the operation of nuclear power plant; and meet the requirements for system information analysis, display, alarm, print, storage, etc.

6.8.2 System Design Principles

The following safety principles, availability principles and human factor engineering principles are mainly taken into consideration during the design of the information display and operation system of marine nuclear power plant:

① The design of control room must ensure that the nuclear power plant is in safe operation under all the operation conditions. In the event of accidents and design basis events, measures can be taken to keep the nuclear power plant in safe state or make it return to safe state. The emergency control points that are independent of the control room shall be set outside of the control room to realize the reactor safe shutdown when the control room cannot be used for implementing the safety functions.

② The control room shall be provided with enough radiation protection measures as well as necessary fire detection, alarm and fire-fighting equipment.

③ When the computerized control system is adopted in the control room, the possibility of losing all the computer information and control systems shall be considered. The bank-up operating panel shall be set in the main control room to maintain the stable operation of nuclear power plant for a period or lead it to the safe shutdown state through the safety display control units, through a few conventional indicating instrument and the hard-wiring control systems on the bank-up operating panel.

④ The influence of various adverse environmental factors on the operators shall be reduced as far as possible to relieve fatigue, stabilize emotion, facilitate operation and ensure the full play of their operation and management.

⑤ The operators shall be able to approach conveniently and rapidly the equipment to be operated and shall have the necessary space for completing operational motions.

⑥ During the man-machine functional assignment, for the operation relating related to nuclear safety, nuclear power plant operation safety and operation with the requirements for quick action, the operators can directly operate according to the corresponding operation operating specifications procedures based on the simple judgment on the observed information and no operation is allowed that requires the operators to perform after calculating on and doing the complex judgment based on the observed information.

⑦ The operation position shall be provided with the amount of visually displayed information, which gives the minimum information content that the operators have to know to complete the specified operation. The display mode shall be bold and clear. The displayed content shall be simple and reasonable to facilitate identification and judgment. The design of color and resolution ratio shall ensure that it is not easy to make the operators get tired after a long-time monitoring.

⑧ Important operating switches and buttons shall be set at the areas that are not easy to touch. The triggers related to safe shutdown and safety drive shall be provided with effective technical measures to avoid human misoperation.

⑨ The factors such as human's physiology, auditory sense, visual sense as well as operation convenience, frequency, habituation, importance system, etc. shall be taken into full consideration in the design of the switch, keyboard, indicator light, indicating instrument and screen display arrangement, luminance, color as well as audible device frequency.

⑩ The color at the top and ambience of the control room, the color of all the equipment arranged in the control room as well as the lighting collocation shall be coordinated with each other to create a quiet and comfortable environment for operators.

6.8.3 System and Equipment Description

(1) Main control room system

The main control room is used for providing a safe, inhabitable and comfortable place for the operators of the nuclear power plant to monitor and control all the process of nuclear power plant. It can be used for providing the centralized man-machine interface resources for nuclear power plant. The typical main control room system of marine nuclear power plant mainly includes the main control console.

The functions of the main control console are as follows: under the normal operation conditions, it is used for starting up of the nuclear power plant, making the nuclear power plant operate at the normal power and meeting the requirements for various control and operation under different loads of the marine nuclear power plant. Under the accident conditions, it can be used for implementing various operation and control to isolate system, slow down the progress of accidents to make the nuclear power plant return to safe state. No matter under the normal operation conditions or the accident conditions, the main control console is used to provide the main operating parameters and the equipment operational state of nuclear power plant for the operators, implement the whole-process operation monitoring and provide the information required for completing various prescribed operations for the operators.

The main control console is composed of several panels, which are respectively used for controlling primary loop, secondary loop and other auxiliary systems of the nuclear power plant. When the analog control mode is adopted, the panels shall be configured with different operators and display instruments according to different control functions. When the computerized operation mode is used as the main operation mode, each panel shall be provided with the screen displays with the identical configuration to realize the operation and control of the nuclear power plant via the graphical interface. Meanwhile, the panels act as the backup for each other.

(2) Emergency control room (point)

The emergency control room is used for implementing safe shutdown at the emergency control point when the main control room cannot fulfill the control functions such as safe shutdown, so that the reactor can enter a safe state.

The emergency control panel is the main equipment of the emergency control room is the emergency control panel. When the main control room is available, the interlocking shall be provided to make the emergency control panel inoperable. When the operators evacuate to the emergency control room due to the non-availability of the main control room, the signals that are from the main control room and conflict with the emergency control panel shall be blocked by using the selector switch on the emergency control panel. The emergency control panel is mainly provided with operating switches, buttons and display instruments that are related to reactor shutdown and residual heat removal. Under normal conditions, the emergency control panel is connected with the field equipment through hardwiring.

6.9 Digitization of I&C System

6.9.1 Technological Development Overview

(1) Technological development of conventional I&C system

For the worldwide development of I&C technology, there are mainly two stages of development (analog technology and digital technology). At present, the analog technology is basically not used anymore, while the digital technology is mainly used to realize the corresponding I&C requirements.

According to the worldwide development of the configuration schemes of I&C system, there are several main stages such as Analog Control System (ACS), Centralized Control System (CCS), Distributed Control System (DCS), and Fieldbus Control System (FCS). Wherein Analog Control System (ACS) is based on the analog technology and mainly composed of the analog combined unit I&C, and it extensively adopted with logic devices such as relays to realize the corresponding interlocking control. Centralized Control System (CCS) is the intermediate phase transiting from analog control system to digital control system and based on the computer centralized control technology. Its complex control is realized by adopting digitalized technology. When compared with Analog Control System (ACS), it has better system performance is relatively good, but it will bring new control problems due to an over over-centralized control, such as the failure of the whole system due to the failure of centralized control equipment. Distributed Control System (DCS) is based on the digitalized technology and composed of field layer, control and protection layer and operation and display layer. The control and protection layer is connected to manipulation control and display layer through the integrated monitoring network to realize a mass data transmission and further improve the system performance. This system scheme can be used for solving the relevant technical problems in Centralized Control System (CCS). Fieldbus Control System (FCS) is developed from Distributed Control System (DCS) and based on fieldbus. This system runs through production field and can be used to realize a bi-directional, serial, multi-node digital communication among measuring instrument, actuating mechanism and control equipment. Compared with DCS, FCS is more aligned with the idea of "centralized information and distributed control" and can be used for improving the system performance further. However, the extensively used fieldbus network will make the system reliability relatively low. A system scheme of combining DCS and FCS is widely used in the market at present. This system scheme combines the advantages of the two and becomes the scheme mainly considered for the configuration of I&C system scheme at present.

(2) Technological development of I&C system of nuclear power plant

I&C system is an important part of the nuclear power plant. Taking the land nuclear power plant as a typical representative, the safety, reliability and economic operation of the unit will, to a great extent, depend on the performance of I&C system. The development of I&C system of nuclear power plant has experienced three stags from the nuclear power projects which have already been already built and are under construction in China. The first stage is the control system mainly composed of the analog combined unit instrument, such as SPEC200 assembled instrument from FOXBORO used in the main control system of 300 MW Qinshan Nuclear Power Plant and Baily 9020 system adopted in the main control system of Daya Bay 2 × 980 MW Nuclear Power Station in operation in China. The analog instrument shall be controlled by using the elements based on the small scale integrated circuit operational amplifier, while the logical instrument shall be controlled by using the hard logic circuit such as relay. Therefore, many instrument controllers are required by the system. The tasks of operation management and maintenance are heavy and most of them are operated manually. The main control room appears relatively large. The second stage is the main control system of combining the analog quantity and digital quantity. By adopting this type of I&C system, the nuclear island system is still controlled by the analog elements based on the small scale integrated circuit operational amplifiers, while a part of conventional island and auxiliary systems adopt PLC automatic control system as well as the software self-diagnosis technology, redundant technology and network communication technology to reduce hardwiring and local control cabinets and improve system operation reliability. I&C system of Ling'ao Nuclear Power Plant (2 × 980 MW) in Guangdong belongs to this type. The third stage is called the fully-digitized I&C system. As the improved and adapted version of the mature Distributed Control System (DCS) of the conventional power plant, it is the new digital I&C system of nuclear power plant and applied in conventional island, BOP and nuclear island (NI). Relatively typical fully-digitized I&C system applied at the present stage include: NUCAMM-90 system developed by Hitachi, etc., N4 control system from Framatome, NUPLEX80+ system from ABB, Eagle21+WDPFn system from Westinghouse as well as TELEPERMXP+XS system from Simens adopted by Tianwan Nuclear Power Plant in China. At present, the fully-digitized I&C system have been fully adopted by the already-built Ling'ao Nuclear Power Plant Phase II and will be adopted by the newly built nuclear power plants, such as Hongyanhe NPP Phase I, Fuqing NPP Phase I, Ningde NPP Phase I, and Yangjiang NPP Phase I.

The technology of I&C system of nuclear power plant plays a crucial role for ensuring safe and reliable operation of the nuclear power plant as well as improving automation, availability and economy of operation. With the development of digital technology, especially the development of microprocessor and network communication technology, the reliability and anti-interference capability of system are greatly improved to realize the modularization, universalization and standardization of equipment and increasingly and greatly reduce the equipment cost. Meanwhile, an open structure is adopted to improve the compatibility of equipment, and the control system also possesses fault-tolerant design, self-check and self-diagnosis functions. Therefore the designers can adopt new digital technology, design and manufacture the technologically advanced digital I&C system by using advanced control, protection, measurement and information processing technology. In the aspect of design means and comprehensive design capability, there is still a certain gap between the technology of I&C system of nuclear power plant in China and the worldwide advanced technology level. As the independent research and development capacity and localization are still insufficient and the digital I&C system of NPP still mainly depend on import, the research and development of the digital I&C technology need to be accelerated.

6.9.2 Technical Schemes of Digital I&C System

(1) Basic composition and structure of digital I&C system

Many digital electronic equipment, digital controllers based on microprocessor, computer systems and network communication technologies have been adopted in the digital I&C system, which can collect and process plenty of field signals and provide digital man-machine interactive interface. The schemes for the typical digital I&C system in nuclear power plant are shown in Fig. 6.14. Compared with Fig. 6.1, the digital I&C system follow the functional classification of the system structure and still include field layer, control and protection layer and operation management layer. As the digital I&C system provide plenty of data and information, an integrated management layer is provided to analyze the data and information comprehensively and assist the system management.

In the field layer, except for the traditional hardwired transmission of electrical signal, intelligent instruments and field bus technologies will be extensively adopted with the advances in equipment technology and the improvement of reliability. The measured and collected parameters or state shall be converted into digital signals and transferred to the control and protection layer through bus network. In the control and protection layer, the digital controllers that are

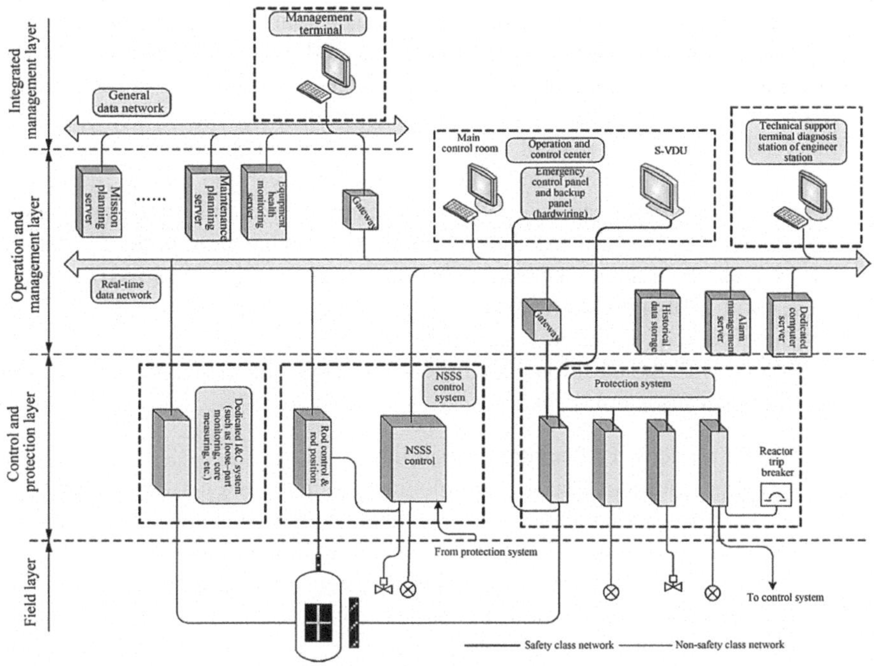

Fig. 6.14 Digital I&C system for nuclear power plants

composed of the standardized modules such as reception, preprocessing, isolation, distribution, calculation, logical processing, and communication are used for processing various types of signals and also returning the control instructions to field layer through hardwiring or bus network. The digital controllers of different subsystems form a distributed digital control node network. Each single control node is used for completing specific control objectives. The information among control nodes is transferred through network. The safety class network and non-safety class network corresponding to the safety classification of system are respectively used to realize the signal interaction of different safety levels, and are isolated by gateway. In the aspect of data processing, the digital technology and network technology shall be fully utilized and the intelligent processing technology, such as self-adaption and self-compensation, shall be used. In the operation and management layer, computers, display equipment and servers with various functions shall be adopted to provide the man-machine interfaces for the operators and the system administrators. The important safety safety-related signals are transferred by using the safety visual display unit (S-VDU) or the emergency (backup) control panel and hardwiring, while other data and information are transferred by using network communications and can be used for providing guidance on procedures and auxiliary decision-making support for the operators through the intelligent processing of software. In the integrated management layer, an integrated data analysis shall be implemented by the computer system according to the data and information provided by the operation and management layer to provide the health management through the whole life cycle, provide the direct instruction for preparing the operation plan and maintenance plan of nuclear power plant and ensure that the nuclear power plant is in a good condition. Within each layer and among the layers, the network communication is a very important information interaction channel and carrier. Generally, the dedicated reliable communication protocol shall be adopted and the redundant communication network shall be deployed to ensure the reliability of signal transmission.

(2) Characteristics of digital I&C system

Compared with the traditional analog technologies, the main characteristics of the digital I&C system include:

① Strong numerical calculation capability.

For the traditional analog technologies, an individual hardware module is required for implementing even an addition operation and the operational capability is extremely limited, therefore, the complex calculation is difficult to realize. However, for the digital I&C system, there is almost no restriction for parameter type, parameter amount and calculation rules. In addition, there are obvious nonlinearities in most of systems and equipment, the traditional analog technologies are almost impossible to be used to deal with these nonlinear calculations, but the digital technologies can be easily used to realize the nonlinear processing.

② Strong judgment capability.

The digital I&C system can be used to realize a series of judgment and processing such as loop, cross call, table look-up, effectiveness judgment, default setting, and logic degradation, to make the system automatically adapt to different external conditions and improve the automatic processing capability of the system.

③ Parameter self-adaptive capability.

With the relatively strong judgment and calculation capability, the digital technologies can be used to realize the adaptive transformation of multiple parameters and multiple algorithms and cover more working conditions with multiple algorithms and parameters.

④ Stronger man-machine interaction capability.

With the computerized display screen, more intuitive overall information can be displayed and the information that the operators care about can be automatically switched with the predetermined mode according to the reactor state. In addition, the operators can implement only the high-level operations through procedures, while the low-level operations shall be automatically implemented by the computer system.

⑤ Distributed arrangement.

Based on the network transmission technology, the real distributed control can be realized to improve the availability and maintainability of the system and reduce the required arrangement space.

⑥ Self-inspection.
The digital I&C system can be used to realize more abundant self-inspection functions to simplify the periodic test and provide more realistic support data for the preventative maintenance.

⑦ Big data processing capability.
The digital I&C system can be used to record the operation data and transient data of important equipment and instrument and analyze the trend thereof. Meanwhile, the data of some relevant systems can also be subjected to correlation analysis to obtain the overall state data and to predict the overall state trend.

(3) Development direction of digital I&C system

① Further digitalized I&C system with more miniaturized and integrated system structure.
With the progress of digital soft and hard technology and digital communication technology, the signal transmission of each layer of I&C system has been developed toward digitalization on a large scale and has been gradually extended to the field sensors and actuators. Meanwhile, the automatic control equipment are also gradually developed and become more miniaturized and integrated. The amount of various signal cables and wiring terminals can also be effectively reduced by using the digital communication technology, which makes it possible for the miniaturization and integration of nuclear power plant I&C system.

② More abundant measured parameters and more diverse measuring methods.
With the progress of sensor technology, the sensor (temperature, water level, neutron flux, etc.) technologies that are more adaptable to the high temperature, high pressure and high irradiation conditions in reactor have been gradually put into the practical use to make the measuring of reactor key parameters possible. With the introduction of measuring methods such as magnetostrictive rod position detection and ultrasonic flow measurement, the measuring means for the parameters of nuclear power plant become more diverse.

③ High automation degree of control systems.
With the improvement of processing capacity of control equipment, it gradually becomes possible to realize a complex control mode gradually. The characteristics of different nuclear power plants shall be reasonably used to adapt to the control requirements under different control modes, to implement the self-adaptation control of various process systems, automatically implement reactor startup, shutdown, power lifting and lowering, special

conditions control, switching of redundant control, complex sequential control, interlocking control and so on, then finally to improve the automation degree of nuclear power plant.

④ Electronic procedures guided operation.
After digitization of I&C system, the signals that can be displayed and operated by the main control room are greatly increased, and the man-machine interfaces become more complex. The systems have been qualified to be used to organize, arrange, process, feedback various kinds of information according to the operating procedures. Therefore, the standardization of the operators' operation can relieve the operators' pressure under complex conditions, thus improve operating efficiency and reduce misoperation.

⑤ More intelligent operation support.
The new technologies of information collection, transmission and storage can be used for improving the requirements for transmission rate and information capacity of operation and control information to implement more precise and more comprehensive monitoring, control, diagnosis and analysis of operating state of nuclear power plant. Supplemented by the progress of operation support methods, the analysis and diagnosis of operating state of nuclear power plant are realized to provide more effective event trending predictions and operation suggestions.

⑥ Spreading of highly reliable and high speed digital communication modes.
The digital communication technology (including bus, network, etc.) is an indispensable element of the digital I&C system. More highly reliable digital communication modes are applied to some severe environment on ocean and ships and their transmission performance is improving. The digital communication modes will facilitate the transmission of plenty of information as long as they have been adequately verified and validated by practices.

⑦ Virtualization of I&C system.
The simulation technology is increasingly used in the design of I&C system. Various characteristics, performance, state, behavior, etc. of I&C system are simulated and virtually implemented by using the simulation models and the relevant computer software to greatly improve the precision and scale. At the early stage of design, the designers are using these technological means to more precisely comprehend the final technical configuration of I&C system to adjust and optimize the design schemes.

References

1. Shuming, Chen. 2010. *Nuclear Power Plant Instrument and Control Basis*. Beijing: Atomic Energy Press.
2. Weiliang, Sang. 1993. *Pressurized Water Reactor Control and Protection Monitoring*. Beijing: Atomic Energy Press.
3. Jianmin, Zhang. 2002. Nuclear Reactor Control. Xi'an: Xi'an Jiaotong University Press.

4. Minjun, Peng. 2009. *Marine Nuclear Power Unit*. Beijing: Atomic Energy Press.
5. Schultz, M.A. 1962. *Automatic Control of Nuclear Reactors and Power Plants*. Beijing: Editorial Board of Nuclear Science Committee of Chinese Academy of Science.
6. Peigao, Yu. 2006. Safety Code of Design for Submarine Nuclear Power Plant—Part 11: Design Criteria for Protection System. Beijing: Military Standard Publication Department of Commission of Science Technology and Industry for National Defense.
7. Harrer, J.M. 1985. *Nuclear Power Reactor Instrument and control system Handbook*, vol. 2. Beijing: Atomic Energy Press.

Chapter 7
Steam Power Conversion System

7.1 Overview

Based on the Rankine cycle, the steam power conversion system (also known as the secondary coolant circuit) is formed by main components such as the secondary side of steam generator, the turbine, the condenser, the condensate pump, the feedwater pump and the feedwater heater as well as steam and water pipes connecting this equipment to realize energy transfer and conversion. Figure 7.1 shows the main equipment of secondary loop forming the Rankine cycle and the ideal energy conversion process [1].

Generated by the fuel fission in the reactor, the heat is taken away by the coolant flowing through the reactor core and transferred to the secondary loop feedwater in the steam generator, which produces the steam with a certain temperature and pressure. Then, the steam is transported to the turbine for power generation or heat-consuming equipment for use through the steam system. The steam exhaust from the turbine is discharged into the condenser and condensed into water by the circulating cooling water from outside of the board, and then is returned to the steam generator for next cycle after being pressurized by the feedwater pump.

The propulsion mode of the marine nuclear power plant can be classified into the direct propulsion and the electric propulsion, and the principle is illustrated in Fig. 7.2.

(1) Direct propulsion

The direct propulsion refers to the mode in which the mechanical energy from the turbine is directly transferred to the propeller via the transmission mechanism and shaft system. It is a traditional propulsion mode and the gear reducer is widely used in the transmission mechanism. This propulsion mode is used for the marine nuclear power plant in most cases.

J. Yu, *Marine Nuclear Power Technology*,
https://doi.org/10.1007/978-981-15-2894-1_7

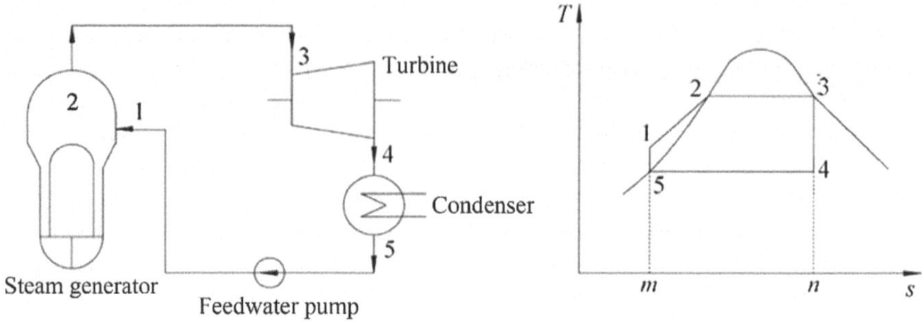

Fig. 7.1 Rankine cycle of nuclear power plant

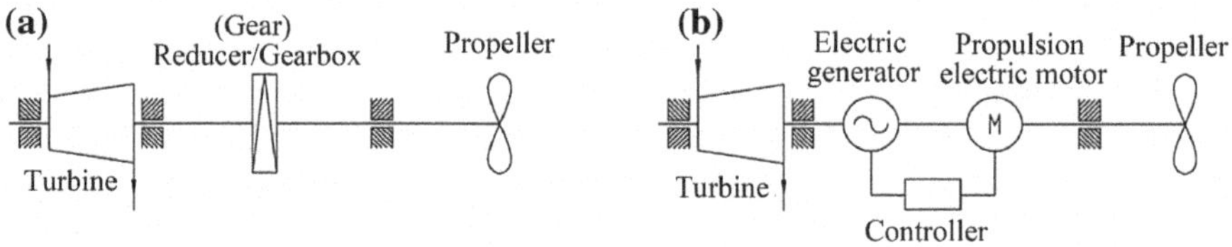

Fig. 7.2 Basic principles of two propulsion modes of ships **a** Direct propulsion; **b** Electric propulsion

(2) Electric propulsion

The electric propulsion refers to the mode that turbine drives the engine and powers the propulsion electric motor, and then the propulsion electric motor drives the propeller. The electric propulsion falls into several categories: DC electric propulsion, AC electric propulsion and AC-DC electric propulsion.

This mode comes with many advantages: it facilitates easy operation and good maneuvering performance; it eliminates the vibration and noise produced by the reducer as no gear reducer is required; it enables selection of the optimum revolution as the rotational speed of propeller is not restricted by the rotational speed of main engine. However, due to heavy weight and large dimensions, the electric propulsion unit brings about higher intermediate losses (generator-propulsion unit) compared with the direct propulsion and its efficiency is relatively low under the rated operating conditions. The efficiency of DC electric propulsion is normally 86–90%; 94–95% for AC electric propulsion and 98% for direct propulsion by gear reducer.

(3) Hybrid propulsion

Hybrid propulsion means that the ships can adopt either the direct propulsion or the electric propulsion based on the requirement, where the nuclear power is the primary power and the electric propulsion is the auxiliary power.

In the PWR nuclear power plant of the ships adopting direct propulsion, the main steam turbine-gear unit converts the thermal energy of the steam into the mechanical energy propelling the ships to sail and the turbo generator set generates the electric power required by the power plants and the whole ship, while the water-making system provides the fresh water for the power plants and the whole ship. In nuclear power plants using electric propulsion, the power generated by the turbo generator set needs to meet the electricity demands of the propulsion electric motor, power plants and the whole ship.

In order to realize the above functions, a series of relevant auxiliary systems are provided for the secondary coolant circuit in addition to such main equipment as the steam turbine-gear unit and turbo generator set, mainly including the steam system, the steam dump system, condensate system, circulating water system and water-making system.

7.2 Steam System

7.2.1 System Description

The function of the steam system is to transport steam. The system transferring steam from the steam generator to the turbine is called the main steam system and the system supplying steam for the auxiliary equipment and auxiliary heat exchanger is called the auxiliary steam system. The system that collects the exhaust steam from auxiliary equipment is called the exhaust steam system. These systems are collectively called the steam system for convenience [1].

The steam system is determined according to the thermodynamic diagram adopted by the secondary loop. The main steam system is working in a tougher condition than that of other steam systems. The design of steam system shall meet the following requirements:

① The service life of the system shall be as long as possible;
② The system shall be with high operational reliability;
③ The flow resistance of steam inside pipes shall be as small as possible;
④ The heat loss of steam in the pipe shall be minimized.

The layout of the steam system is closely related to the number of steam generators and the number and layout of the main turbines and other steam consumers. There are three common layout types of the main steam system: single-line layout, double-line layout and annular layout. The auxiliary steam system is also arranged in three ways: single-line layout, annular layout and independent layout. The exhaust steam system is usually a header pipe running through the compartment in which the exhaust steam from the back-pressure auxiliary equipment is gathered and from which it is distributed.

For the primary loop system equipped with two steam generators, the main steam system usually adopts the double-line layout, as shown in Fig. 7.3. The two steam generators are connected with the primary equipment via one main steam line pipe respectively. Such layout enhances the reliability of steam supply to the primary

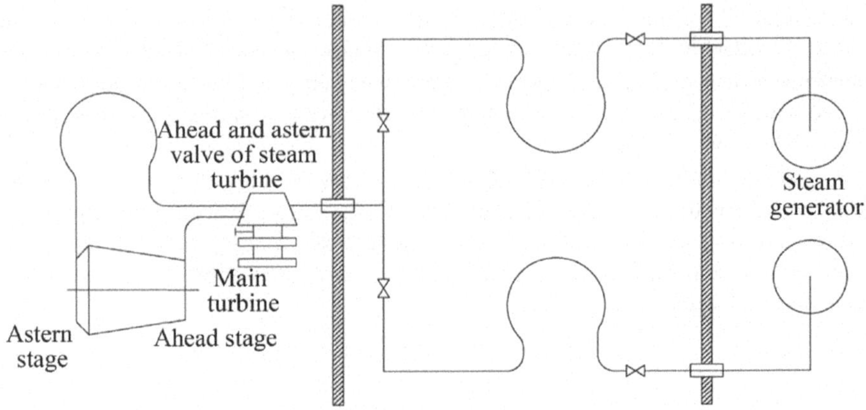

Fig. 7.3 Double-Line layout of main steam system

equipment by the steam generators, and it allows cutting off the damaged pipe. Meanwhile, the thermal stress of pipes is reduced due to the relatively small pipe diameter of main steam lines pipe and thus the reliability of thermal compensation of pipes pipe itself is improved.

For natural-circulation steam generator, the pressure of the main steam changes in a wider range with the reactor loads, resulting from the characteristics of the steam generator, so the working pressure of the main steam system also changes with the loads. Therefore, the design pressure of the main steam system shall be the potential maximum working pressure, whether the natural-circulation steam generator or the once-through steam generator is used.

7.2.2 Equipment Description

(1) Main steam pipeline

The main steam line is used to transport the steam from the secondary side of steam generator to the main turbine. Certain corrosion allowance shall be considered in selecting the wall thickness of main steam lines and the flow velocity in the main steam lines shall be controlled within a reasonable range. In order to protect the steam lines from damage caused by the excessive internal stress arising from the deformation of steam lines due to thermal expansion, appropriate compensation measures against thermal expansion shall be taken during the arrangement of steam lines to avoid the thermal expansion. Measures to compensate the deformation of steam lines are divided into two categories: ① self-compensation, i.e., taking advantage of the bending and torsional deformation of steam lines to make compensation by utilizing the arrangement of steam lines; ② special compensation device, such as the bend compensator.

(2) Main steam isolation valve

The main steam isolation valve is used to limit the discharged steam to the quantity of steam generated by one steam generator in case of the break of main steam line. It will be closed within the stipulated time upon receipt of manual or automatic isolation signal and remain at the fully closed position.

7.2.3 System Operation

During normal operation, the steam system will supply necessary steam for the main turbine, auxiliary turbine and steam consumers. In case of the design basis accident of the large-break of steam pipeline, the main steam isolation valve will close automatically to ensure only one steam generator discharges the steam to the break.

7.3 Condensate and Feedwater System

7.3.1 System Description

The functions of the condensate and feedwater system are as follows:

① Pump out the condensate in the condenser and then transfer it to the steam generator after being heated by the feedwater heater;
② Extract the air in the condenser to guarantee the required vacuum degree;
③ Adjust the water levels of the condenser and steam generator and the water volume of balanced system.

In order to realize the above functions, the condensate and feedwater system is usually composed of the condensate pump, the air extractor, the feedwater pump, the feedwater heater, the feedwater regulating valve and the connecting pipelines and instruments. The system flowchart is shown in Fig. 7.4.

The condensate and feedwater system shall ensure that the marine nuclear power plant can work reliably under all operating conditions. This system shall work more reliably in case the water volume of the steam generator is small and the steam production is large (such as the once-through steam generator). Therefore, the design of the condensate and feedwater system shall meet following principles:

① Main equipment (e.g., condensate pump and feedwater pump) have to be of dual configuration;
② The whole system shall be provided with reliable and sensitive automatic regulating devices for controlling the water levels of steam generator and condenser, etc.;

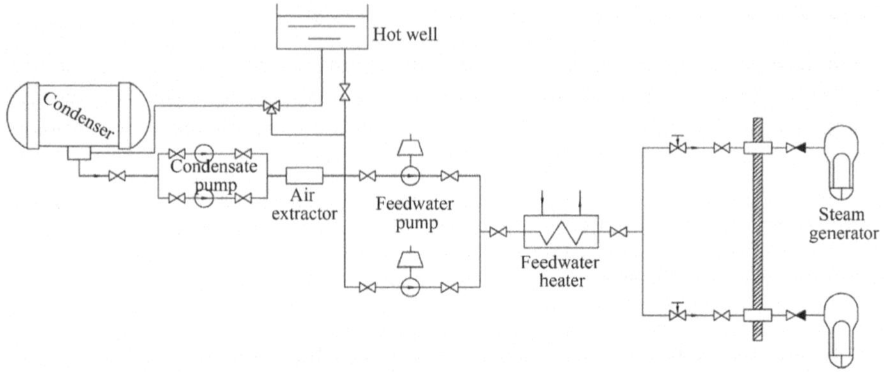

Fig. 7.4 Condensate-Feedwater system process

③ The inconsistency between the condensate quantity and the feedwater quantity shall be taken into account;
④ Technological requirements of all equipment shall be satisfied, such as characteristics of condensate pump and feedwater pump, cavitation characteristic, characteristics of the feedwater regulating valve;
⑤ One emergency suction pipe shall be installed at the inlet of feedwater pump to absorb water from the condensate tank;
⑥ The feedwater quality shall be guaranteed under normal conditions.

7.3.2 Equipment Description

(1) Condensate pump

The function of the condensate pump is to continuously pump out the condensate in the condenser and transfer it to the feedwater system after pressurization. The configuration of the condensate pump is determined by considering the configuration of the condenser and the operation reliability of the system. In general, there shall be at least two condensate pumps. The electric centrifugal pump is usually used as the condensate pump in the nuclear-powered ship, and the pump speed is kept constant. To adapt to the inconsistency between the condensate quantity and the feedwater quantity during operation and to meet the requirements on water level control of the condenser, it is necessary to adjust the condensate flowrate according to the operating status of the system. Figure 7.5 shows two common means to adjust the condensate flow [1].

Figure 7.5a illustrates the reflux regulation method. A return pipe from the outlet pipeline of the condensate pump to the condenser tank is provided. The automatic water level regulator controls the opening of the valve on the return pipe according to the water level of the condenser, and thus adjusts the condensate flowrate by changing the reflux flow.

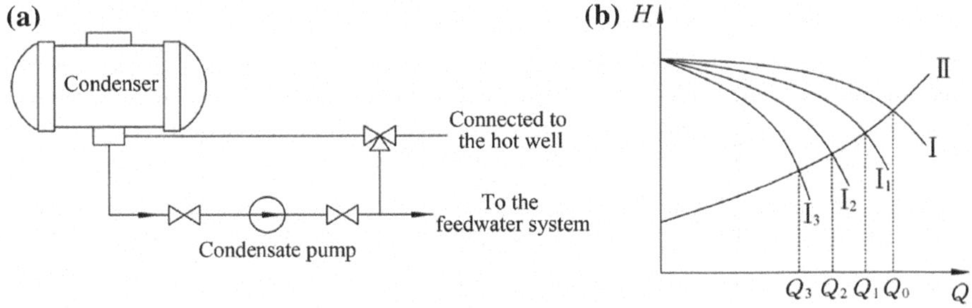

Fig. 7.5 Means to adjust condensate flow **a** Reflux regulation; **b** Cavitation control

Figure 7.5b shows the cavitation control method, i.e., changing the discharge capacity of the pump according to the cavitation performance of the pump. This method can simplify the operation of water pump, but it is required that the pump shall have relatively strong anti-cavitation capacity or pump impeller shall be replaced regularly during repair.

(2) Feedwater pump

The function of the feedwater pump is to continuously supply the steam generator with water to meeting the flow and pressure requirements. Mostly, the feedwater pumps in the marine nuclear power plants are centrifugal pumps driven by the turbines. The configuration of feedwater pump shall ensure that in case of failure of one feedwater pump, the rest pumps can provide rated feedwater flow to the steam generator. The displacement of the feedwater pump shall be consistent with the steam production of the steam generator and a certain displacement shall be reserved in consideration of the blowdown of the steam generator. The outlet pressure of the feedwater pump depends on the steam pressure in the steam generator, the geometric height of water lift, and the water resistance of the pipelines, accessories and heat exchangers. The characteristics of the feedwater pump shall be compatible with the characteristics of the steam generator.

(3) Feedwater heater

The function of the feedwater heater is to heat the feedwater to ensure the feedwater temperature of the steam generator. The requirements on the feedwater and condensate temperatures of steam generators determine whether it is necessary to configure the feedwater heater. The heat source of the feedwater heater is determined by the thermodynamic diagram of the secondary loop, and is generally the exhaust steam. The feedwater heater is provided with a bypass pipeline, and the feedwater is transferred directly to the steam generator through the bypass line without being preheated when the feedwater heater fails.

(4) Feedwater regulating valve

The function of the feedwater regulating valve is to control the feedwater flowrate of the secondary loop to ensure the safe and stable operation of the nuclear power plant.

The feedwater system of the marine nuclear power plant generally adopts the electric control valve with linear regulating characteristics. For the natural-circulation steam generator, the feedwater electric control valve controls the feedwater flow according to the water level of the steam generator, feedwater flow and steam flow so as to adjust and keep the steam generator water level within a specified range. As the water level of once-through steam generators is unstable, the feedwater regulating valve is generally controlled by the steam pressure, the feedwater flow and the steam flow.

7.3.3 System Operation

During normal operation, the condensate is transferred to the cooling air extractor after being pressurized by the condensate pump, and then sent to the feedwater pump inlet. The inlets and outlets of two condensate pumps are connected by bridge pipes and only one condensate pump is under operation while the other one is kept on standby under normal conditions. An automatic water level regulator is installed to maintain the water level required by the condenser. In order to maintain the specified condenser vacuum, the air is evacuated from the main condenser with an air extractor.

Hot wells (balancing tanks) are provided to balance the water volume of the entire system. They are connected with the atmosphere, and the required amount of water is made up by the fresh water system. In order to reduce the oxygen content of the feedwater, the water in the hot well is discharged into the condenser after being heated to a certain temperature with the help of steam exhaust.

The condensate water is pressurized in the feedwater pump, transferred to the feedwater heater, heated to the required temperature in the heater by the steam exhaust, and then supplied to the steam generator through the motor-driven feedwater regulating valve.

7.4 Steam Dump System

7.4.1 System Description

The steam pump system is used to remove the excess steam generated by the steam generator and discharge it into the condenser after temperature and pressure reduction so as to prevent the pressure in the main steam lines from exceeding the allowable value. The serious fluctuation in loads and accident conditions will make the main steam system over-pressured and a portion of the steam shall be discharged to reduce the steam pressure below the rated value in such cases. The steam discharge in the former case is called flexible discharge condition and that in the latter case is called safe discharge condition.

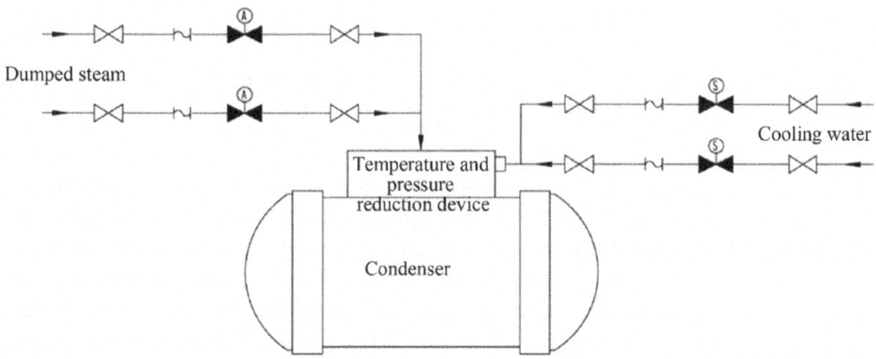

Fig. 7.6 Flow of steam pump system

In the transient process of rapid reduction in the loads of power plants, the steam produced by the steam generator cannot quickly keep up with the change due to the delay of the primary loop, so the steam pressure of the steam generator will increase sharply and even exceed the allowable value of the main steam system. Thus the excess steam shall be discharged into the condenser after temperature and pressure reduction to balance the power consumed by the second loop with the reactor power.

The steam pump system is generally composed of the steam dump valve, the temperature and pressure reduction device, and the connecting pipes and accessories, as shown in Fig. 7.6.

7.4.2 Equipment Description

As a key component of the steam dump system, the temperature and pressure reduction device usually utilizes orifice plates to reduce the pressure and spray water

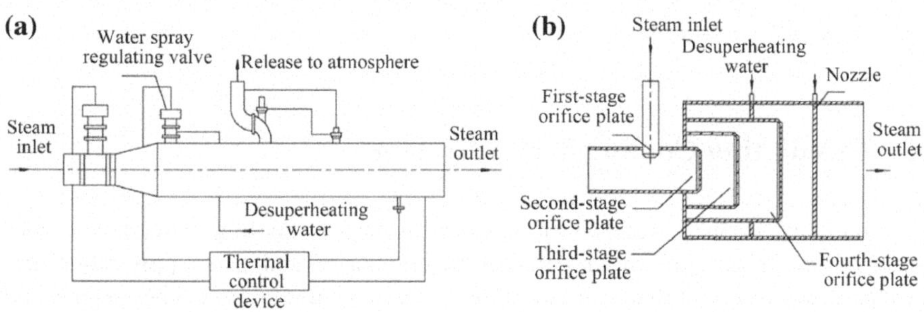

Fig. 7.7 Steam Temperature and Pressure Reduction Device. **a** Temperature and Pressure Reduction Device; **b** Four-grade Temperature and Pressure Reduction Device

for cooling. The Fig. 7.7 shows two types of temperature and pressure reduction devices [1].

The temperature and pressure reduction device shown in Fig. 7.7a uses an orifice plate to throttle and reduce pressure of the steam and reduce the temperature through spraying water after depressurization. The water spray regulating valve is designed with equal percentage characteristics and the valve disk is a cylinder with many circular holes of different diameters arranged on it, controlling the flow and pressure of desuperheating water. Jet orifices are formed by small holes regularly arranged on the cylindrical outer sleeve into which the disk fits. The desuperheating water that reaches the depressurized steam pressure and required flowrate after undergoing multi-stage throttling of valve disks is nebulized before being sprayed into the steam pipeline. The jet regulating valve has multiple functions, such as controlling the flow and pressure of the desuperheating water and injecting the desuperheating water. As orifice throttling with equal percentage operating characteristics is adopted, there are many advantages: the flow rate can be accurately adjusted under various operating conditions; the load range is up to 10–100%; the leakage is low and the entire desuperheating water system is greatly simplified.

The temperature and pressure reduction device shown in Fig. 7.7b is a structure of four-stage depressurization and first-grade one-stage cooling. Steam with high temperature and high pressure is progressively depressurized through the four-stage orifices, and the vapor steam pressure is gradually reduced with no change in the enthalpy value, but and the steam temperature is relatively high. In order to avoid the thermal shock from the high-temperature steam to the condenser components, a first one-stage desuperheating measure is taken after the fourth-stage depressurization. The desuperheating water is sprayed into the steam already throttled and depressurized, and the atomized desuperheating water will have a larger contact area with the steam, which is beneficial to the smooth heat transfer and mass transfer processes and enables the steam to be cooled uniformly. In the design of the temperature and pressure reduction device, the design of the water injection and pressure reduction device shall not only guarantees sufficient cooling water, but also ensures the sufficient full atomization of water. Poor water injection and atomization will influence the temperature decrease decreasing effect, and the airjet flow containing the water drops can easily cause impact erosion.

7.5 Circulating Cooling Water System

The circulating cooling water system is used to supply cooling water to the condenser for condensing the steam entering the condenser, and also supply other auxiliary equipment (e.g., lubricating oil cooler, component cooling water system, and residual heat removal heat exchanger) with required cooling and flushing water. The circulating cooling water is the outboard seawater and Fig. 7.8 shows the principle of the circulating water system [1].

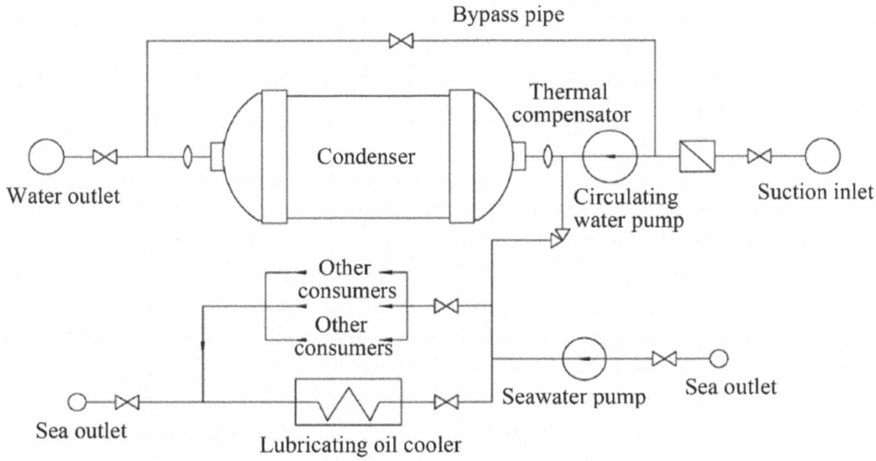

Fig. 7.8 Flow of circulating cooling water system

The outboard seawater enters the system through the suction hole and inlet pipe of the ship. After being pressurized by the circulating water pump, some seawater is transferred into the condenser as the circulating cooling water and discharged to the sea via outlet pipelines and outlet holes after the absorbing of heat; the rest seawater is transferred to the lubricating oil cooler and other equipment and used as the water for cooling and flushing via the branch pipes at the outlet of the circulating water pump and then drained back to the sea through the draining pipes and the sea outlets. In order to guarantee the seawater required for the operation of auxiliary equipment under accident conditions, the emergency seawater pump is installed to introduce seawater through specialized sea outlets and inlet pipes.

Thermal compensators are installed on the inlet and outlet pipes of the condenser to compensate for the expansion deformation of pipes caused by temperature changes. In order to control the subcooling of the condensate, a bypass pipe connecting the inlet and outlet pipes of the main condenser is installed to control the condensate temperature through control of the bypass flow. Filters are installed on both the inlet and outlet pipes to prevent marine life and floating objects from entering the circulating cooling water system to block the pipelines. Gate valves with small flow resistance are installed at different locations of the system pipeline according to requirements to isolate the system and equipment during maintenance and under other special conditions.

The pump-flow circulating cooling water system refers to the system totally depending on the circulating water pump to provide driving head for the flow of seawater in the system. As the required seawater flow is large and flow resistance of system is small, the circulating water pump usually uses the axial flow pump with large flow and low lift and the turbine or motor as the prime mover. For circulating water pumps with large power, it is better to use the turbine as the driving mechanism, which can reduce the overall dimensions of the pump assembly and make it convenient to control the flow of circulating cooling water by adjusting the rotation speed of the circulating water pump. The steam exhausted by the turbine can also be used to heat the feedwater or as the heat source of the evaporating-type water making system, which is easy to realize the heat recovery cycle and improve the economical efficiency of the device.

The self-flow-type circulating cooling water system is widely adopted in the modern marine turbine units. In this system, the seawater flows through the condenser under the dynamic head generated by the relative motion between the hull and the seawater during the sailing of the ship. When the ship speed is not less than 10–12 knots, the seawater can meet the requirements for normal operation of devices. The circulating water pump is only used under low power conditions, so it is convenient for driving by the motor. This system will increase the sailing resistance of ships and consume the propulsion power of the turbine, but it is more economical as the energy utilization coefficient of the main turbine is larger than that of the auxiliary turbine.

The flow of circulating cooling water entering the system will change with the speed of ship. The higher the speed, the greater the flow of circulating cooling water, and vice versa. For the condenser, to a certain extent, this plays the role in automatic adjustment of the flow of circulating cooling water, which is useful for controlling the under saturation degree of the condensate. Compared with the pump-flow-type circulating cooling water system, the gravity-flow circulating cooling water system does not use any circulating water pump in the intermediate or high ship speed, greatly reducing the technical requirements for the circulating water pump and improving system reliability.

7.6 Steam Turbine-Gear Unit

Marine nuclear power mostly adopts the direct propulsion by gear reducer and the steam-turbine-gear unit consists of the turbine, the condenser and the gear reducer. According to different turbine configurations, the steam-turbine-gear units are classified into two types: the single double-cylinder turbine type and the double single-cylinder turbines type, as shown in Fig. 7.9 [1].

7.6.1 Turbines

Turbines use the steam as the working medium and do work through expansion of the steam in the turbine to convert the thermal energy of the steam into mechanical energy, providing power needed to propel the ship.

The turbine consists of 4 basic parts.

① Flow section: steam flows through this section and energy conversion is completed in it and it consists of nozzles and rotating blades;

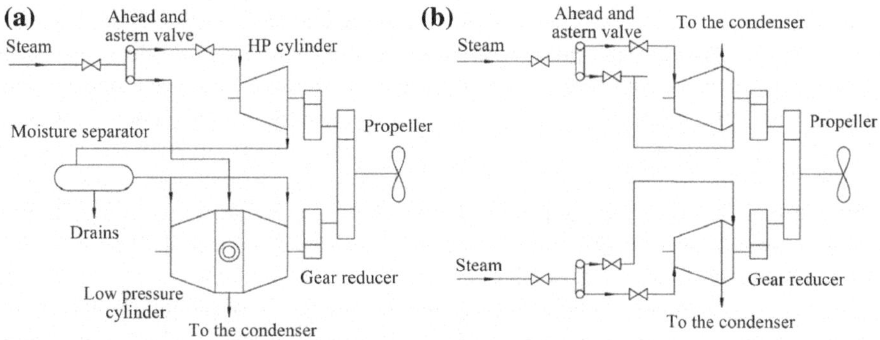

Fig. 7.9 Two types of steam turbine-gear unit **a** Single-machine with double-cylinder type; **b** Double-machine with single-cylinder type

② Stator: the general term for all stationary parts including the cylinder and its fixed parts (e.g., nozzle box and spacer);
③ Rotor: the general term for all rotating parts, including the impeller and the crankshaft;
④ End sections: all parts required to ensure the normal operation of the turbine at both ends of the turbine, including steam seals, thrust bearings, bearings.

Due to the need for maneuvering, turbines shall provide the astern power to ensure the navigation safety of ships. Therefore, turbines usually need to be equipped with an astern turbine or astern stage. The power of the astern turbine is determined according to the requirements, and the astern power of ships is generally about 40–60% of the ahead power.

Most nuclear-powered turbines use saturated steam for operation, but the saturated steam has many negative effects on the design and operation of nuclear-powered turbines. Details are as follows:

① Low steam parameters: and the steam driving the turbine is the saturated steam or slightly-superheated steam with low pressure;
② Large steam volume flow;
③ High steam humidity.

Therefore, corresponding dehumidification measures shall be taken in the design of nuclear-powered turbines to ensure the safe and reliable operation of turbines.

7.6.2 Gear Reducer

As an effective measure to increase the operating efficiency and reduce the weight and dimensions, the higher rotational speed of turbines is of great significance for improving the ship's endurance. On the premise of satisfying the reliability requirements, the operating speed of the turbine is generally within the range of 3000–9000 r/min.

Limited by operating characteristics, if the rotating speed of the high-power propeller is too high, the "cavitation" phenomenon will occur, which will erode the surface of the propeller blade and reduce the thrust and efficiency of the propeller or even cause the propeller unable to work. Therefore, the propeller speed is generally within 100–400 r/min to ensure high propulsion efficiency and working stability. In order to ensure both the turbine and the propeller can normally operate at their respective favorable speeds and the output power of the turbine is effectively transmitted to the propeller, an intermediate transmission device, i.e., a gear reducer shall be installed between the turbine and the propeller.

The main performance indicator of the gear reducer is the reduction ratio, namely the ratio of the input rotating speed to the output rotating speed. The transmission ratio of single-stage reducer can be up to 14–20, while that of two-stage reducer can reach 40–100.

7.6.2.1 Condenser

As an important part of the turbine unit, the condenser is used to receive the steam from the turbine and the steam dump system, recover the drainage water at startup and other draining, condense the steam into water, and establish and maintain the back pressure required for the turbine operation.

The marine condensers are of surface-type and can be divided into the single-pass and double-pass condensers based on the cooling water flow path. Most main condensers for ships are of single-pass dual-channel design structure, which can sufficiently cool down the tube bundles of condenser, and the circulating cooling water has relatively small resistance but large volume, generally 50–80 times of the condensate volume. The dual-channel structure can improve the endurance of ships and ships can operate normally with single channel at low power.

The gas extractor is used to extract the non-condensable gas out of the condenser so as to maintain the condenser vacuum. The two-stage booster ejector using steam as the working medium is generally adopted in the marine condensing equipment. Although it is uneconomical, it is characterized by simple structure, reliable operation, and compact configuration and convenient for use and maintenance.

7.7 Turbo-Generator Set

The function of the turbo-generator set is to convert the heat energy of the steam generated by the steam generator into electrical energy. The turbo-generator set consists of the turbine, the condenser, the generator, etc. A reducer may be installed between the turbine and the generator according to the revolving speeds of these two devices.

At present, most marine nuclear power plants adopt direct propulsion. The turbo-generator set is mainly used to provide electricity for the whole ship due to its relatively low power and its configuration and capacity are determined according to the electricity demand of the whole ship.

With the development of technologies, more applications will be available for marine electric propulsion. For nuclear-powered ships using electric propulsion, the propulsion electric motors shall be added to the electricity consumers, through which the energy is transferred to the propeller and converted into the mechanical energy for sailing. Therefore, the performance requirements for turbo-generator sets are enhanced, such as higher unit power, higher dehumidification efficiency and higher reliability of high-speed motors.

Reference

1. Minjun, Peng. 2009. *Marine Nuclear Power Unit*. Beijing: Atomic Energy Press.

Chapter 8
Source Term and Radiation Protection

This chapter will discuss the radiation protection of marine nuclear power plants. It starts with the concept, objectives, principles of the radiation protection and the characteristics of the radiation protection of the marine nuclear power plants; secondly, the design of source term, which is an important data basis for establishing the radiation protection measures; and finally introduces radiation protection measures taken to realize the objectives of the radiation protection from the perspectives of radiation protection facilities and management.

8.1 Concept and Principles of Radiation Protection

8.1.1 Concept of Radiation Protection

Radiation refers to the outward diffusion of the energy in the form of electromagnetic waves or particles, and it falls into two types, ionization radiation and non-ionization radiation based on whether it can directly or indirectly cause the ionization of substances. The non-ionization radiation mainly refers to electromagnetic field, infrared ray, ultraviolet ray and laser, while the ionization radiation refers to the energy diffusion of charged particles such as α-particles, β-particles and protons, or uncharged particles like neutron, X-ray, and γ-ray.

The fundamental characteristics of particles or rays involved in ionization radiation are as follows:

① The α-ray is a stream of charged particles and is emitted by some heavy atomic nucleus during the nuclear decay. The alpha ray has strong ionizing capacity, with energy range of 4 MeV–8 MeV. Because of large mass, its penetration capability is poor and it can be blocked by one piece of paper or healthy skin.

② The β-ray is also a stream of charged particles and is the electrons emitted by the atomic nucleus. The daughter of β-emitter can also emit β-ray in most cases. The energy range of beta energies ray is between 5 KeV and 3.5 MeV, and the ionizing

J. Yu, *Marine Nuclear Power Technology*,
https://doi.org/10.1007/978-981-15-2894-1_8

capacity is much smaller than that of α-ray, but its penetration capacity is greater than that of α-ray. However, β-ray is easy to be absorbed by the aluminum case and organic glass.

③ X-ray and γ-ray are uncharged electromagnetic waves, and the former is emitted by inner layer of electrons of atoms after being excited and the latter by atomic nucleus. The γ-ray has higher energy and stronger penetration capacity than the X-ray. The energy of X-ray ranges from several KeV to hundreds KeV; while γ-ray is generated during nuclide decay, such as ^{60}Co and ^{137}Cs. The range of X-ray and γ-ray is long and is not influenced by magnetic field, hence generally can be shielded with a lead plate of a few centimeters thick or concrete of a few meters thick.

④ Neutron ray, typically with energy below 50 MeV, mainly comes from the spontaneous fission of some elements, and may also result from the nuclear reaction due to the bombardment of γ-ray, α-ray, neutrons and protons towards some nuclides. Neutrons have no electric charge and have strong penetrating capacity. Though they cannot perform ionization directly, they can cause the ionization of substances indirectly by producing secondary particles through interaction with substances, just as γ-ray does. It should be noted that neutron ray is generally mixed with γ-ray and this shall be taken into account in the protection.

The ionization radiation can release energy in body tissues, causing cell damage or death. Little damage occurs under small dose, however cells become abnormal, instead of dead, and even develops into the cancer cells in some cases. Extensive cell death occurs under large dose. The ionization radiation affects the human body in two ways, i.e., external exposure and internal exposure. The external exposure is caused by the action of γ-ray and X-ray with long range, fast speed and small ionization density on the organism from the outside of the body, and the degree of hazard is related to the dose, position subject to radiation and ray types, graded irradiation method and irradiated area. The internal exposure is mainly caused by the radionuclide entering the body through inhalation, ingestion, skin or wounds. The hazard caused by internal exposure is more complex than that of external exposure, because the radiation source of the internal exposure is in the body and all the radiation emitted effects the body; due to the chemical properties of the radiation source, the radiation accumulates in special organs or tissues; the radiation from the radiation source effects the body all the time unless it is removed from the body.

Limited to the early historical conditions and technological level, people paid a price in the practice of studying and applying the nuclear energy and ionization radiation technologies. With deepening understanding of radiation hazards to human body and rapid development of relevant technologies, researches and measures for radiation protection and safety have been greatly developed and strengthened. At present, radiation protection has become an important branch of atomic energy science and technology. It is a comprehensive interdisciplinary subject studying how humans are protected from or less exposed to the ionization radiation, which centers on radiation dosimetry, radiation protection standard, radiation protection technology, assessment of radiation protection and radiation protection management.

The basic tasks of radiation protection are to protect the safety of people engaged in radiological work and their descendants as well as the general public and the humankind, and protect the environment, while those practices that might produce radiation but are necessary for the benefit of humankind can be implemented.

Radiation protection work aims to prevent the harmful deterministic effect and limit the probability of stochastic effect to an acceptable level.

8.1.2 Ionization Radiation Source of Nuclear Power Plants

Taking the PWR as an example, fissile materials (like ^{235}U) are used as the fuel and the nuclear energy is mainly converted to the other energies mainly thermal energy through the controllable self-sustaining chain fission reaction. These thermal energies energy are carried by coolant and then converted to mechanical kinetic energy or electric energy through primary and secondary loop systems. Structurally, the reactor is mainly composed of the core, vessel, pipes, steam generators and pressurizer. During the operation, the cooling water, flowing from the vessel inlet, is heated while passing the core, and flows into the steam generators from the vessel outlet to the reactor coolant piping and then reaches the main pumps through the reactor coolant piping, finally returns to the inlet via the main pump, completing a cycle. The heat of primary loop is transferred to the secondary loop in the steam generators.

In the nuclear power plants, the original ionization radiation comes from the fission reaction of the reactor core and the nuclide decay in the nuclear power plants. Neutrons and γ-rays with high energy and strong penetrability will escape from the vessel after passing through the fuel cladding, coolant and vessel, causing harm to personnel and equipment outside the vessel; In addition, there are many radiation sources in the coolant, including fission products released due to the failed fuel cladding, activated corrosion products and ^{16}N resulting from the capture of neutrons by oxygen atoms in the coolant, and the generation, migration, accumulation and leakage of these radiation sources will cause harm to personnel. In the event of nuclear reactor meltdown and reactor disruption, massive radiation sources will be released into the environment, posing serious hazards to the mankind and environment.

8.1.3 Basic Principles of Radiation Protection

In order to protect the health and safety of workers and the public as well as the environment, the ionization radiation produced by nuclear power plants shall be shielded to prevent the non-stochastic effect and reduce the probability of stochastic effect to an acceptable level.

The three basic principles of radiation protection of nuclear power plants are:

① Legitimacy of radiation practices. A radiation practice is regarded as legitimate only when the benefits that the radiation practice brings to the exposed persons or the society can compensate for the radiation hazards it may cause, based on comprehensive consideration of social, economic and other factors and full demonstration with tradeoffs.
② Optimization of radiation protection and safety. On the premise that the individual dose and potential exposure hazard resulting from the radiation sources (including radiation facilities) used during radiation practices are lower than the dose constraint and potential exposure constraint respectively, the individual exposure dose, number of exposed personnel and exposure possibility shall be maintained at a level as low as reasonably achievable (ALARA) with full consideration of economic and social factors.
③ Limit of individual dose. Since benefits and costs are not evenly distributed among different people, and though the radiation practices have satisfied legitimacy requirements and the protection and safety have been optimized, this fact does not necessarily mean that sufficient protection can be provided for each and every person. Therefore, the normal exposure to each person shall be limited, to ensure that the total effective dose to individuals and total equivalent dose to related organs or tissues resulting from combined radiation of approved radiation practices do not exceed the corresponding dose limits specified in relevant standards.

8.1.4 Dose Limit for Radiation Protection

The estimated average doses to the public caused by the practices shall not exceed the following limits [1]:

① Annual effective dose: 1 mSv;
② In special cases, the effective dose can be up to 5 mSv in one single year only when the annually averaged dose is less than 1 mSv for five consecutive years;
③ Annual equivalent dose to the eye lens: 15 mSv;
④ Annual equivalent dose to the skin: 50 mSv.

The exposure of any practitioner in radiation practices shall be limited within the following limits:

① Average annual effective dose in five consecutive years: 20 mSv (but not averaged retrospectively): 20 mSv;
② Effective dose in any single year: 50 mSv;
③ Annual equivalent dose to the eye lens: 150 mSv;
④ Annual equivalent dose to the extremities (hands and feet) or the skin: 500 mSv.

In case of accidents of nuclear power plants, the dose is limited as below [2]:

① In case of a major accident, the effective dose to anyone of the public shall not exceed 5 mSv, and equivalent dose to the thyroid shall not exceed 50 mSv;
② In case of a serious accident, the effective dose to anyone of the public shall not exceed 0.1 Sv, and equivalent dose to the thyroid shall not exceed 1 Sv.

8.1.5 Design Principles for Radiation Protection of Marine Nuclear Power Plant

Radiation shielding is crucial in the design of marine nuclear power plants. A reasonable design can not only guarantee the normal operation of nuclear power plants, but also ensure the safe radiation dose to workers. The shields of a nuclear power plant are generally divided into two parts: the primary shield and the secondary shield. The former is the most important and most direct shield against the core radioactivity and its main task is to reduce the radiation dose level around the reactor to a reasonably acceptable level. More specifically, it is used to prevent excessive exposure of equipment and instruments around the reactor; meawhile, it is applied to reduce the burden of the secondary shield, and to minimize the shield weight to avoid the loss of maneuverability of nuclear-powered ships. The main task of the secondary shield is to shield the radiation rays leaking from the primary shield to ensure the radiation safety of workers outside the secondary shield.

(1) Basic principles of shield design
The shield shall be designed as an integral part of the entire reactor system; the adjacent parts of the shield with the same design standard shall be designed to have the same performance; the reactor shield is a safety facility and shall be considered as an integral part of the reactor safety system; the shield design shall satisfy the requirements on mechanical properties of the reactor, supporting structures, and various structures of its ancillary system without sacrificing the integrity of the reactor and the shield, the consistency of the shielding performance and the safety of the shielding, and the design costs shall be as low as possible; the shields structure shall adopt a design as simple as it can be.

(2) Requirements of shielding performance
① Performance requirements for the primary shield of reactors: the primary shield shall reduce the neutron current emitted from the reactor to a level that does not cause significant activation of secondary-loop coolant and minimize the γ radiation caused by capturing the neutrons inside the shield of primary primary-loop coolant; reduce the remaining radiation emitted from the active region to a low level, allowing personnel to stay in the area between the primary shield and secondary shield for a certain time after reactor shutdown; reduce the neutron current emitted from the reactor to a low level to avoid the accumulation of activation activity of structures and equipment that prevents personnel from approaching the reactor when necessary.

② Performance requirements for the secondary shield of reactors: the secondary shield shall reduce the gamma radiation and neutron flux current on it to the extent that personnel working near the shield will not suffer from radiation damage; the excess radiation streaming shall be prevented from passing through areas with poor shielding performance; sometimes it shall play the role of sealing the radioactively contaminated air; when the reactor is in operation, it shall also prevent workers from directly approaching areas with strong radioactivity.

(3) Shield design procedures
The following steps are recommended for the design of primary shield:

① Determining the intensity and distribution of various radiation sources generated in the active area;
② Considering the secondary radiation sources generated outside the active area;
③ Selecting the material of the additional shield arranged at the periphery of the reactor based on the analysis of materials in the active area of the reactor, materials of the reflector and the structural materials; and
④ The above three points shall be satisfied at full power operation or shutdown.

The design of the secondary shield shall take into account the contribution of all nearby sources and the dose rate outside the secondary shield shall be reduced to allowable limit values.

8.1.6 Characteristics of Radiation Protection for Marine Nuclear Power Plant

Compared with that of nuclear power plants, the radiation protection of the marine nuclear power plants has the following characteristics:

① As limited by the space, weight and dimensions, the radiation protection technology featured by small size and light weight shall be adopted as far as possible. For example, special shielding materials and radiation monitoring instruments are used.
② Marine reactors are characterized by variable operating conditions, which need to be considered in the source term analysis and radiation zoning. The nuclear-powered ships have a variety of operating conditions, and the change in the flow rate of the primary coolant caused by the power increase and decrease has a significant effect on the radiation field intensity, thus affecting the variation of the radiation field intensity in the same zoning.

③ Electromagnetic compatibility of corresponding radiation protection monitoring instruments shall be considered because the space in the ship is limited and the equipment is densely arranged.

④ The design of shielding structures and the equipment installation shall take into account the bearing capacity under the tilting, swinging or impact conditions since ships are operating in the marine environment.

8.2 Source Term Design

8.2.1 Overview

The study of the intensity, distribution, and migration of radiation sources in the radiation protection work is helpful for the understanding of the damage extent, and provides the design inputs for the deployment of radiation protection facilities like shields, and the design of radiation protection management measures such as operation procedures and emergency plans. The methods of source term design for nuclear power plants are described as follows.

Based on the operating conditions of reactors, the source term design is divided into three categories: source term under normal operation, shutdown source term and accident source term.

8.2.2 Source Terms Under Normal Operation

When the reactor is operating normally at certain power, the radiation source terms in the reactor core, reactor coolant and nuclear auxiliary systems are the major considerations, including neutrons and gamma radiation sources during reactor operation, ^{16}N and ^{17}N radiation sources generated by oxygen activation in reactor coolant, and radiation sources of fission products and corrosion products.

(1) Radiation sources during normal full-power operation

① Neutrons and gamma radiation sources

The calculation of neutrons and gamma radiation sources for marine nuclear power plant can be divided into the primary shield and secondary shield calculations. The former mainly calculates the fast neutron and thermal neutron fluence rates on the outer surface of the primary shield, γ-ray energy fluence rate, and heat release ratein the shield; the latter is performed on the basis of the former.

Internationally, computer codes related to Monte Carlo and SN methods are generally used to conduct the calculation of neutrons and gamma radiation sources.

② ^{16}N source term in the reactor primary coolant

The ^{16}N in the primary coolant is mainly generated from the $^{16}O(n, p)^{16}N$ reaction and it is a high-energy gamma radioactive source with a half-life of T = 7.13 s and the energy of γ-ray released from the decay of 6.129 MeV (68.8%) and 7.115 MeV (4.73%). The yield of ^{16}N is directly proportional to the reactor power, i.e., the higher the reactor power, the higher the yield. During the power operation of the reactor, ^{16}N is generated in the primary coolant, which can enter the secondary loop or even be released to the outside of the primary and secondary loops through the damaged steam generator tube. Therefore, ^{16}N source term is usually the main radiation source term considered in the shield design of primary loop equipment rooms.

③ Source terms of fission products and corrosion products in the primary coolant of the reactor (^{16}N excluded)

In case of the failure of fuel elements, fission products in fuel elements will be released into the coolant and migrate in the coolant system. These fission products mainly include inert gases such as ^{85m}Kr, ^{133m}Xe and ^{133}Xe, iodine isotopes such as ^{135}I, cesium isotopes such as ^{134}Cs, and other isotopes such as ^{89}Sr, ^{90}Sr, ^{95}Nb and ^{131}Te.

In the primary system, the radiation sources of corrosion products will be formed through oxidation corrosion, erosion release after erosion, and activation of the parent metal, mainly including isotopes like ^{51}Cr, ^{54}Mn, ^{59}Fe, ^{58}Co, ^{60}Co, ^{100m}Ag, and ^{124}Sb.

The production and migration mechanisms of fission products and corrosion products are relatively complex, and specific computer codes are required to calculate the quantitative results. During the calculation of fission products, major considerations include the influences of the production, release and dispersion of fission products, and operating parameters related to the reactor coolant system. The calculation of corrosion products mainly considers the oxidation, corrosion and release after the erosion, the dissolution, the crystallization and the precipitation of the parent metal; the morphology of the corrosion products and the corresponding purification mechanism; the migration, activation, decay, and leakage in the loop.

④ Calculations of source terms in key systems

Based on the different characteristics of each system in the reactor, the calculations of the source terms shall be conducted for key systems such as the pressurizer and the residual heat removal system.

(2) Radiation source under shutdown condition

Under shutdown condition, some major radiation sources shall also be considered in the shield design, including source terms from spent fuel assemblies and residual heat removal systems.

① Source terms from spent fuel assemblies

Shutdown radiation sources from spent fuel assemblies include γ-ray radiation sources and neutron radiation sources.

Internationally, the ORIGEN code is generally used to calculate the spent fuel source term. The ORIGEN code, with its database covering more than 1,700 nuclides, is mainly used to calculate the accumulation and decay of radionuclides, and the change in the nuclide composition after various processes. This code can compute and give out the source terms of core fission products, actinides, and activated products of structural materials, including the mass, activity, toxicity, neutron absorption rate, fission rate, spontaneous fission neutrons, neutrons from (α, n) reaction, photon spectrum and core decay heat.

② Source terms from residual heat removal system

i During reactor shutdown, when the temperature and pressure of the reactor coolant system are below a certain threshold, the residual heat removal system begins to operate in order to remove the heat of the reactor. After reactor shutdown, the radionuclides in the reactor coolant decay exponentially.

ii During the calculation, the degassing factors of gaseous radionuclides, the purification efficiency and purification flow rate of non-gaseous nuclides shall be considered comprehensively to compute the γ source intensity of radionuclides in the residual heat removal system.

8.2.3 Source Terms in Accidents

According to the design principle of "defense-in-depth", nuclear power plants adopt three leak-tight barriers to protect the public against the hazards of radioactive release. These barriers include fuel claddings, reactor coolant pressure boundaries and containment. In case of an accident, one or more barriers may fail, resulting in the release of radioactive fission products in the reactor into the environment and causing radiation harm to the environment or the public. The accident source term analysis addresses the generation, migration and release routes of radioactive substances after an accident, to obtain the amount of radioactive substances released to the environment.

Typical accidents considered in the calculation of the accident source terms include loss of coolant accident, control rod ejection accident and steam generator tube rupture accident.

(1) Source terms due to loss of coolant accident

After the break of the loop pipe in the nuclear power plant, the radioactive substances like fission products in the primary coolant are released to the reactor compartment and other compartments through the break. The calculation shall consider the fuel failure and the release of the radionuclides such as corrosion products accumulated in the coolant into the reactor compartment following the flow passing the break. The radioactive intensity and concentrations of aerosol nuclides in the primary loop and each compartment at different times after the accident are obtained through calculations. These concentrations are evaluated in comparison with the derived concentrations corresponding to the radiation safety limits.

(2) Source terms due to control rod ejection accident

The control rod ejection accident is defined as the ejection of rod cluster control assemblies due to the mechanical damage of the pressure housing of the control rod drive mechanism. After the control rod ejection accident, the reactor power rises sharply, resulting in the sharp rise of the pressure and temperature of the primary coolant in the primary loop and the quick openning of the pressure relief valve of the pressurizer. The primary coolant can also be released to the reactor compartment from the break of the pressurized housing of the drive mechanism. During the calculation, the amount of radionuclides accumulated in the primary coolant shall be considered according to the fuel failure. The concentration of radionuclides in the atmosphere of each compartment shall be calculated based on the relevant leakage paths, and compared and evaluated with the derived limits.

(3) Source terms due to steam generator tube rupture accident

In case of steam generator tube rupture accident, the primary coolant leaks to the secondary side through the broken tubes. It is assumed that the core fuel elements are damaged to a certain degree; the primary loop purification and the steam generator blowdown terminate at the early stage of the accident; the inert gas in the secondary side vapor is completely released to the rear auxiliary engine compartment. The concentration of radionuclide in the rear auxiliary engine room after the accident shall be calculated, and compared and evaluated with the derived limits.

(4) Source terms due to severe accident

Severe accident refers to the accident condition which is more severe than a design basis accident and causes significant core degradation. In the analysis of the severe accident source terms, the accident process and the migration of radioactive substances are generally simulated based on the severe accident phenomena and then the amount of radioactive substances released to the environment is obtained and used as the input for off-site dose analysis.

Under severe accident conditions, the radionuclides in the core are classified into the following groups based on the physical and chemical properties of elements: inert gases, halogens (elementary and organic), alkali metals, antimony, alkaline earths, precious metals and refractory oxides. The release fraction and migration behavior of each group shall be considered respectively.

The escape fraction of the primary coolant system is based on the following considerations: the release of vapors and aerosols of fission products outside the primary coolant system by huge air flow; the deposition of fission products on the surface of reactor internal components; and the absorption of fission products by emergency core cooling (ECC) water. The behaviors of fission products in the containment are: natural migration and deposition (natural deposition, gravity setting, thermal migration, diffusion migration); the migration of fission products through water ejection; the recirculation of filtration system; once-through filtration system; pit flushing; leakage and discharge from the containment to the outside atmosphere.

The radiological consequences outside containment under severe accident are mainly determined based on the following three factors: the failure mode of the containment, including modes such as containment bypassing, containment isolation failure, severe damage to the containment, or intact containment; containment failure time from the start of release of a large number of fission products from the core corium; the removal mechanism of fission products in the containment.

8.3 Radiation Protection Facilities

Radiation protection facilities can effectively prevent harmful radiation in the reactor compartment from entering the operating compartment and minimize the radiation level in the operating compartment. In addition to the primary and secondary shields, radiation protection facilities mainly include relevant systems used for controlling gaseous, liquid, solid and other radioactive substances, systems related to radiation monitoring, and personal protective equipment.

The related systems used for controlling gaseous, liquid, solid and other radioactive substances include systems for controlling airborne radioactive materials through air purification, external exhaust and other devices; and systems for collection, temporary storage and transfer of liquid and gaseous radioactive wastes.

The relevant systems for radiation monitoring include radiation monitoring systems of each compartment to ensure personnel safety, process monitoring systems for ensuring the safe operation of equipment, and effluent radiation monitoring systems for ensuring the environmental safety.

Personal protective equipment includes personal protective equipment such as respiratory tract protective equipment, special protective underwear, masks, shoes, socks, gloves and work hats, and they are used for personnel radiation protection in case of maintenance and inspection in the control area and contaminated area and accidents.

8.4 Management of Radiation Protection Work

8.4.1 Control Through Radiation Zoning

Radiation zoning can effectively control the normal radiation, prevent the spread of radioactive pollution, and prevent potential exposure or limit the scope of potential exposure, facilitate radiation protection management and occupational exposure control, so that the annual effective dose of personnel is lower than the specified limit, and kept as low as reasonably achievable.

Internationally, workplaces with radioactive sources are generally divided into the controlled area and the supervised area. Measures or safety regulations are needed or may be required within the controlled area to control the normal exposure under normal conditions or prevent the spread of contamination and prevent or limit the range of potential exposure. The supervised area is an area where operating conditions shall be continuously checked but special procedures are normally not required.

8.4.2 Emergency Plan

Prior to the service of nuclear-powered ships, emergency plans shall be formulated for severe accidents of the release of a large amount of radioactive substances to the environment, including establishing an organizational leadership system for emergency plan activities, defining the responsibilities of organizations at all levels and various types of personnel, and determining the principles and measures for controlling the release of radioactive substances and reducing radiation hazards.

8.4.3 Radiation Protection Requirements for Nuclear Power Plant at Each Stage

(1) Radiation protection requirements for engineering design

Nuclear-powered ships shall be designed in accordance with the principle of optimization of radiation protection to ensure that the exposure of workers and the public is kept as low as reasonably achievable under normal and accident conditions.

It is necessary to establish redundant safety barriers to prevent the radioactive release and guarantee their integrity, to ensure that the impact can be controlled below the corresponding design limits? in the event of an accident.

The compartments shall be divided into different radiation limited areas according to the radiation level and the permitted retention time for workers, and marked with obvious signs.

The design of equipment, instruments and special tools shall meet the requirements of reliable performance, convenience for dismantling, replacement and decontamination & maintenance. The compartments shall be equipped with sensitive and reliable radiation monitoring systems.

(2) Radiation protection requirements during operation

During the operation of nuclear-powered ships, practical measures shall be available and taken to ensure that the radiation exposure to workers meets the prescribed dose limits.

The following shall be specified clearly: radiation protection of workers; instruments and equipment for personal monitoring; radioactivity monitoring of nuclear-powered ships and the surrounding environment; decontamination of workers, equipment and nuclear-powered ships; collection, processing, storage and disposal of radioactive wastes.

The applicable procedures shall be formulated according to the specific characteristics and conditions of nuclear-powered ships, so that the radiation exposure to the workers is as low as possible. The workers should be familiar with the procedures related to radiation protection.

(3) Radiation protection requirements during maintenance

Radiation protection requirements shall be considered during the development of maintenance program to ensure that the dose received by workers is kept as low as possible.

The radioactive operation card system is implemented for the maintenance of radioactive equipment, which includes radiation monitoring results and radiation protection measures.

An environmental impact report shall be submitted when a large amount of radioactive substances are likely to be released to the environment during the maintenance.

During the operation of components of high radioactivity, specific shielding and remote control technologies shall be adopted, or working duration time shall be limited to reduce the exposure to workers. Systems with potential contamination shall be designed in a manner that water flushing or remote chemical cleansing shall be completed prior to the operation needing contact.

In order to reduce operational errors, workers shall be skillful in the operating techniques so as to shorten the exposure time. For the operation of components of high radioactivity, exercise shall be conducted on the simulator before the operation needing contact. The guardianship system is applied for major operations so as to correct the misoperation in time.

(4) Radiation protection requirements for decommissioning

A detailed decommissioning plan shall be prepared for nuclear-powered ships, in which radiation protection measures shall be particularly considered. The decommissioning plan shall be reviewed and approved by the competent authority before implementation.

An environmental impact report shall be prepared during the development of a decommissioning plan. Radiation monitoring measures after decommissioning shall be formulated and implemented according to the decommissioning method. The radioactive solid wastes generated during the decommissioning shall be stored and disposed in accordance with the relevant regulations.

References

1. Ministry of Environmental Protection of the People's Republic of China. 2002. Basic standards for Protection against Ionizing Radiation and for the Safety of Radiation Sources (GB18871-2002).
2. National Environmental Protection Agency. 1985. Regulations for Environmental Radiation Protection of Nuclear Power Plant (GB6249-S6_).

Chapter 9
Vibration and Noise Reduction

9.1 Overview

The vibration noise influences the ships mainly in two ways: ① worsening the working and living environments, and even the health of sailors; ② affecting the normal and high-precision performance of relevant instruments and equipment. Therefore, vibration and noise reduction has drawn more and more attentions in the ship design.

The mechanical noise, among of ship noise sources, mainly comes from the pumps, electrically-operated valves and other power equipment installed in the marine nuclear power plants and vibration of fluids in the pipelines.

9.2 Sources and Transfer Paths of the Vibration Noise

During the normal operation of the ship, the main pump, purification pump and cool water pump of equipment in the primary loop system of the marine nuclear power plant, and the steam turbine-gear unit, turbo-generator set, feedwater pump and condensate pump in the steam power conversion system are the rotary or reciprocating machinery under continuous operation, which constitutes the major vibration noise sources in the marine power system. During the condition transition, the short-time operation equipment such as check valves, isolation valves, control valves constitute the transient noise source in the marine power system. The fluid excitation caused by the flowing fluid in pipes and the discharge of safety valves - the noise sources partially.

The vibration of power equipment operating for a long period, valves with short-term actions and their connected pipes are transmitted through equipment supports and pipe hangers and supports to the ship structure, thus forming the compartment noise and radiation noise.

J. Yu, *Marine Nuclear Power Technology*,
https://doi.org/10.1007/978-981-15-2894-1_9

9.3 Control Measures for Vibration Noise

The control of vibration noises involves controlling the vibration sources, transfer paths and protection of receivers. A brief introduction to the first two aspects is given in this book.

9.3.1 Control Measures for Vibration Noise

The control of vibration noise source is a measure to control the "sources" by reducing the number of operating equipment or lowering the vibration noise of operating equipment itself, mainly including following aspects:

(1) Configuration and design of low noise mechanical equipment

The noise output power of stand-alone mechanical equipment shall be reduced by structural development, such as usage of materials beneficial to reduce sound source, improvement of machining precision and assembly quality of moving parts, control of dynamic balancing precision of rotators, adoption of effective lubrication systems and improvement of the streamline of water (air) intake and drainage components [1].

With regards to the configuration of low noise equipment, measures like "integration improvement, redundancy reduction and pipeline simplification" and "combination of pumping equipment of low and high capacity" can be taken to control the vibration noise source of mechanical equipment system.

(2) Control of fluid noise sources

When fluids flow in the pipelines equipped with bends, valves and throttling elements under the drive of the pumps and blowers, unbalanced flow field excitation and pressure fluctuation will cause the vibration of pipelines and loud flow noises. The fluid noise source can be controlled by optimizing the piping layout (direction and bend, etc.), reducing the throttling loss of valves, or using special noise-abatement device.

(3) Dynamic vibration absorption technology

When the mechanical equipment vibrates under excitation, an auxiliary system (composed of mass block, elastic component and damping component) can be added to this equipment. As the original main system vibrates, the auxiliary system will also vibrate and produce the power offsetting that of the main system so as to control the vibration of the main system.

Other measures to control vibration sources include the visco-elastic damping technology and granular vibration damping technology, which can be referenced in professional books related to vibration control.

9.3.2 *Vibration Isolation of Transfer Paths*

Used in the isolation of pumps, blowers, pipelines and other equipment producing vibration as vibration control measures in its transfer paths widely, Vibration isolation aims to isolate or weaken the vibration energy transmitted to the mounting base.

The vibration isolation design can be divided into single-stage, two-stage and floating raft vibration isolations based on different vibration isolation systems; active and passive vibration control technologies based on whether there is external energy supply.

(1) Single-stage vibration isolation

Only one group of vibration isolator is installed between the power equipment and the pedestal in the single-stage vibration isolation equipment, and the main disadvantage is that the effect of isolating high-frequency vibration is poor (see Fig. 9.1). It is currently used for systems or equipment with not very strict equipment on vibration isolation but strict alignment requirement.

(2) Two-stage vibration isolation

Two groups of vibration isolators are installed between the equipment and the pedestal with the intermediate mass block [2] inserted between the two layers of isolators, utilizing the intermediate mass to attenuate some vibration from the upper vibration isolator so as to improve the vibration isolation effect.

This technology is now widely applied in the generating units, ventilation units and pumping equipment of naval ships in the world.

(3) Floating raft vibration isolation

The floating raft vibration isolation is a kind of two-stage vibration isolation and the difference is that floating raft vibration isolation corresponds to multiple vibration sources and the two-stage vibration isolation corresponds to one vibration source

Fig. 9.1 Single-stage vibration isolation system

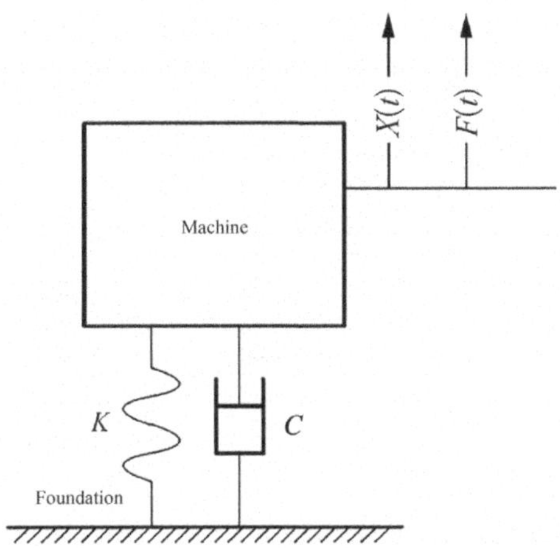

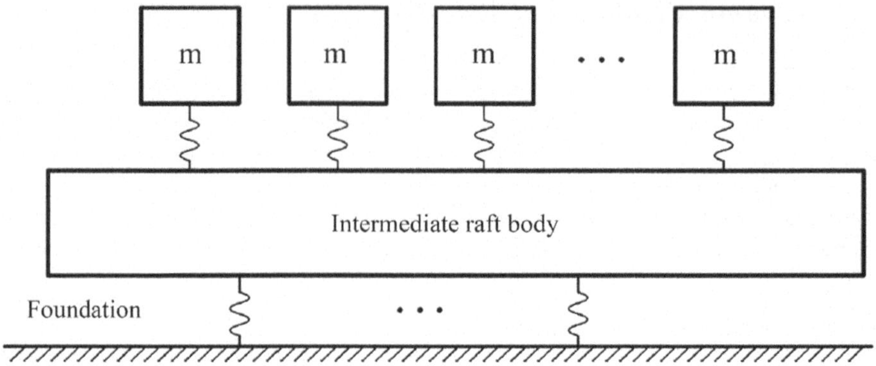

Fig. 9.2 Floating raft vibration isolation System

(see Fig. 9.2). The intermediate mass of floating raft unit produces large mechanical impedance and has better isolation effect than the single-stage and two-stage vibration isolation.

(4) Active vibration control technology

The active vibration control technology, namely the active vibration reduction or active vibration isolation, can be applied in the aforesaid single-stage and two-stage as well as floating raft vibration isolations [2].

As computer technology, control theory, materials science and other subjects develop rapidly, the active vibration control technology becomes an advanced technology. It provides energy to support the operation of vibration reduction devices based on additional power and is characterized by good effect of isolating low-frequency vibration and effective control of equipment line spectrum, becoming an important way to control the vibration.

Common active vibration control systems often adopt the closed-loop control, mainly composed of the controlled object, sensor, actuator, controller and external energy, in which the external energy, such as power supply, air supply and hydraulic oil supply is used to support the actuator operation, as shown in Fig. 9.3.

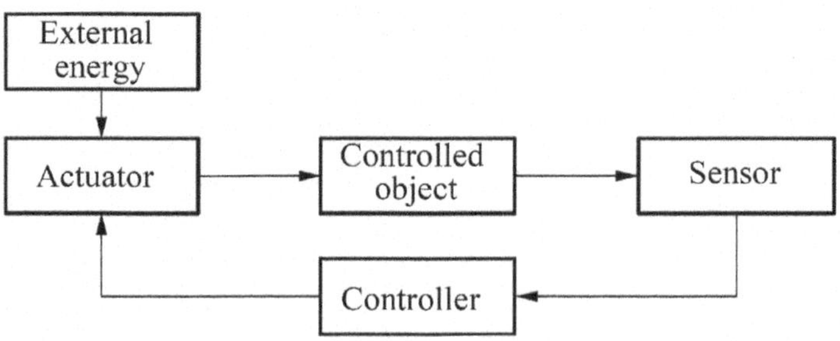

Fig. 9.3 Active vibration isolation system

First, the sensor picks up the vibration response signal of the controlled object and inputs the signal into the controller; and then the controller designs necessary control procedure based on different control methods and outputs the control signal to control the actuator to apply an acting force or moment on the controlled object. In this way, the vibration isolation is controlled.

The actuators for active vibration isolation are divided into those of servo air-cushion, electro hydraulic servo, electrical magnetic, magnetic suspension, electrorheological and magnetorheological fluid, magnetostriction material, shape memory material and piezoelectric material [3].

References

1. Xiaojian, Chen. 2013. *Submarine Noise Control Technology*. Shanghai: Shanghai Jiao Tong University Press.
2. Chengjun, Wu. 2008. *Engineering Vibration and Control*. Xi'an: Xi'an Jiaotong University Press.
3. Lin, He. 2007. Development of submarine acoustic stealth technology. *Ship Science and Technology* 9 (5): 29–42.

Chapter 10
Mechanical Analysis and Evaluation

10.1 Overview

The reactor system (including related systems) is an important part of the marine nuclear power plant and the power source of the ship, and also called the nuclear power plants in engineering. The reactor system shall not only meet the relevant functional requirements, but also be able to withstand the loads generated during its operation, such as temperature, pressure, environmental loads, tilt and swing caused by storm, accident loads and shock blasts generated by non-contact bomb explosions. The task of mechanical analysis and evaluation, i.e. mechanical design, is to ensure that the mechanical behaviors of the reactor system under the above-mentioned load conditions satisfy requirements specified in related regulations. The mechanical design mainly includes three aspects: design inputs, analysis methods and evaluation criteria. The design inputs are used to define the load conditions to be withstood by the reactor system, the analysis methods are used to predict the mechanical behaviors of the reactor system under various load conditions, and the evaluation criteria are used to determine the acceptability of the mechanical behavior occurred in the reactor system under various load conditions.

As an important branch of engineering mechanics and applied mechanics, the reactor structural mechanics is deeply involved in the reactor system mechanical design. The widespread application of nuclear power technology poses many requirements and also provides good opportunities for the development of the reactor structural mechanics. In the design of marine nuclear power plants, the reactor structural mechanics plays an important role in design optimization, including optimization of overall and detailed structural layout, reduction of arrangement space and MNPP weight, and the improving of safety and reliability. During the research and development of new types of marine nuclear power plants, the reactor structural mechanics is applied to help determine structural material structural layout, performance parameters and dimensions [1].

J. Yu, *Marine Nuclear Power Technology*,
https://doi.org/10.1007/978-981-15-2894-1_10

After decades of engineering practice and relevant technological development, reactor structural mechanics has covered the design and analysis on relevant systems, equipment, pipes, supports and other accessory structures of nuclear power plants. The research domain of reactor structural mechanics falls into two levels: the analysis of the system level load distribution and that of the component (local) level stress assessment. The former is to determine the distribution of the working loads and environmental loads within the system (various components and locations). The latter is to determine stress-related mechanical behaviors of components according to the load distribution information. The analysis of load distribution includes static force (including dead weight, pressure, thermal expansion and tilt) analysis, swing analysis and shock resistance analysis. The analysis of component (local) level stress includes stress fatigue, thermal ratcheting, fast fracture, sealing performance, stability, and active components operability and functional capacity. Reactor structural mechanics is an interdisciplinary field involving mechanics of materials, theoretical mechanics, structural mechanics, elastic theory, elastic-plastic mechanics, fracture analysis, vibration, fluid dynamics, materials science and heat transfer theory.

10.2 Main Theories of Mechanical Analysis

10.2.1 Analysis Theory for Shock Resistance of System and Equipment

The shock [2] refers to the physical phenomenon in which the system and equipment experience drastic change in position, velocity/acceleration and stress/strain in a rather short time due to violent excitation effect. The shock has the following features: a. extreme short load action with energy release, transfer and transmission completed very shortly; b. excitation function is aperiodic; c. motion response of the system is directly related to shock duration and its intrinsic structural dynamic properties; d. both excitation and response are rich in high frequency components.

There are three main types of shocks that ships may experience during service [3–5]: a. contact explosion; b. non-contact explosion in water; and c. shock caused by recoil during the firing of on-board weapons. The shock environment of a ship during a non-contact explosion in water is the fundamental premise for the evaluation of ship damage degree and vitality when the ship is attacked by various underwater weapons. Meanwhile, the marine nuclear power plant is installed in the reactor compartment. Normally this compartment will not take direct contact explosion. Under the recoil effect of the on-board weapons, the shock load transferred to the foundation position of the nuclear power plant is far smaller than the shock load of non-contact explosions in water. Therefore, for the shock resistance design of marine nuclear power plants, the primary consideration is the shock effect caused by non-contact underwater explosions applied to systems and equipment.

The source of shock load in the shock resistance design of a marine nuclear power plant is the equivalent explosion. Equivalent explosion is associated with the distance between explosion center and the ship. The shock load generated by the underwater equivalent explosion is transferred via the fluid to the ship structure, and then through the ship structure to the supports of systems and equipment of nuclear power plant. In the analysis of the shock resistance of system and equipment, the shock loads transferred to the supports are taken as the inputs. Proper analysis methods such as finite element calculation are used to obtain the shock responses (including force, torque, acceleration, velocity and displacement) at various components of the system and equipment. Such responses are further used as inputs for structural design and component stress assessment.

Based on the engineering design experience and the in-depth defense safety requirements of nuclear powered ships, shock loads are classified into two categories. I: under certain level of shock, nuclear power plants should meet the requirements for normal operation, and shall not cause any harm to the health and safety of personnel Level II: under certain level of shock, the nuclear power plants shall meet the nuclear safety requirements to safely shutdown the reactor and maintain its shutdown state. Structural integrity should be maintained to ensure reactor coolant pressure boundary, and to prevent or mitigate the consequences of nuclear accidents. Radiation level outside the reactor compartment should meet the requirements specified in relevant laws, regulations and specifications.

There are mainly three analysis methods for the shock resistance of systems and equipment in marine nuclear power plants [6]: shock design factor method, response spectrum analysis method, and transient time history shock analysis method.

(1) Shock design factor method

The shock design factor method is also called the equivalent static method. In this method, the total weight of the system or equipment is multiplied by the design shock acceleration to obtain the force applied at the center of gravity of the equipment. The selection of design shock acceleration shall meet the conservative principle. At that time when the computer hardware and software technologies were not well developed or before the application of the large finite element software, this was the main method for the shock design of systems and equipment. This method is also used to determine the dimension of structural supports and fixing bolts [7].

(2) Response spectrum analysis method

A shock spectrum is a curve with each point indicating the relation between maximum absolute value of response of a series single degree of freedom spring-mass-damper oscillators with varying resonant frequencies subject to the same shock motion. The response can be displacement, velocity or acceleration.

The shock resistance analysis by response spectrum method includes two steps: modal analysis and spectrum analysis.

① Modal analysis

Considering the degree of freedom of the analyzed system or equipment as n, the undamped free vibration equation of the system is as follows:

$$[M]\{\ddot{u}\} + [K]\{u\} = \{0\} \tag{10.1}$$

where $[M]$ is the $n \times n$ mass matrix, $[K]$ the $n \times n$ stiffness matrix, $\{\ddot{u}\}$ the n nodal acceleration vector, $\{u\}$ the n nodal displacement vector, and $\{0\}$ the n zero element vector.

Let $\{u\} = \{\Phi\}\sin(\omega t + \alpha)$, where $\{\Phi\}$ is the n amplitude vector. Substitute it into Eq. (10.1),

$$\left([k] - \omega^2[M]\right)\{\Phi\} = \{0\} \tag{10.2}$$

According to linear algebra, having Eq. (10.2) to have non-trivial solution must have the following condition being satisfied:

$$\left|[K] - \omega^2[M]\right| = 0 \tag{10.3}$$

Solving the characteristic Eq. (10.3) can obtain n eigenvalues $\omega_i^2 (i = 1, 2, \ldots, n)$. Resonant frequencies of the system can be written as follows:

$$f_i = \frac{\omega_i}{2\pi} \tag{10.4}$$

Substituting ω_i into Eq. (10.2) can obtain corresponding amplitude vector $\{\Phi\}_i$, which is called the ith order mode shape vector.

The mode shape vector depicts orthogonality conditions, namely, when $i \neq j$:

$$\{\Phi\}_i^T[M]\{\Phi\}_j = 0 \tag{10.5}$$

$$\{\Phi\}_i^T[K]\{\Phi\}_j = 0 \tag{10.6}$$

When $i = j$:

$$\{\Phi\}_i^T[M]\{\Phi\}_j = M_i \tag{10.7}$$

$$\{\Phi\}_i^T[K]\{\Phi\}_j = K_i \tag{10.8}$$

For subsequent response spectrum analysis, $\{\Phi\}_i$ is normalized to satisfy the following conditions:

$$\{\Phi'\}_i^T[M]\{\Phi'\}_i = 1 \tag{10.9}$$

$$\{\Phi'\}_i^T [K]\{\Phi'\}_i = \omega_i \tag{10.10}$$

$\{\Phi'\}_i$ is called the normalized mode shape.

② Spectrum analysis

A response spectrum is a curve indicating the relation between the maximum absolute response and the resonant frequency of the structure. Response spectrum is used for linear analyses of structures. Based on different response forms, the response spectra can be divided into displacement, velocity and acceleration response spectra.

For the single point response spectrum method, the excitation is applied to all supporting locations.

For a given excitation direction, the modal participation factor is computed by the following equation:

$$\gamma_i = \{\varphi\}_i^T [M]\{D\} \tag{10.11}$$

where, γ_i is the modal participation factor of ith order, $\{\varphi\}_i$ the normalized mode shape, and $\{D\}$ the vector describing the excitation direction.

When using acceleration response spectra as the shock input, modal coefficient of the ith order is calculated as:

$$A_i = \frac{A_{ai}\gamma_i}{\omega_i^2} \tag{10.12}$$

When using velocity response spectra as the shock input, modal coefficient of the ith order is calculated as:

$$A_i = \frac{V_{ai}\gamma_i}{\omega_i} \tag{10.13}$$

When using the displacement response spectra as the shock input, modal coefficient of the ith order is calculated as:

$$A_i = U_{ai}\gamma_i \tag{10.14}$$

where, A_{ai}, V_{ai} and U_{ai} are the spectral acceleration, velocity and displacement corresponding to the ith order of mode respectively (each computed spectra value corresponds to the frequency at f_i).

Nodal displacement vector of the structure under each order of mode can be calculated according to the modal coefficient:

$$\{u\}_i = A_i\{\phi\}_i \tag{10.15}$$

Structural response under excitation at a specific direction is consist of various orders of modal displacement responses $\{u\}_i$. There are multiple combination methods to compute the spectrum response from the modes, appropriate combination method can be selected according to the structural modal frequency distribution. Mode distribution is categorized into dense and sparse according to their corresponding resonant frequencies. If the frequency difference between two adjacent modes is smaller than 10%, these modes are dense modes. Otherwise, they are sparse modes. For sparse modes, the square root of the sum of the squares (SRSS) method is used for mode combination. For dense modes, the grouping method, ten percent method, double sum method or complete quadratic combination (CQC) can be used for combination [8].

Response spectrum method is used in associate with eigen-problem solving for shock resistance analysis. According to the mode analysis results of system and equipment, the appropriate cut-off frequency is selected to ensure that the effective mass of the selected mode accounts for more than 90% of the total mass of the analytical model.

The shock resistance analysis by response spectrum method requires "design shock spectrum", that is, the curve indicating the relation between the shock acceleration (velocity or displacement) and the frequency. The design shock spectra can be obtained empirically. Such spectra should be comparable with other suitable equipment. Alternatively, the spectra can also be determined according to the analysis of full-scale ship explosion test results and the performance requirements.

(3) Transient shock analysis method

For the design of marine nuclear power plants, the shock loads are usually given in the form of shock response spectra. For transient analysis, shock response spectra should be converted into the time history [9]. The relevant technologies are briefly introduced below.

The shock response spectra are usually mixed spectra. Shock response at low-frequency band depicts constant displacement feature, at middle-frequency band shock depicts constant velocity, and at high-frequency is constant acceleration. Before shock spectra are converted into the accelerogram, they shall be converted into the same type of spectra, i.e. displacement spectra, velocity spectra or acceleration spectra. The conversion relationship is as follows:

$$S_a(\omega) = \omega^2 S_d(\omega) \tag{10.16}$$

$$S_v(\omega) = \omega S_d(\omega) \tag{10.17}$$

where, $S_a$is the acceleration response spectra; S_v the velocity response spectra; S_d the displacement response spectra; and ω the angular frequency.

Shock process usually occurs in the order of magnitude of millisecond. Structure damping can be neglected over this short period of time, hence small damping ($\xi << 1$) can be assumed. The response of a small damping system of a single degree of freedom is a narrow-band process with maximum response occurred near its resonant frequency. Therefore, in the narrow-band range, the power spectral density can be approximately replaced by the power spectral density at the resonant frequency. This relation can be expressed as follows:

$$G_{in}(\varpi_n) \cong \frac{2\xi_t \omega_n^3 S_d^2(\omega_n)}{\pi \eta^2} \tag{10.18}$$

where, ω_n is the resonant frequency, ξ_t the virtual damping; and η the peak factor.

According to the Pasval principle, power spectral density equals 1/2 power of the Fourier spectrum modulus. The amplitude $|C_x(k)|$ of the Fourier spectrum is calculated by the following equation:

$$|C_x(k)| = \sqrt{G_{\text{in}}(\omega_k)\omega_0} \qquad k = 1, 2, 3 \cdots \tag{10.19}$$

where, $\omega_0 = 2\pi / T_0$; $\omega_k = k\omega_0$; and T_0 the time span.

The above Fourier spectrum is discrete. According to the sampling theorem, in order to avoid aliasing, the number of terms of the Fourier series, N, should meet the following condition:

$$N \geq 2 f_c T \tag{10.20}$$

where, f_c is the cut-off frequency, and T the duration.

Shock process is a random process. The Gaussian equiprobable random phases are generated in the range of [0, 2π], and the amplitude of the Fourier spectrum obtained by above operation is adopted to perform Fourier inverse transformation to obtain the initial stationary Gaussian random shock accelerogram. Because of the non-stationary nature of the shock, the non-stationary envelope shall be performed for the obtained shock accelerogram. The strength envelope equation is as follows:

$$\ddot{x}_0(t) = I(t)\ddot{x}(t) \tag{10.21}$$

where, $\ddot{x}(t)$ is the stationary shock accelerogram, $\ddot{x}_0(t)$ the non-stationary shock accelerogram, $I(t)$ the non-stationary envelope function, which is also known as the shape function. $I(t)$ is defined as follows:

$$I(t) = \begin{cases} \dfrac{t}{T_1} & 0 \leq t < T_1 \\ 1 & T_1 \leq t < T_2 \\ e^{-\alpha(t-T_2)} & T_2 < t \leq T_0 \end{cases} \tag{10.22}$$

where, T_1 is the stationary starting time, T_2 the stationary ending time, T_0 the duration of shock process, α the shape function coefficient.

Commonly used envelope functions include trapezoid window, Hanning window and Laplace–Gauss window.

In order to characterize the correlation between the stationary time and the strength envelope function, the equivalent stationary time, i.e. the stationary time calculated by T_1, T_2, Ts and specific shape function, is usually used.

The peak value of shock accelerogram obtained by the above process may not equal to the spectral value at cut-off frequency (zero period acceleration) of the shock response spectrum. In this case, it is necessary to adjust the peak value of shock accelerogram. Two cases are considered herein.

In the first case, the maximum value in the shock accelerogram is smaller than the spectral value at the cut-off frequency of the shock design response spectrum. The maximum value in the shock accelerogram value is directly set equal to the spectral value at the cut-off frequency of the shock design response spectrum. The shock accelerogram values at other time instants remain unchanged. Of course, the above method is for the case where the time history values for adjacent time points in the shock accelerogram have opposite signs. Otherwise, the acceleration values aside the maximum with the same sign should be adjusted as well. This adjustment ends until the maximum time history value has opposite sign of its aside accelerations. In addition, if there are several absolute maximum values in the shock accelerogram, the above process shall be performed for every maximum.

In the second case, the maximum value in the shock accelerogram is larger than the spectral value at the cut-off frequency of the shock design response spectrum. For this case, the following equation is used for correction.

$$\ddot{x}_1(t) = \ddot{x}_0(t) + C_0 + C_1 t + C_2 t^2 \tag{10.23}$$

where, $\ddot{x}_0(t)$ is the initial shock accelerogram, $\ddot{x}_1(t)$ the adjusted shock accelerogram, and the selection principle of C_0, C_1 and C_2 is to minimize the mean square velocity. The principle for time series adjustment is consistent with the principle used for the first case.

During the artificial synthesis of shock accelerograms, shock displacement time history usually drifts, making it inconsistent with the actual situation. Therefore, appropriate drifting processing should be performed to correct the shock accelerograms. The correction should ensure zero displacement, velocity and acceleration at starting and ending time instant. In the meantime, the displacement over the entire time span should have zero mean.

The shock accelerograms obtained are converted into shock response spectra, via the Duhamel integral:

$$S_{\mathrm{d}}(\omega) = \left| \frac{1}{\omega_{\mathrm{d}}} \int_0^t \ddot{x}(\tau) \mathrm{e}^{-\xi\omega(t-\tau)} \sin[\omega_{\mathrm{d}}(t-\tau)] \mathrm{d}\tau \right|_{\max} \tag{10.24}$$

where, $S_{\mathrm{d}}(\omega)$ is the shock displacement response spectrum, $\ddot{x}(\tau)$ the shock accelerogram, ω the angular frequency, and $\omega_d = \omega\sqrt{1-\xi^2}$.

Once the spectrum is generated, check on the compatibility between generated and design spectrum should be carried out. This check ensures the error lies within the allowable engineering range. At the same time, the compatibility between the statistical characteristic parameters of shock accelerogram (mean value, Pearson coefficient, mean square error and so on) and the actual parameters corresponding to the shock accelerogram should be reviewed. If the reviews yield satisfactory results, the calculation ends with ensured quality of generated spectrum. Otherwise, the spectrum amplitude or phase should be adjusted. The amplitude of Fourier spectrum of the shock accelerogram is adjusted according to the following equation:

$$\left|C_{\mathrm{x}}^{*}(k)\right| = \left|C_{\mathrm{x}}(k)\right| \times \frac{S_{\mathrm{v}}^{T}(\omega_k)}{S_{\mathrm{v}}(\omega_k)} \tag{10.25}$$

where $|C_{\mathrm{X}}(k)|$ is the amplitude of Fourier spectrum remain to be adjusted, $\left|C_{\mathrm{X}}^{*}(k)\right|$ the adjusted amplitude, $S_{\mathrm{V}}^{\mathrm{T}}(\omega_{\mathrm{k}})$ the shock design velocity response spectrum, and $S_{\mathrm{V}}(\omega_{\mathrm{k}})$ the shock velocity response spectrum [10].

The main analysis techniques of the time history methods include the direct integration and modal superposition. Direction integration can be further divided into the implicit and explicit integration scheme. Implicit scheme includes the Newmark β method and the Wilson θ method. Explicit scheme includes the central difference method, the double-loop iteration method and the Runge-Kutta method [11].

The basic idea of the modal superposition method is to use the undamped mode shape (mode) of the system as the spatial variable and decouple the original dynamic equation by coordinate transformation. The n mutually independent equations are solved to obtain the modal displacements. System response can be obtained by the superposition of various orders of modes.

For the modal superposition method, there are two methods for nonlinear problems. One is the incremental method. In this method, the stiffness matrix of the system is modified in each time step according to the nonlinear states of the system, and the frequency and mode shape are re-modified accordingly. The other is the fictitious force method. When this method is used to solve the nonlinear problems, the mass and stiffness matrix of the system are calculated only once, and the frequencies and mode shape obtained from the updamped system free from external load are used for modal superposition. This state is exactly the original state for the dynamic analysis of the system. In the process of time history integration, the fictitious force vector is only modified according to the nonlinear states of the system. The govern equation of the modal superposition method is as follows:

$$[M]\{\ddot{u}(t)\} + [C]\{\dot{u}(t)\} + [K]\{u(t)\} = \{P(t)\} \tag{10.26}$$

where, $[M]$ is the mass matrix, $[C]$ the damping matrix, $[K]$ the stiffness matrix, $\{\ddot{u}(t)\}$ the acceleration vector, $\{\dot{u}(t)\}$ the velocity vector, $\{u(t)\}$ the displacement vector, and $\{P(t)\}$ the load vector.

The displacement vector is expressed as follows:

$$\{u(t)\} = \sum_{i=1}^{n} \xi_i(t)\{\phi_i\} = [\Phi]\{\xi(t)\} \tag{10.27}$$

where, $\{\xi(t)\}$ is the general displacement vector, which is also known as the modal displacement vector, $[\Phi]$ the mode shape vector (also known as the master modal) matrix.

The response of each order of modal is obtained by integrating n single degree of freedom vibration differential equations, and the response of the system is obtained by coordinate transformation, i.e. the superposition of responses from all orders of vibration.

For the modal superposition method, eigenvalue problem is solved first to obtain the resonant frequency and the corresponding main mode shape (master modal). It is impossible to obtain those of all orders. Generally, only several low orders of modes. The number of modes extracted is determined by the cut-off frequency. If the number of modes extracted for superposition is too few, there will be a rather large error, leading to incredible results. The amount of error depends on whether the loads applied on the system would excite the abandoned high orders of modal.

There are some limitations on the use of modal superposition method. For example, the modal eigenvector of the system shall be obtained, and this method only applies to linear systems. However, linear method is relatively low cost computationally, and there is no convergence problem. Therefore, this method is widely used in linear or some weakly nonlinear systems.

The direct integration methods are divided into the implicit explicit integration schemes. Among implicit schemes Newmark βand Wilson θ method are widely used in engineering. Direct integration method solves the differential equation for system movement by the step-by-step integration of the time domain excitation, and solve the state vectors $\{u_{n+1}\}$, $\{\dot{u}_{n+1}\}$ and $\{\ddot{u}_{n+1}\}$ at t_{n+1} by using the structural state vectors (displacement $\{u_n\}$, velocity $\{\dot{u}_n\}$ and acceleration $\{\ddot{u}_n\}$) at t_n. The basic idea of direct integration methods is that the displacement vector $\{u\}$, governed by the dynamic equilibrium equation at any time instant t can be solved via specific assumptions of the relations between displacement, velocity and acceleration at the intervals between two consecutive time instants. For structures with geometric nonlinearities, the stiffness matrix is related to the deformation of the structure, and thus the equilibrium is dependent on deformations. Assuming that the displacement vector $\{u_i\}$ is known, it is possible to obtain $\{u_{i+1}\}$ by solving the $\{\Delta u_i\}$ as per the dynamic equation. This iterative process starts from known initial condition $\{u_0\}$ of the system and repeats until the convergence occurs.

The most commonly used Newmark-β method is selected for introduction. The dynamic equilibrium equation for a structural system with multiple degrees of freedom is as follows:

$$[M]\{\ddot{u}\} + [C]\{\dot{u}\} + [K]\{u\} = \{F_a\} \tag{10.28}$$

where, $[M]$ is the mass matrix of the structure, $[C]$ the damping matrix of the structure, $[K]$ the stiffness matrix of the structure, $\{u\}$ the displacement vector, $\{\ddot{u}\}$ the acceleration vector, $\{\dot{u}\}$ the velocity vector, and $\{F_a\}$ the external acting force vector.

The basic idea of direct integration methods is that the displacement vector $\{u\}$, which is governed by Eq. (10.28) at any time instant t can be solved via specific assumptions characterize the relations between displacement, velocity and acceleration at two consecutive time instants. The basic solving process of Newmark-β method is as follows:

Within the time interval Δt from t_n to t_{n+1}, Lagrange's mean value theorem is applied to express the velocity vector at t_{n+1} as follows:

$$\{\dot{u}_{n+1}\} = \{\dot{u}_n\} + \left[(1-\delta)\{\ddot{u}_n\} + \delta\{\ddot{u}_{n+1}\}\right]\Delta t \tag{10.29}$$

Similarly, Taylor expansion for displacement is applied to express the displacement vector at t_{n+1} as follows:

$$\{u_{n+1}\} = \{u_n\} + \{\dot{u}_n\}\Delta t + \left[\left\{\frac{1}{2} - \alpha\right\}\{\ddot{u}_n\} + \alpha\{\ddot{u}_{n+1}\}\right]\Delta t^2 \tag{10.30}$$

where, α and δ are Newmark integral parameters, $0 \le 2\alpha \le 1, 0 \le \delta \le 1$, $\Delta t = t_{n+1} - t_n$, $\{u_n\}$ is the nodal displacement vector at t_n, $\{\dot{u}_n\}$ the nodal velocity vector at t_n, $\{\ddot{u}_n\}$ the nodal acceleration vector at t_n, $\{u_{n+1}\}$ the nodal displacement vector at t_{n+1}, $\{\dot{u}_{n+1}\}$ the nodal velocity vector at t_{n+1}, and $\{\ddot{u}_{n+1}\}$ the nodal acceleration vector at t_{n+1}.

The dynamic equilibrium Eq. (10.28) at t_{n+1} states:

$$[M]\{\ddot{u}_{n+1}\} + [C]\{\dot{u}_{n+1}\} + [K]\{u_{n+1}\} = \left\{F_{a,n+1}\right\} \tag{10.31}$$

Equations (10.29) to (10.31) are the fundamental equations for Newmark-β method. If state vectors $\{u_n\}$, $\{\dot{u}_n\}$ and $\{\ddot{u}_n\}$ at t_n are known, the above three equations can be solved to obtain the state vectors $\{u_{n+1}\}$, $\{\dot{u}_{n+1}\}$ and $\{\ddot{u}_{n+1}\}$ at t_{n+1}. With known initial conditions at $t = 0$, the step-by-step integration can be used to obtain the system state vector at any time.

Newmark integration parameters are selected as:

$$\alpha = \frac{1}{4}(1+\gamma)^2 \tag{10.32}$$

$$\delta = \frac{1}{2} + \gamma \tag{10.33}$$

where, $\gamma = 0.005$. Under this case, the Newmark-β method is unconditionally stable.

When applying the Newton-Raphson method for a structure with geometric nonlinearities, the stiffness matrix is related to the deformation of the structure. The equilibrium equation can be written at the deformed position. The equilibrium equation of structure reads:

$$[K]\{u\} = \{F_a\} \tag{10.34}$$

Newton-Raphson method is a linear iterative method for solving the above nonlinear equations. The iterative formula is as follows:

$$\left[K_i^T\right]\{\Delta u_i\} = \{F_a\} - \left\{F_{nr,i}\right\} \tag{10.35}$$

$$\{u_{i+1}\} = \{u_i\} + \{\Delta u_i\} \tag{10.36}$$

where $\left[K_i^{\mathrm{T}}\right]$ is the tangent stiffness (Jacobian) matrix, and $\left\{F_{nr,i}\right\}$ the elemental restoring force vector.

With known displacement vector $\{u_i\}$, the tangent stiffness matrix $\left[K_i^{\mathrm{T}}\right]$ and the elemental restoring force vector $\left\{F_{nr,i}\right\}$ can be calculated, $\{\Delta u_i\}$ is calculated by Eq. (10.35), and $\{u_{i+1}\}$ is calculated by Eq. (10.36). This iterative process starts from a known initial condition $\{u_0\}$ of the system and repeats until convergence occurs.

The calculation time and requirements for software and hardware are quite varying for nonlinear dynamical analysis methods, so does the calculation accuracy. Usually, a method with a higher accuracy requires a longer calculation time and higher software and hardware requirements.

10.2.2 Analysis Theory for Structural Stress

For the stress theory, except for the derivation of some formulas, the components in the coordinate axis directions are not used, because such components do not have direct relationship with the deformation or material strength of the structure. The components directly related to the deformation and strength of the structure are normal and tangential components, i.e. the normal stress and shear stress [12]. The maximum shear stress at a point is equal to half of the difference between the algebraic maximum and the algebraic minimum of the three principal stresses at this point. The concept of stress intensity is commonly used in stress analysis. The stress intensity, as the equivalent intensity of combined stress, is twice the maximum shear stress. That is, the stress intensity means the difference between the algebraic maximum principal stress and the algebraic minimum principal stress at a given point. The tensile stress is positive and the compressive stress is negative.

Based on the nature of stresses and the resulted failure mode, the stresses can be divided into primary stress, secondary stress, peak stress and total stress.

The primary stress means any normal stress or shear stress generated by the applied load that is necessary to meet the equilibrium rule of external force, internal force and moment. The basic characteristic of primary stress is non-self-limited. When the primary stress exceeds the yield strength of the material, it will lead to failure or at least overall excessive deformation. The primary stress shall be limited to prevent any excessive deformation failure. The primary stress is divided into primary membrane stress and primary bending stress. The primary membrane stress is further divided into overall primary membrane stress and local primary membrane stress. The overall primary membrane stress will not cause the redistribution of the loads because of yielding. The local primary membrane stress is the membrane stress generated by pressure or other mechanical loads, and related to the excessive deformation that is generated by the unlimited discontinuity effect when the load is transferred to other parts of the structure.

The secondary stress is the stress caused by the constraints of adjacent materials or the constraints of the structure itself. Its basic characteristic is self-limited. Local yield and micro-deformation can cause the stress partially. This stress does not cause the expected failure if applied only once. The overall thermal stress and the bending stress at the discontinuous points of the overall structure belong to the secondary stress.

The peak stress means the stress increment added to the sum of primary stress and secondary stress because of local discontinuity or local thermal stress including stress concentration effect. Its basic characteristic of peak stress is that it does not cause any serious deformation and, instead, it is harmful only when it is the source of fatigue crack or brittle fracture. The stress that is not within the highly concentrated local region and does not cause obvious deformation also belongs to peak stress.

The total stress means the sum of primary stress, secondary stress and peak stress. It is important to identify each type of stress that acts alone for determining appropriate stress limits.

The design by analysis method is mainly used to analyze and evaluation for the structures of nuclear power plants. The analysis and evaluation is mainly to evaluate the equivalent stress intensity and the equivalent stress. The stress analysis and evaluation of nuclear power plant are based on the Tresca theory. The equivalent stress based on Tresca theory in the specifications is called the stress intensity.

The stress state at a certain point is represented by a set of components containing three normal stresses and three shear stresses. On a set of principal axes, the stress components are principal stresses σ_1, σ_2 and σ_3.

$$S_{12} = \sigma_1 - \sigma_2 \tag{10.37}$$

$$S_{23} = \sigma_2 - \sigma_3 \tag{10.38}$$

$$S_{31} = \sigma_3 - \sigma_1 \tag{10.39}$$

The stress intensity S at this point is the maximum absolute value of S_{12}, S_{23} and S_{31}.

In a given cross section, the membrane stress intensity and the membrane with bending stress intensity shall be determined by the average value or linear value (i.e. the linearization) along the cross section for stress analysis.

The stress distribution on the cross section is obtained by the stress calculation of the structure. The stress distribution is represented by the local coordinates on the cross section.

The bending stress is obtained by subcontracting the membrane stress for each component in the linear stress.

For the specified cycling stress obtained by steady state and transient analysis, the stress alternate amplitude shall be calculated at any time point.

In the design analysis of marine nuclear power plants, besides the intensity analysis according to conditions and stress classification, the fatigue analysis shall also be performed for the structures since the components are exposed to transient conditions for various high temperature and high pressure variations. The fatigue refers to a process in which a material undergoes a local permanent structural change at a certain point or points under the cyclic stress or cyclic strain, thus forming a crack or fracture after a certain number of cycles. The analysis methods mainly include infinite lifetime design method, finite lifetime design method, local stress-strain design method, damage tolerance design method and fatigue reliability design method [13].

In the design of nuclear power plant, the nominal stress fatigue design method is usually used. The nominal stress fatigue design method takes the nominal stress as the basic design parameter and takes the *S-N* curve as the main design basis. This method can be divided into infinite lifetime design method and finite lifetime design method. The infinite lifetime design method requires that the components do not suffer from any fatigue failure within the infinite service time, that is, the nominal stress is less than the endurance limit of the material. The finite lifetime design method requires that the components do not suffer from any fatigue failure within a certain service time. The nominal stress and cycle index in all possible loading cycles within the service life and the *S-N* are used to calculate the fatigue usage factor, and judge whether the total fatigue usage factor is less than 1 and whether the structural design meet the fatigue usage requirements.

The *S-N* curve reflecting the fatigue strength characteristic of material is obtained from cyclic data of uniaxial strain. The stress value is obtained by multiplying the forced strain amplitude (semi-range) by elasticity modulus. The least square method is used to process the logarithm of stress value to obtain a smooth experimental curve. This experimental curve can only represent the fatigue performance of standard smooth specimens. The dimensions, shapes, and surface conditions of the actual components are various and quite different from those of the standard specimens. In order to consider the factors such as shape factor, size effect, surface machining effect, average stress effect and load condition effect, the *S-N* curve obtained from the test shall be processed as follows: a. to divide the stress value by a coefficient of 2; b. to divide the cycle index by a coefficient of 20 and perform the lower envelope before the fatigue evaluation. It should be noted that these coefficients are used to consider various effects such as environment, size or result dispersion, not safety margins. The *S-N* curve in the specifications has usually been processed accordingly and can be used directly in the design.

The fatigue strength reduction coefficient is used to consider the influence of local structural discontinuity (stress concentration) effect on the fatigue strength and shall be considered in the fatigue analysis. The specifications usually give the fatigue strength reduction coefficient under certain special conditions. The stress concentration factor may be used in case of the lack of experimental data.

Calculation of cumulative fatigue damage coefficient:
The capability of a structure to withstand a specified utilization and operation cycle without generating fatigue failure shall be determined according to the specified method and procedure. The allowed number of the stress cycle shall be sufficient for the specified working load at each point. It is only necessary to consider the stress difference caused by the utilization and operation cycle specified in the design.

For each *S* alternation 1, *S* alternation 2,... and *S* alternation *n* values, if this type of cycle works individually, the applicable design fatigue curve shall be used to determine the allowable maximum times of repetition, which are called N_1, N_2, ... and N_{n} respectively.

For each type of stress cycle,$U_1 = n_1/N_1$, $U_2 = n_2/N_2$ ... and $U_n = n_n/N_n$ are used to calculate the damage coefficients U_1, U_1, ... and U_{n} of the cycle. The cumulative damage coefficient *U* shall not exceed 1.0.

When the vessel shell is subject to a constant internal pressure and an alternating temperature difference, there shall be a progressive accumulation of plastic strain under the combined effect of mechanical stress and alternating differential temperature stress. This effect is called thermal ratcheting effect. As a result of the thermal ratcheting effect, there may be large deformations, that is, the increase of deformation per cycle is almost equal. In order to prevent the periodic increase of the diameter, the maximum cyclic thermal stress shall be limited in a partially ax symmetric shell under a stable internal pressure load.

If the ratcheting effect terminates after several load cycles and the structure tends to be stable, the subsequent structural response is elastic and there is no increasing non-elastic deformation. The elastic stability refers to the situation in which the subsequent response is elastic.

The structural stress analysis is mainly divided into elastic analysis, simplified elastic-plastic analysis, elastic-plastic analysis and experimental stress analysis. The analysis cost and technical difficulty of these methods increase in the order given above, so the elastic analysis method is preferred.

Elastic analysis theory: In the analysis and calculation of stress, it is usually assumed that the material is characterized by continuity, complete elasticity, isotropy, homogeneity, infinite divisibility and unchanging performance, and that the material follows the Hooke's law. Under this assumption, the stress is directly proportional to the strain. Meanwhile, it is assumed that the displacement and deformation are minor, that is, the displacement at any position of the entire structure is far smaller than the dimensions of the structure itself, and the strain and the rotation angle are far less than 1. In this way, the dimensions of the structure before the deformation can be used as the dimensions after deformation without causing significant errors. During the observation of the deformation or displacement of the structure, the second power or product of rotation angle and strain can be neglected [14]. These assumptions simplify the equations in elastic mechanics into linear equations. In the actual structures, these assumptions are not absolutely accurate. The structural materials are usually an aggregate of crystals, fibers or bonding particles. The arrangement may be either irregular or regular. In case of irregular arrangement, the material is anisotropic if the part studied is rather large compared to the constitutional unit. In addition, when the part studied contains only a part of single crystal, fiber or other unit, its performance is likely to be different from the performance of a larger part of the aggregate of these units during the further subdivision of this part. Finally, very careful experiments indicate that all materials may be inconsistent or deviate from the Hooke's law under the effect of any stress, no matter how small the stress is [15].

Although these actual conditions impose some limitations to the traditional stress analysis methods and thus shall often be considered, the stress and strain formulas derived by mathematical methods according to above assumptions can used to obtain satisfactory results for most mechanical problems of marine nuclear power plants. Especially within the proportional limit, the Hooke's law can be considered accurate. Although the proportional limit often cannot be clearly defined, it can be specified rather accurately for most materials. Therefore, a fairly specific elastic limit can also be determined. In most cases, the elastic limit is very close to the proportional limit, so it is not necessary to make a distinction. It can be generally assumed that the elastic properties of most metal materials used in the nuclear power plants do not change with stress when the stress is lower than the nominal proportional limit [16].

Simplified elastic-plastic analysis theory: In the elastic analysis, the elastic-plastic strain correction factor is introduced to correct the calculation results. The elastic-plastic strain correction factor is the ratio of the actual strain amplitude to the strain amplitude determined by the elastic analysis theory. According to the forms of loads, the elastic-plastic strain correction factors are divided into the one caused by mechanical loads (including dead weight, pressure, shock, tilt and swing) and the one caused by thermal loads. At the same time, the factor is related to the magnitude of stress change, the material types, and the maximum operating temperature.

Elastic-plastic analysis theory: The limit analysis indicates that when the load applied on a structure composed of perfectly plastic materials reaches a certain limit load value, the deformation of this structure will increase limitlessly. The limit analysis method can estimate this load. For a given load, if the stress distribution satisfies the equilibrium law and the stress at any point does not exceed the yield strength of the material, the given load is less than or equal to the limit load. This is the lower bound theorem for limit analysis. This theorem can determine the lower bound of the limit load. It is necessary to analyze the excessive deformation, the elastic, elastic-plastic or plastic instability, the progressive deformation and fatigue. The actual (elastic and plastic) properties of the material are considered in the analysis, and the overall properties of the part under discussion are analyzed by cyclic elastic-plastic calculations. In addition, if the plastic adaptability conditions are met in sufficient number of cycles, the design meets the requirements. In addition, the deformation before the adaptation shall not exceed the given limit [17].

Experimental stress analysis: When the theoretical stress analysis method is not sufficient to determine the components subject to load or stress corresponding to the damage types considered, or when there are no available design criteria, the experimental stress analysis shall be performed to determine the load and stress. The experimental stress analysis mainly includes the control stress determination test, the limit load determination test and the fatigue test and destructive test for evaluating cyclic loading. It mainly includes loading, strain measurement and result analysis. The strain measurement requirements include the strain meter requirements and the determination of strain measurement or deformation measurement model. If the model can show the effect of external loads on the structures, both 2D and 3D photo elastic experiments are acceptable. The loading includes the gradual application of pressure, temperature and mechanical loads to simulate the enveloping conditions of loads that the structure may bear during the entire lifetime. The analysis of results includes the determination of the stress, the determination of the scope of stress analysis and the determination of the limit load criteria.

10.3 Main Methods for Mechanical Analysis

10.3.1 Theoretical Analysis

The variation method for the problems in elasticity, also known as the energy method [18], is divided into the displacement variation and the stress variation. It is closely related to the deformation potential energy of the elastic objects. The starting point of the displacement variation method aims at using displacement variation equation to derive the principle of minimum potential energy, and the starting point of the stress variation method aims at using stress variation equation to derive the principle of minimum complementary energy. The principle of minimum potential energy states that under a given external force, for all displacement distribution possibilities that meet the boundary conditions, the actual displacement is the one that minimize the potential energy. If the displacement component expressions are set to meet the boundary conditions, there are several parameters to be determined. The process of solving these parameters by displacement variation equation is called the displacement variation. The principle of minimum complementary energy states that in all stresses meeting the equilibrium differential equation and stress boundary conditions, the actual stress combination is the one that minimizes the complementary energy. If the stress component expressions are set to meet the differential equation and boundary conditions, there are several parameters to be determined. The process of solving these parameters by stress variation equation is called the stress variation.

Because of the complexity of mechanical analysis of marine nuclear power plants, the energy method is often used in the theoretical analysis. That is, according to the principle of minimum energy of the system, the displacement functions field is assumed to analyze various types of energy of the system, the total potential energy is determined expressed via the assumed form of displacement field., the Then variation is performed for the total potential energy, and relevant parameters are obtained according to the extremum principle. When time parameter is introduced into the displacement function, the momentum conservation can be used to study the response of the structure at any moment. The use of energy method for marine nuclear power plants mechanical analysis usually depends on experiments to determine proper failure modes. Mathematical models and energy equations are established based on these tests.

Analytical method: Based on the basic theory and equations of mechanics, analytical method solves the mechanical problems with known initial and boundary conditions.

Empirical method: This method is based on dimensional analysis of mechanics and statistical analysis of experimental results. The researches with empirical approaches are generally limited to specific kinds of mechanical problems under certain conditions. The research purpose aims at giving the govern law in the form of empirical formulas that characterize the relations between loads and target characteristic parameters (such as material parameters and geometric parameters). Empirical method can only present the final results of various mechanical problems while unable to reveal the intermittent derivation process.

10.3.2 Finite Element Method

In engineering practices, due to the complexity of differential equations and the difficulty in determining the accurate boundary and initial conditions, it is generally impossible to obtain exact solution. In order to solve this problem, it is often necessary to find an approximate solution by means of numerical methods. The numerical solution is the analytical solution sought at discrete points (nodes). The numerical method is divided into two kinds: finite difference method and finite element method. In contrast, the finite element method uses the integration method rather than the differential method to establish the algebraic equation set of the system. Moreover, finite element method uses a continuous function to approximately describe solution within an element. Since the boundaries of the internal elements are continuous, the solution to the entire problem can be connected or assembled by individual solutions. Therefore, the finite element method is widely used to solve the practical problems in complex structural engineering [19].

The finite element method is an effective method for solving the static and dynamic problems of complex structures. A large number of large-scale finite element software has been available for engineers. Since excessively coarse mechanical models will lead to serious loss of accuracy, numerical computation often unable to compensate for the loss. To cope this, the finite element modeling technology is constantly improving. With the development of computer technology, the scale of the model is increasing [20].

The main way for the mechanical evaluation of marine nuclear power plants is the design by analysis, and the finite element numerical simulation method is the core of design by analysis. When applying the finite element method, a continuous elastic object is divided into finite elements, displacement method, force method or the mixed method can be used for the solution. The finite element method is a comprehensive discipline combining the theories of mechanics, mathematical modeling, computational techniques, computer programming and software and hardware technology. With the rapid development of computer software and hardware technology, it is possible to solve large-scale problems quickly and accurately. Hence finite element method has become an indispensable method for marine nuclear power plants design. The accuracy of numerical calculation mainly depends on the following aspects: a. element type and size of the physical model, b. boundary, interface and coupling setting of the finite element model; c. accuracy of constitutive equation of materials, which also includes the conformity between the material constitutive equation used and the actual material constitutive equation; d. failure criteria selection, which includes the goodness of fit between the failure criteria and the actual structural failure conditions; and e. the effectiveness and accuracy of solving algorithms. At present, many kinds of finite element algorithms have been developed. In early stages, such algorithms were based on the variation principle, and the widely used methods were the Laplace method and the Eulerian method. Later, the Galerkin method and the least square method were developed. Different algorithms result in various calculation accuracies. The analysts should reasonably select the appropriate algorithm according to the structure, load type and accuracy requirements.

Numerical simulation can theoretically solve all complex mechanical problems, but this statement is premised on available mathematic description in advance. Otherwise, programming cannot be realized without mathematical descriptions. At the same time, the scale of the mechanical analysis model and the complexity of the problems definitely rely on the computation capacity of the computer used to perform the numerical simulation. Currently, existing analysis software can solve most transient nonlinear problems with the functionalities to calculate the transient stress and strain fields, instantaneous displacement, velocity and acceleration of the structure in different media, to analyze and determine the failure of strain, stress and stability, and to analyze the fracture failure and simulate its propagation process.

10.3.3 Experimental Research Methods

Experimental research methods study the mechanical quantities of the structure such as stress, strain and displacement by using the fundamental physical principles of light, electricity and so on. Experimental approaches combine the application of measuring techniques, mechanics theories and mathematical derivations [21]. Experiment research can obtain relatively close-to-real results. It is the basis of theoretical analysis and numerical methods. Most laws of mechanics can be summarized from a large number of experimental data. Meanwhile, it is necessary to use the experiments to verify whether the results obtained by applying the theoretical analysis meet the reality. Therefore, in principle, the importance of the experiments shall not be underestimated. However, on the one hand, the scale of the experimental models is often limited, full scale experimental research is usually impossible for large-scale systems. In addition, there are some conditions that are difficult to materialize experimentally. On the other hand, the experiments often consume a lot of manpower and material resources, technical level of instruments and the measuring techniques often limit the experiments. These difficulties hamper experimental research [22].

For the mechanical problems of marine nuclear power plants, the experimental research is one of the important means. It can effectively simulate the response of the system and equipment, such as operating conditions, stress-strain, displacement, velocity and acceleration, and intuitively judge whether the design of system and equipment meets the relevant requirements. However, experiments are costly with the entire project cycle is quite long, and there are some loads or load combinations cannot be simulated in the actual operation. Therefore, this method is usually used only when the theoretical methods cannot solve the problem.

The experiments in mechanics mainly include structural characteristics experiments, structural integrity experiments and functional verification experiments.

Structural characteristics experiments are set to measure the certain performance indicators of the system, equipment or components, such as stiffness characteristics, resonant frequencies, mode shapes, damping parameters and mass parameters. The stiffness characteristics are usually measured in the actual structure in room temperature. When applied to finite element analysis or structural design, computational parameters are derived from test results follow the principle of similarity. The resonant frequencies, mode shapes, damping parameters and added mass parameters shall be measured separately in air and water according to the actual working state. In addition, appropriate theoretical analysis and calculation shall be performed to determine the relevant parameters.

Structural integrity experiments include the validation of structure strength under shock, flow-induced vibration, and drop during transportation. These experiments also include fatigue examine and hydrostatic tests. Shock resistance experiments are performed with light shock machine, medium shock machine or floating shock platform. Flow-induced vibration experiments mainly involve examine of reactor internal, fuel assemblies, steam generator heat transfer tubes and main pipes (including pumps and valves). Flow-induced vibration experiments mainly simulate the response of the structures under the fluid effects, and judge whether the structure will fail within the design lifetime according to the calculation and analysis of experimental results. Flow-induced vibration experiments are usually performed on scaled models under normal and representative high pressure and temperature. The parameters to be measured mainly include amplitude, acceleration, strain and pulsating pressure. In order to test the fatigue performance of the structures under the effect of flow-induced vibration, durability experiments shall be performed according to the specified time (number of times). The vibration experiments of piping system are usually performed at the mooring state with the measurement of vibration acceleration, velocity and displacement at the key positions. The experimental results are analyzed to judge whether the piping system design meets the relevant requirements. The drop during transportation experiments are set to simulate the drop situations that may occur during the transport. The relevant safety requirements, such as no exposure of radioactive substances, shall be met during a drop event. The fatigue experiments are set to test the limit use cycles under the design load conditions for the parts of which the fatigue analysis obtained from computational approaches do not meet the requirements in the specifications. Experiments are therefore set to test the actual bearing capacity of the structures and judge whether the structures meet the fatigue use requirements. The hydrostatic experiments include the hydrostatic experiments for equipment and system. The purpose is to test whether the equipment or system meets the relevant stress, deformation and sealing performance requirements under corresponding experimental pressure.

For equipment with functional capability and motility requirements, such as control rod drive mechanism, pumps and valves, the functional tests shall be performed to verify that the equipment has the capacity to perform actions under operating and accident conditions specified in the design.

10.4 Main Content of Mechanical Analysis

10.4.1 Load Distribution of Systems and Equipment

The purpose of load distribution computation of systems and equipment in marine nuclear power plants is to obtain the response behaviors of systems and equipment under internal or external loads, thereafter provides the load inputs for structural integrity and functional evaluation of equipment and pipes in the systems. It also provides the input loads for structural design and tests. Load distribution computation model usually contains a very large number of equipment and components, so the model is relatively rough and it is not necessary to consider the influence of the local structures, but the connection relationship between the components should be carefully simulated.

The response of equipment and components at equipment nozzle, support, connecting part are of primary concern. Main physical variables under investigation at these locations include local displacement, overall dynamical response, displacement time history, velocity time history, accelerogram, etc.

The load distribution mainly includes load distribution of reactor coolant system, and the load distribution of primary equipment and internals.

For the load distribution of reactor coolant system, the focuses are the interaction between pipes and supports and the interaction between supports and ship structure. Therefore, the equipment is sharply simplified in the models. Generally, it is only necessary to consider the mass and stiffness characteristics of the structures, and the mutual connection relationships, the support stiffness and the support clearances shall be accurately modeled. In the support stiffness calculation and simulation, there are various types of supports in the reactor coolant system, and the characteristics of these supports shall be accurately carefully simulated for each type during the system load distribution. In the analysis, usually, 3D solid model or plate-shell model is should be established according to mimic the support structure, and the element load method is used to calculate the stiffness value of the support structure. For system-level model, spring, beam, rod and abstract stiffness matrix elements are used for simulation. If the mass characteristics are not negligible for the system load distribution, equivalent density, lumped mass or mass matrix elements shall be used to simulate the mass characteristics. Pipes are simulated by straight pipe segments and elbow elements. Pipe or beam elements are usually used to simulate the stiffness characteristics of attachments such as valves, and the lumped mass elements shall be used to simulate the mass and calibrate the center-of-gravity. After the establishment of models for all equipment and pipes, these models are assembled to create a complete system load distribution computational model.

For load distribution of equipment and internals, the purpose of analysis is to obtain the loads on the internal structures of the equipment. The simulation of connection relationship between the internal structures of the equipment is particularly important. Pipes connected to the equipment can be simplified or decoupled by related principles. The model establishment for equipment and internals mainly includes the following aspects: use the beam and pipe elements to simulate the stiffness and mass characteristics of relatively geometrically regular structures; calculate the contact stiffness of possible contact parts under various loads; use the spring elements to simulate the simplified contact relations; calculate the stiffness and mass matrix of the analyzed equipment in the system loops; use the spring elements or stiffness matrix elements to simulate the supporting effect of the loops against the equipment, and use the mass matrix elements to take the mass effect of the loops into account; include the elements that can accurately simulate the equipment support characteristics in the analysis model, and generally use the stiffness matrix elements, the spring elements and the rod-beam elements for simulation; use the lumped mass elements, the equivalent density elements and the fluid-solid coupling elements to simulate the fluids in the equipment, generally use the lumped mass elements or the equivalent density elements to simulate the weakly coupled areas, and use the fluid-solid coupling elements to simulate the strongly coupled areas; and assemble such elements into a complete load distribution model for equipment and internals.

Loads under consideration mainly include dead weight, inner pressure, transient pressure, transient temperature, shock, tilt and swing. The analysis methods are as follows: static method for dead weight, thermal expansion and pressure load analysis; response spectrum method for shock analysis, and, if necessary, equivalent static method or time history method for simplified shock analysis; harmonic response method or load factor method for swing analysis; and static method for tilt analysis. The elastic analysis method is usually used for the load distribution of systems, equipment and internals.

10.4.2 Stress Analysis for Structures and Components

The purpose of equipment stress analysis is to confirm that the equipment will not fail under the loads or load combinations of different loading conditions. In order to meet this requirement, the specified stress criterions are defined for different loading conditions. In this process, some important parameters are determined by analysis, and compared with the allowable values under the corresponding loading conditions. Thus, whether the design of structure meets the mechanical requirements can be determined.

For the equipment structure and components, the stress analysis mainly includes structural stress analysis, fatigue analysis, thermal ratcheting analysis, sealing analysis, instability analysis, fracture mechanical analysis and deformation analysis. For the pipes, the stress analysis mainly includes pipe stress analysis, fatigue analysis, thermal ratcheting analysis, displacement calculation of special position, support load calculation, valve load calculation, nozzle load calculation, valve acceleration calculation, operability evaluation of active component and functional capability evaluation of pipe.

The elastic stress analysis method is normally used. If the elastic analysis cannot meet the requirements, the elastic-plastic stress analysis method or the limit analysis method is used. The experimental stress analysis method can also be used, if necessary.

The stress analysis of structures and components mainly includes finite element modeling, application of boundary conditions and loads, analysis and calculation, post-processing of results and evaluation according to the codes.

① Finite element modeling: considering the characteristics of the structure and load, the 2D or 3D finite element models can be adopted in the stress analysis of the structures and components,. The models shall simulate the actual conditions of the structure in detail, especially the structural discontinuities that are likely to cause stress concentration, so as to avoid stress distortion. For the areas with the loads applied, the extension shall be performed according to the Saint-Venant's principle in order to eliminate the influence of boundary effect on the evaluated areas under the premise of meeting the static equivalent conditions.

② Application of boundary conditions and loads: the application of boundary conditions is after the modeling. The boundary conditions shall accurately represent the relationship between the analyzed zone and the zone connected with this area. The load conditions shall accurately represent the actual load conditions of the analyzed zone.
③ Analysis and calculation: after the application of boundary conditions and loads, suitable methods shall be selected for the analysis and calculation. Thus, the physical values such as stress, strain and displacement under different loading conditions can be obtained.
④ Post-processing of results: during the post-processing of results, the evaluation path shall be reasonably determined based on the stress conditions in the analyzed zone. The selected evaluation path shall include the most dangerous section in the analyzed zone. The stresses under different loading conditions shall be linearized along the evaluation path to determine the maximum stress intensity under different loading conditions.
⑤ Evaluation according to the code: after the determination of the maximum stress intensity, according to the code class, loading condition and stress intensity category, the maximum value shall be compared with the corresponding limitation to determine whether the analysis satisfied the corresponding requirements.

The analyzed loads mainly include internal or external pressure, weight of components and contents, static and dynamic loads due to the fluid under the analyzed conditions, loads at the outside zone but acting on the boundary of the area, shock loads, tilt loads, swing loads, reactive forces of supports and loads due to the pressure and temperature transients.

The analyzed conditions include design conditions, operating conditions and hydrostatic test conditions.

The loads to be considered under different conditions are different. The loads to be considered under the design conditions include design pressure, design temperature, dead weight, external mechanical loads and shock loads. The loads to be considered under operating conditions include pressure transients, temperature transients, dead weight, external mechanical loads, loads induced by hull tilt and swing, vibration-induced loads and shock loads. The loads to be considered under hydrostatic test conditions include test pressure and test temperature.

Different conditions shall meet their corresponding criteria.

(1) Support stress analysis

According to the structural forms, the supports are categorized into plate-shell supports, linear supports and standard supports. According to the classes of equipment supported, the supports are categorized into Class 1, Class 2, Class 3 and Class MC supports.

The support analysis methods include 3 types: analytical method (including maximum shear stress theory and maximum stress theory), experimental stress analysis method and rated load method. The Class 1 plate-shell supports are analyzed by the elastic analysis method under the maximum shear stress theory. The Class 2 and Class

3 plate-shell supports are analyzed by the elastic analysis method with the maximum stress. The Class 1, Class 2, Class 3 and Class MC linear supports are analyzed by the elastic analysis method of the maximum stress theory. If the components and their connections are subject to high-cycle fatigue effect, the limit analysis method shall be used for analysis. The standard supports are analyzed by corresponding methods for plate-shell supports or linear supports and the rated load method.

In the analysis and evaluation, besides the stress limits under various conditions, it is also necessary to consider critical yield, anchor displacement, stability and stress caused by the movement at anchor point. The analysis methods include elastic analysis method and the limit analysis method. Corresponding standard supports are usually evaluated by comparing the bearing load with the allowable load.

(2) Sealing analysis

In the sealing analysis, it is mainly required to simulate small elastic-plastic deformation, elastic-plastic contact, steady and transient temperature field calculations and thermal contact calculations. The incremental forms of the mixed contact finite element method are usually used to solve the contact problems, and the Newton-Raphson method is used for the iterative solution of material nonlinearity (i.e. small elastic-plastic deformation problems). The coupling problems for material nonlinearity and boundary nonlinearity are solved under the same incremental loading. The plastic correction iteration and contact iteration loop is nested to solve the elastic-plastic contact problems. In the analysis, it is necessary to simulate the bolt preloading loads, solve the bolt deformation, and calculate the deformation, clearance, resilience force and bolt load distribution of nut bottom and flange, so as to obtain the loads and deformations for the actual state of the structure. In the analysis of transient temperature fields, the finite element discretization shall be performed for spatial dimensions, the same set of element forms and interpolation functions for the elastic-plastic analysis shall be used, and the Galerkin method in the weighted residual methods shall be used to establish the finite element formulation. The finite difference discretization shall be performed for the time dimensions, and the control parameters shall be added for calculation and analysis. Considering the nonlinearity of the contact heat transfer, the difference in the orders of stress functions will result in errors. Generally, the temperature field and the stress field are coupled for solution, so that the influence of the transient temperature field on the performance of the sealing structure during the temperature increasing and decreasing can be fully considered.

The separation quantum or specific seal pressure of the sealing surface is calculated by the finite element and compared with the corresponding limit. When the separation quantum is used as the limit, if the calculated separation quantum is smaller than the limit, the sealing performance is considered to meet the requirements. When the specific seal pressure is used as the limit, if the calculated specific seal pressure is larger than the limit, the sealing performance is considered to meet the requirements.

(3) PTS analysis

Pressurized thermal shock (PTS) analysis of the reactor pressure vessel: there is a thermal shock at the inner wall of the reactor pressure vessel when the coolant is injected into the vessel. At the same time, the inner wall of the vessel is subject to high pressure. Therefore, this transient is called the pressurized thermal shock (PTS). The content of PTS analysis is as follows:

First, the analyzed area shall be determined by the analyzed object. There are usually three (3) areas, i.e. the safety injection nozzle, the shell weld and the core shell. The initial crack shape, length and depth are assumed according to the code.

Determination of the fracture toughness: the core shell and the shell weld are exposed to the neutron irradiation during the lifetime, and their toughness will decrease with the increase of irradiation. The nil ductility temperature RT_{PTS} of the material shall be calculated.

In order to obtain the crack size at the end of lifetime, the fatigue crack propagation analysis shall be performed with the assumed crack. The analysis includes material fracture toughness calculation, stress intensity factor calculation, fatigue crack propagation calculation and fast fracture mechanical analysis. The normal stress calculation is first performed for the stress intensity factor calculation. The normal stress refers to the stress perpendicular to the plane of the crack. It is the stress generated by all loads under the analyzed conditions. The analysis can be divided into two parts: thermal stress analysis and pressure stress analysis. The thermal stress analysis includes the calculation of transient temperature gradient according to the temperature transient curve, the temperature of the mixed zone and the heat transfer coefficient distribution, and the calculation of the transient thermal stress distribution according to the obtained temperature gradient. The pressure stress analysis is to calculate the pressure stress distribution according to the transient pressure curve. The stress vectors of the thermal stress and the pressure stress changing with time are superimposed to obtain the total stress.

For vessels and pipes that may be subject to external pressure, the instability analysis or buckling analysis shall be performed. The eigenvalue method is usually used to calculate the critical load and the buckling mode of the structure, so as to evaluate whether the structure will suffer from buckling instability.

(4) Pipe stress analysis

The pipe stress analysis can be categorized into primary pipe stress analysis and high-energy auxiliary pipe stress analysis.

For the primary pipe stress analysis, the load distribution of reactor coolant system is used to obtain the loads, and the equations for class 1 pipe are used to evaluate the stress. If the evaluation results cannot meet the requirements according to the criteria of class 1 pipe, the stress evaluation can be performed according to the method of equipment stress evaluation.

For the high-energy auxiliary pipes, special software can be used to generate a model containing the analyzed pipes, pipe accessory, pipe supports and other pipes and equipment that are connected to them but do not meet the decoupling criteria.

During the evolution of the model, efforts shall be focused on the positions and types of welds, because they directly affect the stress index and calculation results. The length of the element in the model shall not be too short or exceed the limits determined according to the elasticity modulus, bending modulus, the line mass of pipe and the cut-off frequency. Various kinds of loads are applied on the analyzed pipe. The analysis and calculation are performed, and the responses under different loads are combined by the specified method.

According to the method for corresponding pipe level and load conditions, the stress calculation results under different conditions are evaluated. For the active components in the piping system, the operability evaluation shall be performed. The active components are actuated by external inputs such as trigger, mechanical motion, power source and can actively influence the working process of the system. It is necessary to ensure the integrity and the operability of active components, especially the operability during and after the accident condition. For the active safety class 1 pumps and valves, the stress limit level shall be increased. For the pipes with functional capability requirements, the functional capability evaluation shall be performed for the pipes by the methods similar as the operability evaluation of active components, and the stress limit level shall be increased.

In order to meet the requirements of the analysis for equipment nozzles, supports and valves, the nozzle loads, support loads, valve loads and valve accelerations are outputted after the analysis of high-energy pipes. If necessary, the displacements of special nodes are outputted as the design input as well.

10.5 Analysis and Evaluation

10.5.1 Load Distribution of Systems and Equipment

Due to the complexity of structures and the diversity operating conditions of marine nuclear power plants, it is necessary to establish an overall finite element model for the systems and equipment consisting of rods, beams, springs and lumped masses, and select proper methods and tools to calculate the load distribution under the effect of loads such as shock, tilt, swing, dead weight, pressure and temperature transients.

(1) Model establishment: establishing the finite element model is time-consuming, model feasibility directly relates to the result correctness, calculation time and post-processing workload. Therefore, modeling the system is the key to the entire load distribution calculation. The job name, analysis title and unit definition shall be given, and the preprocessor shall be used for the definition of element types, real constants of elements, material characteristics and geometric models. As the first part of the file name generated in the analysis, the job name is used to identify the job and ensure that the file will not be overwritten. The analysis title will be included in the graphic display and all solution output.

(2) Unit definition: a lot of software does not specify the system units for analysis. Before the analysis, the users can define the units according to specific needs. It is better to use the SI units in the calculation of load distribution of systems and equipment, so as to avoid confusion of the units.
(3) Definition of element types: in the finite element software, there are usually many types of elements available, and each element type has a specific number and a prefix for identifying this element type. The element type determines the number of degrees of freedom of the element, two-dimensional space or three-dimensional space of the element and the scope of application of the element. The element types commonly used in load distribution calculations are beam, rod, pipe, spring, lumped mass, stiffness matrix and mass matrix elements.
(4) Definition of real constants of elements: the real constants of elements depend on the characteristics of the element type, such as the cross section characteristics of beam and rod, the stiffness of spring and the concentrated mass of lumped mass elements.
(5) Definition of material characteristics: material characteristics are classified into linear, nonlinear, isotropic, orthotropic, inelasticity, time-invariant or with time-dependent categorizes. The elasticity modulus, Poisson's ratio, density and coefficient of thermal expansion are usually used in the calculation of load distribution within linear range.
(6) Establishment of a geometric model: during the establishment of a geometric model, it is usually necessary to determine the coordinate system, working plane, coupling relation, constraint equations and so on. For the calculation of load distribution of systems, equipment and internals, it is recommended to prepare such modeling information in the form of command files,which facilitates the modification and inspection.

During the establishment of load distribution models for systems, equipment and internals, it is recommended to adopt the modular modeling technology. The essence of modularization is to decompose equipment and pipes from the system into several modules and perform the segmentation management of nodes, elements, material characteristics and real constants. All modules are assembled into a complete system-level model once establishment of all modules are complete. The advantage of this method is that if the design status of the equipment changes, it is only necessary to modify the module subject to the change, thus improving the efficiency and quality.

(7) Loading and solving: during this phase, it is first necessary to define the analysis type and analysis options. The analysis type can be selected according to load conditions and responses to be calculated. Typical analysis options include calculation method selection, stress hardening on/off and Newton-Raphson settings. The loads mainly involve constraint equations, forces, surface distributed loads, body loads, inertial loads, coupled field loads and response spectra. Most of these loads can be applied to solid models (key points, lines and planes) or nodes/elements. The load step option shall be defined, which refers to the option to change the load step, such as the number of substeps, the constraint time of load step, and the output control. During the solution, the program obtains the

model and load information from the database and performs the calculation. The calculation results are written into the result file.

(8) Post-processing of results: once the calculations are completed, the postprocessor can be used to view and process the calculation results. The post-processing of results is divided into steady post-processing and transient post-processing. The steady post-processing can view the results at each substep (time step) of selected parts from the entire model, obtain contour line display, deformation shapes and list data for inspecting and interpreting the analysis results. Moreover, error estimation, load condition combination, path operation for stress evaluation can be carried out during post-processing. The post-processing of transients enables to view the results for a specific point of the model at all time steps, obtain the relation curve and list of result data versus time (or frequency), and perform various operations on the result data, such as addition, subtraction, multiplication, division, derivation and integration. During the post-processing, it is better to generate a command file to record the process, so that the post-processing of results can be performed again when design status changes and that the design flaw can be discovered and treated timely.

Once load distribution calculation model is established for the systems, equipment and internals, static analysis and modal analysis shall be performed for the model to validate the modeling. The results of static and modal analysis are used to judge the rationality of the model and, if necessary, model adjustment should be carried out until the model is validated.

Before load distribution calculation under thermal expansion, temperature transients should be converted to temperature load groups used in thermal expansion analysis. All transients given in the thermal analysis are coolant transients, but the thermal expansion analysis considers the temperature transients of the system pipes and equipment structures. Therefore, the one-dimensional thermal conduction analysis shall be performed to obtain the temperature transients of the structures under all transients.

For the shock resistance load distribution calculation, if it is necessary to perform a shock time history analysis while the shock inputs are given in the form of shock response spectra, then shock response spectra should be converted into shock accelerograms firstly.

The loads to be considered in the design are applied to the model. According to the factors such as load type, model conditions and output results, an appropriate method is selected to calculate the load distribution. For the dead weight, thermal expansion and tilt load distribution, static method is selected. Effect of dampers can be ignored for static analysis. For swing load distribution, harmonic response method or load factor method is selected, still with damper ignored. For the shock load distribution, three (3) methods are available, i.e. response spectrum method, equivalent static method and time history method, the equivalent static method can be selected for very simple systems or equipment with single degree of freedom, the response spectrum method is selected for linear or weakly nonlinear systems and equipment, and the time history method is selected for strongly nonlinear systems or equipment.

After the analysis, corresponding post-processing scheme is selected according to the analysis type (steady analysis or transient analysis) for the post-processing of calculation results to obtain the loads required for stress analysis and structural design. Such loads shall normally be given in the coordinate system facilitating the stress analysis loading. Such loads include support loads, pipe loads, component loads, interaction between components, equipment nozzle loads and external loads of valves. Such loads are usually given in the form of extrema of force and moment. The maximum acceleration is also required for the valve analysis. For some positions with special requirements, the displacements at key points are also required. If components and equipment are verified by test, the acceleration and displacement time history of the relevant positions should be provided.

The load distribution calculation of systems, equipment and internals usually gives the inputs required for component-level stress analysis and structural design. The analysis results are mainly the forces and moments applied on various parts, the displacements, velocities and accelerations of various parts. Resonant frequency, modal shape, modal participation and modal mass are analyzed for the system and equipment to determine its general dynamical properties. It is necessary to verify the analysis results by comparison against theoretical solutions, the existed engineering design analogy or experimental data. If limitations are set for displacement and acceleration responses, then comparison between computed and specific requirements should be performed to verify the structural design.

10.5.2 Stress Analysis of Structures and Components

The stress analysis is to prove that the components in systems and equipments will not suffer from any form of failure under corresponding loads. To meet this requirement, each load and load combination shall comply with the corresponding stress criteria. In this process, some important quantities are calculated using applicable analysis methods and compared with allowable values to judge whether the design of systems and equipments meets the requirements of relevant specifications. If the design does not meet the requirements, the structure shall be modified and the analysis shall be considered until the requirements are met through continuous optimization and iteration.

Determination of calculation areas: Due to the limitations of hardware and software technologies, it is necessary to divide the systems and equipments into several calculation areas. In order to determine the boundary conditions of the analysis area, an overall analysis (i.e. the load distribution calculation) must be performed on the systems and equipments. In the analysis of different areas of a single component, the boundary conditions shall be determined according to the structure and load conditions, so as to obtain conservative results that can ensure the capability of preventing various types of failures.

The stress analysis and evaluation of structures and components are performed by classifying their own levels based on the standard classes of the equipments or systems. The analysis mainly focuses on the prevention against excessive deformation, plastic instability, elastic instability or elastic-plastic instability, progressive deformation under alternating loads, fatigue failure and fast fracture. The corresponding stress limits for different conditions are used to evaluate various types of stresses under possible loads. The basis for the determination of stresses is the maximum shear stress theory.

In order to meet the special application requirements, the specific stress criteria that are different from the basic stress criteria are specified. If the specific stress criteria and the basic stress criteria are inconsistent, the specific stress criteria shall prevail.

For the sealing structures, the separation quantum or the specific seal pressure of sealing surface shall be analyzed and the calculation results shall be compared with the design limits to judge whether the sealing structure meets the requirements for sealing performance. According to the actual conditions of sealing structure and the evaluation criteria, an analysis model shall be established after proper simplification. A case study of the sealing analysis of reactor pressure vessel: considering the symmetry of structure and load as well as the economic efficiency and accuracy, the calculation model can be established only containing half of a bolt. The key to the analysis is the accurate simulation of the content of calculation and analysis such as bolt preloading, frictional contact analysis, elastic-plastic analysis, and thermal and structural deformation coupling analysis. After the completion of all analysis calculations, the separation quantum or the specific seal pressure are obtained through post-processing of calculation results, then the obtained results are compared with the limits to judge whether the sealing requirement are met.

After neutron irradiation, materials in some positions will suffer from irradiation embrittlement. In order to prevent the fast fracture in such positions, the fracture mechanical analysis shall be performed to confirm that the components will not suffer from fast fracture under the given loads.

Based on a complete set of analysis processes for reactor structural mechanics, reliable analytical tools and methods are used to determine some important mechanical parameters in nuclear power plants. These parameters are evaluated according to a complete set of laws, regulations, and standards, so as to ensure the mechanical safety of nuclear power plants. In addition, this process also requires a set of formal and standardized review processes.

10.5.3 Analysis Example

In the following, a case study of the load distribution calculation for reactor primary system and the stress analysis of primary pipes and equipment nozzles is introduced to illustrate the mechanical analysis and evaluation process for nuclear power plants.

The analysis model for primary system of reactor includes reactor pressure vessel, steam generator, main pump, valves and supports for such equipment and pipes.

Load distribution calculation for primary system of reactor:

In the analysis model, the reactor pressure vessel and its internals are modeled by dividing them into straight pipe elements and lumped mass elements. The reactor supports and the primary shielding structures are modeled by dividing them into beam elements and lumped mass elements. The steam generator and its internals are simulated by dividing them into straight pipe elements and lumped mass elements. The upper supports are modeled by dividing them into beam elements and spring elements, and the lower supports are modeled by dividing them into spring elements. The main pump and its internals are simulated by dividing them into straight pipe elements, and the main pump supports are simulated by dividing them into spring elements. Straight pipe sections and elbows of the main pipe are simulated by dividing them into straight pipe elements and elbow elements. The valves are simulated by dividing them into straight pipe elements and lumped mass elements, and the distance from the center of gravity to the centerline of the pipe is simulated by dividing them into rigid elements. The main pipe supports, including dampers and spring hangers, are simulated by dividing them into spring elements. The dampers will not function under static loads and the spring hangers will not function under dynamic loads.

The analyzed loads include dead weight, thermal expansion, tilt, swing and shock.

Dead weight, thermal expansion and tilt analysis are performed by static method. The gravitational acceleration is directly applied to the model during for dead weight analysis. The tilt analysis is based on the transverse and longitudinal inclination angles, gravitational acceleration is decomposed and applied on the model. For the thermal expansion analysis, the coefficient of linear expansion is entered into the element material characteristics, and the temperature load acts directly on the element nodes.

Swing analysis is carried out using the harmonic response analysis. The magnitude of dynamic response of primary system of reactor under swing conditions depends on the swing angle and cycle. Since the fundamental frequency of the structure is much higher than the maximum frequency of the excitation, it is rather far away from the resonance frequency. In this case, the shorter the swing cycle is, the greater the dynamic response of the structure would be. Therefore, in the swing analysis, the swing angle is taken as the maximum specified for the ambient conditions, and the swing cycle is taken as the minimum.

The shock load inputs are provided in the form of frequency domain response spectra. A special program for converting shock response spectra into accelerograms is used to generate the time history acceleration sequence. Figure 10.1 illustrates the resultant accelerogram curve. As can be seen from Fig. 10.2, the response spectrum converted from the resultant accelerogram time history agrees well with the original shock spectrum.

The above-mentioned system load distribution calculations can obtain the loads of structures at main pipes, equipment inlet/outlet nozzles, and equipment/pipe supports under the effect of dead weight, thermal expansion, tilt, swing and shock.

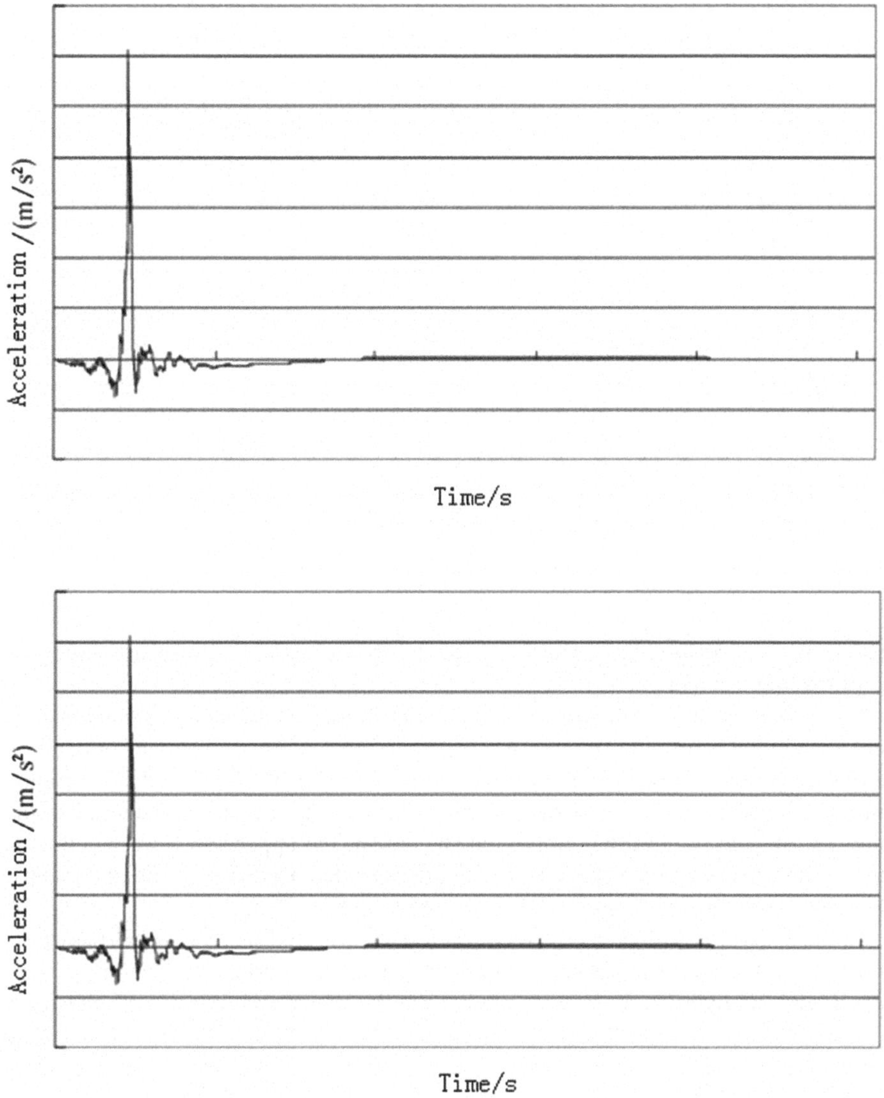

Fig. 10.1 Shock accelerogram curve

Stress analysis and evaluation of main pipes: the main pipe consists of elbows, welds and tee joints. The loads considered in the analysis include pressure, temperature, dead weight, thermal expansion, tilt, sway and shock. The stress exponents of pipe fittings in the main pipe are calculated. The calculated stress exponents and the considered loads are substituted into corresponding equations to analyze and calculate the stresses under various conditions. In addition, the evaluation is performed

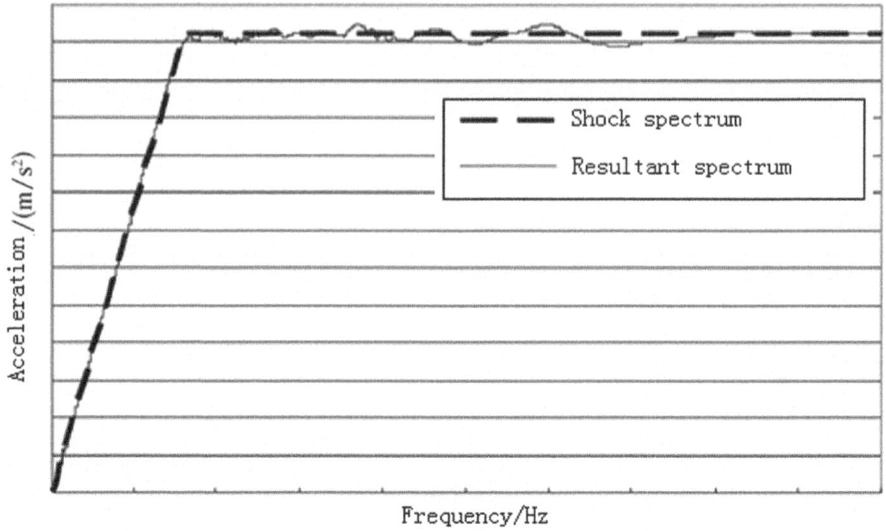

Fig. 10.2 Correlation curve for resultant spectrum and original shock spectrum

according to corresponding criteria to check whether calculation results meet the specified requirements.

Stress analysis and evaluation of equipment nozzles: a case study of reactor pressure vessel inlet is provided herein. According to the detailed dimensions of the inlet of reactor pressure vessel, the 3D finite element model is established. The contribution of welding layer on surface to the overall strength is not considered in the analysis. In order to eliminate the boundary effect, the upper and lower shell ends and the nozzles of reactor pressure vessel are extended according to Saint-Venant's principle. The finite element model is illustrated in Fig. 10.3.

In the analysis, all degrees of freedom of the lower end face of support skirt are constrained, the uniform pressure is applied on the inner surface of reactor pressure vessel, and the hydrostatic force is applied on the upper and lower shell end faces

Fig. 10.3 Finite element model for reactor pressure vessel inlet nozzle

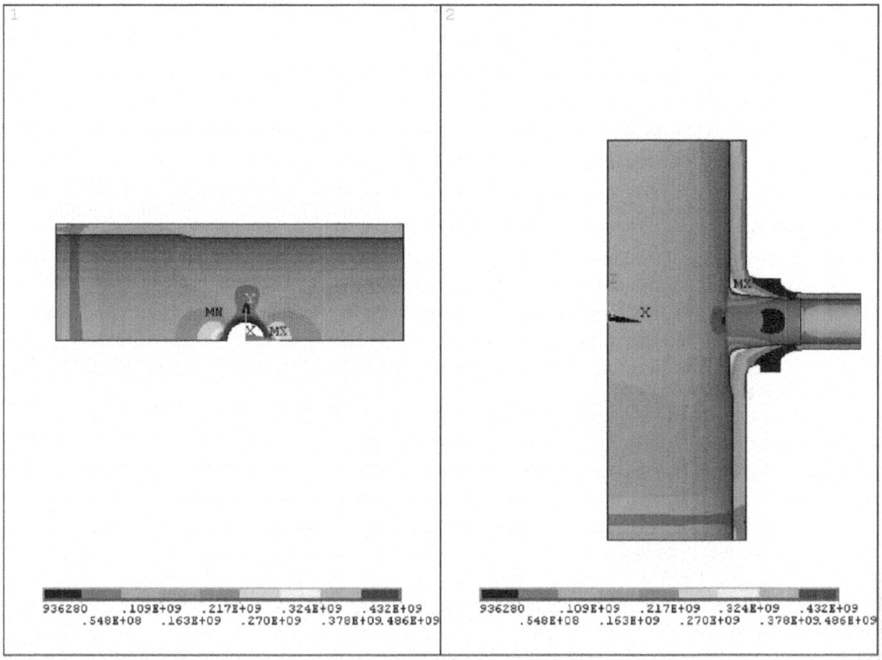

Fig. 10.4 Stress distribution of inlet nozzles of reactor pressure vessel

and the nozzle extension cross section of reactor pressure vessel. The load obtained in the system load distribution is applied on the extended end of nozzle.

The analyzed loads include dead weight, pressure, thermal expansion, tilt, swing and shock. The analyzed conditions include design conditions, operating conditions and hydrostatic test conditions.

The stress distribution is illustrated in Fig. 10.4. The stress evaluation of representative paths indicates that the inlet nozzles of reactor pressure vessel meet the requirements in the specifications.

References

1. Gu, Fangyu, and Sun Jia. 2001. Structural Mechanics in Reactor Technology Facing New Century. *Nuclear Power Engineering* 22: 146-14-9.
2. Xi, Zhu, Zhang Zhenhua, Mei Zhiyuan, et al. 2013. *Damage Mechanics of Warship Structure Subject to Explosion*. Beijing: National Defence Industry Press.
3. Wang, Yu. 2006. *Shock Response Analysis Technology for Naval Ships and Equipment*. Beijing: Haichao Press.
4. Wang, Yu. 2007. *Underwater Explosion Test and Shock Analysis Technology for Naval Ships*. Beijing: National Defense Industry Press.
5. Wang, Yu. 2010. *Underwater Explosion Shock Test and Protection Technology for Real Ships*. Beijing: National Defense Industry Press.

6. Translated by Liu Shulin. 2008. *Shock and Vibration Handbook*, 5th ed. Beijing: China Petrochemical Press.
7. Translated by Zhou Kang. 2006. *Naval Shock Analysis and Design*. Harbin: Harbin Engineering University Press.
8. Translated by Xie Lili. 2009. *Dynamics of Structures: Theory and Application in Earthquake Engineering*, 2nd ed. Beijing: Higher Education Press.
9. Liu Wenjin, Wu, Yang Yu. Wanjun, et al. 2011. Research on Shock Resistance and Earthquake Resistance Analysis Technology for Main Class 1E Equipment. *Nuclear Power Engineering* 32: 150–153.
10. Wanjun, Wu, Liu Wenjin, and Zhang Yixiong. 2009. Study on Generation of Artificial Accelerogram from Shock Response. *Nuclear Power Engineering* 30: 70–71.
11. Translated and Proofread by Wang Guangyuan. 2006. Structural Dynamics (2nd Edition, Revised Edition). Higher Education Press.
12. Translated by Yue Zhufeng. 2005. *Roark's Formulas for Stress and Strain*, 7th ed. Beijing: Science Press.
13. Shunming, Li. 2006. *Mechanical Fatigue and Reliability Design*. Beijing: Science Press.
14. Zhilun, Xu. 1986. *Elasticity*. Beijing: Higher Education Press.
15. Yiquan, Xie, Lin Zhongxiang, and Ding Haojiang. 1988. *Elasticity*. Beijing: Higher Education Press.
16. Mingwan, Lu. 2001. *Foundation of Elasticity*. Beijing: Tsinghua University Press.
17. Zhuhua, Xiong. 1986. *Foundations of Plastic Mechanics*. Beijing: Higher Education Press.
18. Hongwen, Liu. 1995. *Mechanics of Materials*. Beijing: Higher Education Press.
19. Yongyan, Wang. 1999. *Theory and Application of Dynamic Substructure Methods*. Beijing: Scicnce Press.
20. Translated by Wang Song. 2009. *Finite Element Analysis—Theory and Application with ANSYS*, 3rd ed. Beijing: Publishing House of Electronics Industry.
21. Qingcheng, Zhao. 1996. *Photomechamics Tutorial*. Beijing: Higher Education Press.
22. Jiakun, Zhu. 1985. *Computational Fluid Mechanics*. Beijing: Higher Education Press.

Chapter 11
Reliability and Maintainability Design

11.1 Overview

During the development of nuclear power plant, systematic efforts shall be made to guarantee its reliability and maintainability, including the preparation of the reliability and maintainability program, the definition of the objectives of the efforts and the specific work items and work requirements, to ensure that the nuclear power plant can satisfy the requirements for reliability and maintainability.

The reliability of a product refers to its capacity of completing the required functions under the specified conditions within the required time. The probability measure of reliability is called the degree of reliability. This definition includes the products, the specified conditions, the required time and the specified functions, and indicates that the reliability is a capacity of a product. The reliability of the nuclear power plant refers to the unit is capacity of completing the required functions under the specified applicable conditions within the required time, indicating the stability of operation and time of a system, machine or equipment, etc. Therefore, the reliability of the nuclear power plant is based on that of various systems, equipment and machineries [1].

The maintainability of a product refers to its capacity of maintaining or restoring to the specified state according to the specified procedures and methods and at the time of the routine maintenance under the specified conditions within the required time. The maintainability belongs to a quality of products, namely the inherent characteristic endowed by the product design to make its maintenance simple, quick and economical.

The reliability and the maintainability are all the inherent attributes of products, which are established by the design and ensured by the production and management.

The reliability and maintainability activities mainly include: ① reliability and maintainability management; ② design and analysis of reliability and maintainability; ③ reliability test and evaluation.

J. Yu, *Marine Nuclear Power Technology*,
https://doi.org/10.1007/978-981-15-2894-1_11

11.2 Reliability and Maintainability Management

The reliability and maintainability management of the nuclear power plant mainly includes the establishment of reliability and maintainability work system, the development of reliability and maintainability work plans, the preparation of the general design specification for reliability and maintainability, and the management of the information related to the reliability and maintainability.

(1) Establishment of the work system for reliability and maintainability

To establish the work system for the reliability and maintainability, and develop corresponding management regulations.

(2) Development of reliability and maintainability work plans

To develop the reliability and maintainability work plans during the development of the nuclear power plant, including:

① To establish the guideline for the reliability and maintainability requirements;
② To set the objectives of the reliability and maintainability work;
③ To prepare the general specifications for reliability and maintainability design.

According to the reliability and maintainability work demands, to prepare the general design specification or guidance relating to the reliability and maintainability design, analysis, etc., or combining with the characteristics of the nuclear power plant, to modify the existing general design specification or guidance. These general specifications or guideline for reliability and maintainability design include:

① Reliability design criteria;
② Maintainability design criteria;
③ Guide to failure modes, effects and hazards analysis;
④ Guide to fault tree analysis;
⑤ Guide to reliability modeling, estimation and allocation.

(3) Management of reliability and maintainability information

During the development of the nuclear power plant, the accurate, effective and timely delivery of information such as the reliability and maintainability design, and analysis shall be emphasized.

The information database about reliability and maintainability shall be established to provide for designers with the data and technical information required in the reliability and maintainability design and analysis.

The reliability and maintainability data are the basis for implementing the reliability and maintainability works. These data can be classified into the development process data, the test data, and the application data by the data sources. Among these data, the most important are the failure information and data that are related to the devices and the equipment composed of the devices, and are identified in the tests.

The purposes to collect the reliability and maintainability data are:

① To evaluate the reliability of the nuclear power plant and improve the design and manufacturing process to enhance the inherent reliability of plant;
② To evaluate the test results and establish the required corrective measures to improve the maintainability;
③ To implement the design and analysis for the reliability and maintainability of the unit and reduce the failure probability of the unit;
④ To provide information for the application of new technologies and the development of new products;
⑤ To provide data items for the analysis of guarantee and recognition.

11.3 Reliability Design and Analysis

As the most important phase in such process, the reliability design and analysis refers to the process of engineering the reliability requirements into the design of the technical documents and drawings of the products through a series of technical activities for reliability such as reliability allocation, estimation, analysis, and improvement to achieve the inherent reliability of the products [2].

The reliability design and analysis of the nuclear power plant includes:

① Establishment of the requirements for the reliability of the nuclear power plant;
② Establishment of the reliability models of the nuclear power plant, and allocation and estimation of the reliability indexes;
③ Implementation of various kinds of reliability design and analysis (such as failure modes, influence and hazards analysis, fault tree analysis, potential analysis, finite element analysis, and durability analysis) to find and confirm the vulnerabilities and determine the key parts for the reliability of the products to select elements, components and raw materials and improve the reliability of the products by improving the design;
④ Adoption of various effective design measures (or technologies) on reliability, for example, the development and implementation of the design criteria for reliability (involving the derating design, the redundant design, the simplified design, etc.) and combination of the design methods with the performance design of the nuclear power plant to reduce the failures of products.

11.3.1 Reliability Requirements

The purposes of establishing the requirements for the reliability of the nuclear power plant are to obtain the reliable and supportable products to realize the required reliability.

The reliability requirements are classified into the qualitative requirements, the quantitative requirements and the requirements for work items. Wherein the reliability work items are usually the reliability design measures or the reliability analysis to be adopted in the development of the nuclear power plant to ensure and improve the reliability. These work items are all general design measures and analysis works and shall be detailed according to the characteristics of actual situations and designed analysis methods at various stages of development. The main work items usually include the reliability modeling, allocation and estimation, the reliability design requirements, the development and implementation of the design criteria for reliability, the confirmation of key parts and important parts, the failure modes, effects and hazards analysis, the fault tree analysis, the reliability tests, etc.

(1) Qualitative reliability requirements

The purposes of the qualitative reliability requirements are to obtain the product reliability. The reliability qualitative requirements are the non-quantitative requirements for design, process, software and other aspects.

The qualitative design requirements for reliability mainly include the simple qualitative requirements as well as the qualitative requirements for redundancy, derating, adoption of mature technologies, environmental suitability and man-machine engineering.

(2) Quantitative reliability requirements

The quantitative reliability requirements refer to the selection and confirmation of reliability parameters, indexes, verification time and verification methods of the products, to evaluate or verify the reliability level by using the quantitative methods in design, production, test verification and products application.

During the reliability design and analysis of the nuclear power plant, the parameter indexes relating to reliability usually include the design lifetime, the failure rate, the mean time between failures and design lifetime, etc.

11.3.2 Methods for Reliability Design

The reliability design aims to dig and confirm the hazards (and the vulnerabilities) in the design processes and effectively eliminate the hazards (and the vulnerabilities) by adopting the design preventive and improvement measures. The quantitative calculation and the qualitative analysis are mainly used to evaluate the current reliability level of the products or find the vulnerabilities. However, the inherent reliability of the products can be improved only by adopting various specific reliability design methods. The specific requirements are as follows.

(1) Establishment of reliability model

The reliability model refers to the reliability block diagram and the mathematical model established for allocating, predicting, analyzing or estimating the reliability of the products. The reliability model is divided into the overall level and the system level. The logic block diagram of reliability and the reliability mathematical model of the corresponding levels shall be established based on the functional block diagram of the products. The reliability model shall be consistent with the technical state of the nuclear power plant.

(2) Reliability allocation

The reliability allocation refers to the allocation of the reliability quantitative requirements to various components according to the given criteria. For the development of the nuclear power plant, the allocation idea and range shall be determined according to the technical characteristics of the power plant. The reliability allocation shall be implemented by adopting the appropriate methods according to the reliability quantitative requirements of the nuclear power plant.

The reliability allocation methods mainly include the equal allocation method, the proportion combination method, the reliability reallocation method, and the scoring allocation method. The specific allocation method may vary with the progress of the nuclear power plant in the development phase.

During the reliability allocation, the previous product experience shall be made full use of and a comprehensive consideration shall be implemented by combining with the factors such as functions, reliability level, and product complexity of the analysis object. The reliability allocation shall be implemented with the progress of the design, or the original allocation results shall be changed and rectified according to the design.

(3) Reliability prediction

The reliability prediction refers to the predication of the reliability of the products under the given working conditions. For the development of the nuclear power plant, the prediction of system and equipment reliability shall be implemented at different levels.

For the reliability prediction of the mechanical products in case of lacking reliability data and mature methods, the prediction can be implemented by adopting the professional evaluation method and making full use of the empirical data.

The reliability prediction is usually implemented by adopting the component counting and the reliability block diagram or assessed through the reliability tests. The component counting is applicable to the electronic equipment, while the reliability block diagram is applicable to system and equipment (excluding the electronic equipment).

(4) Failure modes and effects analysis

Failure modes and effects analysis: in the process of product design, the adoptable preventive and improvement measures are proposed through the analysis of various potential failure modes of each component of the product and their effects on product functions as well as the classification of each potential failure mode according to the severity to improve the reliability of products. Refer to Fig. 11.1 for the working procedures of failure modes and effects analysis [3].

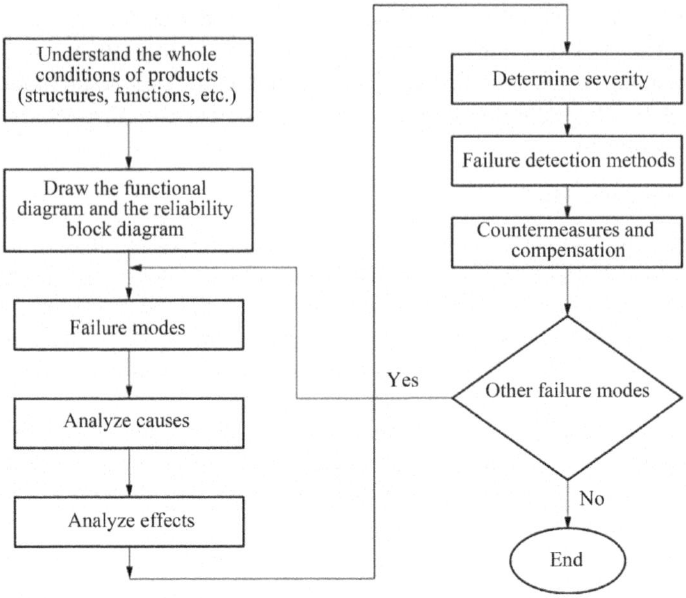

Fig. 11.1 Working procedures of failure modes and effects analysis

Failure modes, effects and hazards analysis: another task is to judge the lethality of this kind of failure to make the analysis quantitative based on the failure modes and effects analysis.

The failure modes and effects analysis is as shown in Table 11.1. Among the headers, the "initially agreed level" refers to the overall products corresponding to

Table 11.1 Failure modes, effects and hazards analysis

Products at initially agreed level ____________ Tasks _____________

Products at agreed level

S/N	Identification of products or functions (names)	Functions	Failure modes	Failure causes	Task phases and working modes	Failure effects			Severity categories	Failure detection methods	Basic maintenance measures	Failure probability level	Failure rate or failure probability data sources	Remarks
						Effects on themselves	Effects on the upper level	Final effects						

the analyzed objects; the "agreed level" refers to the products of the previous level adjacent to the products being analyzed; the "task" refers to the general basic tasks to be completed by the product.

Based the requirements and the actual conditions, only the failure modes and effects analysis is implemented without the hazards analysis.. The failure modes and effects analysis, etc. should be implemented at the early stage of the design to provide the basis for the design review and the arrangement of improvement measures.

At the different phases of the development of the nuclear power plant, the failure modes and effects analysis shall implemented repeatedly on systems and equipment. For the design vulnerabilities found in the analysis, the improvement measures shall be adopted in time.

For the failure modes, effects and hazards analysis of systems, the equipment and accessories shall be regarded as the lowest agreed level; for the failure modes, effects and hazards analysis of equipment, the minimum replaceable unit shall be regarded as the lowest agreed level.

The failures with a severity of Class I (catastrophic) and Class II (serious) confirmed in the failure modes, effects and hazards analysis shall be eliminated by adopting measures.

(5) Fault tree analysis

Fault tree analysis: in the system design process, various factors that may cause system failure (including hardware, software, environment, human factors, etc.) shall be analyzed and the logic block diagrams (i.e. fault trees) shall be drawn to determine various possible combinations of the causes of system failure and their probability. The system failure probability shall be obtained and the corresponding corrective measures be adopted to improve the system reliability. Refer to Fig. 11.2 for the working procedures of fault tree analysis [3].

In the design phase, the fault tree analysis can be used to distinguish the potential system failure modes and catastrophic hazards and find the reliability and safety vulnerabilities to improve the design.

During the development of the nuclear power plant, the fault trees shall be continuously improved and repeatedly iterated with the progress of the development phase. When the design is changed, the corresponding modifications shall be made to the fault trees. As a supplement for the failure modes, effects and hazards analysis, the fault tress mainly aim at the catastrophic and fatal failure modes that affect safety

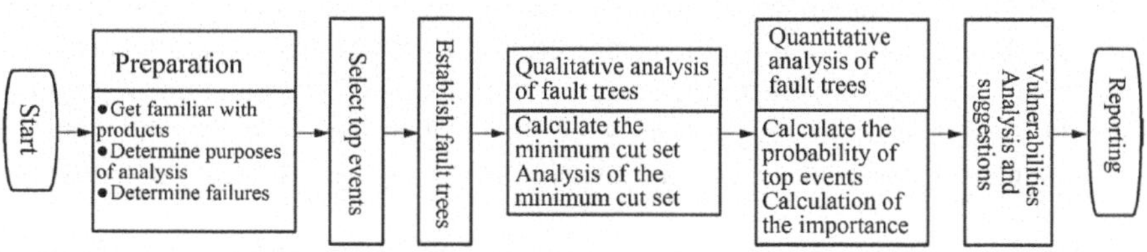

Fig. 11.2 Working procedures of fault tree analysis

and tasks and are used as the top events for implementing the fault tree analysis. Generally, the important safety systems shall be subjected to the quantitative analysis of the fault trees and a part of systems and equipment are only subjected to the qualitative analysis of the fault trees according to the demands and the acquainted data conditions.

The fault tree analysis shall be implemented for the important safety systems and equipment to implement the detailed analysis of various failure causes that cause the top events of the product and find the hazards in the design.

The analysis of the system fault trees shall be performed up to the equipment and accessories, while the analysis of the equipment fault trees shall be performed up to the minimum replaceable unit. The single point failures that cause Class I and Class II failures shall be eliminated through the fault tree analysis.

During the quantitative fault tree analysis, the quantitative analysis of the relevant reliability data of basic events can be implemented to emphasize the relative comparative significance of the results. During the analysis, if there is a lack of the relevant data of next level, the assumptions and treatment can be made to achieve the accuracy within the allowable range.

(6) Development and implementation of design criteria for reliability

Practices prove that the failures can be avoided through the development and implementation of the design criteria for reliability, which is of great importance for improving the inherent reliability of the nuclear power plant [4]. The so called design criteria for reliability refers to the summarizing of the engineering experience of the existing and similar products to make the engineering experience well organized in a systematic and scientific manner and become the criteria and requirements that the designers must comply with in the reliability design.

The relevant reliability requirements are converted into the specific requirements that are operable in the design of hardware and software through the formulation of the reliability design criteria of system and equipment of nuclear power plant to satisfy the qualitative design requirements of reliability. The designers shall comply with the relevant design criteria for reliability and shall implement each criterion in the design of the nuclear power plant by combining with the specific characteristics of the designed nuclear power plant to ensure that each design criterion is provided with the corresponding design guarantee measures. After the completion of the design, the designers shall submit the reports about the implementation of the design criteria for reliability.

The design criteria for reliability are usually applicable to a specific product. However, the common points of the design criteria for reliability of various products can also be combined into the design criteria for reliability of a certain type of product.

In the design criteria for reliability of the nuclear power plant, the detailed requirements for different aspects are usually proposed according to the following provisions.

① Adoption of mature technologies and process; ② simplified design; ③ control of selection of components and raw materials; ④ derating design; ⑤ redundant design; ⑥ environmental design; ⑦ design of human factors; ⑧ design of electromagnetic compatibility; ⑨ design of circuit tolerance; ⑩ design of anti-transient overstress ⑪ reliability design of software.

(7) Determination of critical parts and important parts for reliability

The list of the critical parts and important parts for the reliability design of the nuclear power plant shall be determined according to the results of the failure modes and effects analysis and other reliability analysis, and the relevant standards.

During the development of the nuclear power plant, the critical parts and important parts shall be emphasized for the reliability design and the effective measures are adopted to improve the reliability.

11.4 Design and Analysis of Maintainability

The maintainability design refers to the process of implementing the maintainability requirements in the product design. Based on the maintainability indexes, it will ensure that the technical status of the finally designed products meet the requirements for the product maintainability by adopting a series of effective design measures. After the product fails, the products can be restored to the specified technical status with the shortest maintenance time, the minimum maintenance hours and expenses and the fewest resources consumption in case of failures.

The purposes of the maintainability analysis of the nuclear power plant are to: ① in the development phase, analyze and evaluate whether the design of the nuclear power plant meets the specified maintainability quantitative requirements or not and provide the basis for the maintainability design schemes of comparing and selecting the products; ② determine the vulnerabilities of the nuclear power plant in the maintainability design and propose the measures for improving the maintainability and prevention as well; and ③ determine the key guarantee resources required for maintenance through analysis to provide the basis for the guarantee decisions.

It shall be noted that, the results of the maintainability analysis are the important basis for the maintainability design, but cannot cause direct influence on the maintainability. In order to improve the maintainability and satisfy the specified maintainability quantitative requirements, the maintainability design must be relied on, i.e. the measures of preventing and improving the vulnerabilities that affect the maintainability and the measures of minimizing the maintenance time shall be adopted in the design.

11.4.1 Maintainability Requirements

The determination of the maintainability requirements is the primary condition for a reliable application of the nuclear power plant. The maintainability requirements can be classified into the qualitative requirements, the quantitative requirements and the requirements for the maintainability work items. Wherein the requirements for the maintainability items are important for the maintainability works and the purposes of these requirements are to realize the specified maintainability requirements through implementing the minimum and effective work items.

The maintainability work items of the nuclear power plant usually include the development and implementation of the design criteria for reliability, the preliminary analysis of maintainability, the design requirements for maintainability, the virtual maintainability verification and analysis, the maintenance task analysis, etc. The selection of the specific work items depends on the factors such as the maintainability level, the complexity and criticality, the proportion of new technologies, etc. of the developed or improved nuclear power plant. The major objectives are to satisfy the maintainability qualitative and quantitative requirements.

(1) Qualitative maintainability requirements

The qualitative maintainability requirements refer to the non-quantitative requirements proposed for product design, process, software and other aspects to make the product maintenance fast, simple and economical.

The qualitative maintainability requirements mainly include: good maintenance accessibility, simplified design and maintenance, improved maintenance standardization and interchangeability, perfect error prevention measures and identification marks, guaranteed maintenance safety, accurate, fast and convenient text and diagnosis, appropriate repairability of valuable objects, qualified man-machine engineering in maintenance.

(2) Quantitative maintainability requirements

If the nuclear power plant meets the qualitative requirements in the maintainability design, the maintainability will be greatly improved. However, the maintainability degree cannot be judged directly. Therefore, the quantitative requirements (i.e. the maintainability index requirements) shall be proposed. As there are many kinds of characteristic quantity used for timing maintenance (also called the maintainability parameters), the maintainability indexes can be expressed in many ways. For the nuclear power plant, the involved maintainability parameters and indexes usually include the mean time to repair, the longest repair time, the preventive maintenance time, etc.

11.4.2 Qualitative Maintainability Design

(1) Development and implementation of design criteria for maintainability

The design criteria for maintainability for the system and equipment that the nuclear power plant belongs to shall be developed by referring to the common criteria for the maintainability design and combining with the characteristics of the nuclear power plant as well as the requirements for the maintainability of systems and equipment to convert the relevant maintainability requirements into the specific requirements that can be implemented directly in the hardware and software design and continuously improve these specific requirements in the development process.

The designers shall carefully analyze and summarize the historical experience according to the design criteria for maintainability and combining with the specific characteristics of the nuclear power plant to develop the detailed design requirements for maintainability and determine the corresponding design measures that shall be implemented one by one.

The maintainability qualitative design gives a specific description of the maintenance simplicity, rapidity and economy. The qualitative design requirements can be summarized as the following aspects:

① The simplified design and maintenance; ② the reduced maintenance content and requirements for maintenance skill; ③ the good maintenance accessibility; ④ the perfect error prevention and identification marks; ⑤ the guaranteed maintenance safety; ⑥ the accurate, rapid and convenient detection and diagnosis; ⑦ the qualified man-machine-environment engineering of maintenance; ⑧ the maintainability design of software; ⑨ the consideration of the preventive maintenance and non-operating state's effects on maintenance.

(2) Maintainability analysis of arrangement of major equipment and reactor compartment

The maintainability analysis of arrangement is mainly implemented for the maintenance accessibility of the important equipment according to the characteristics of arrangement of the equipment and pipes in the nuclear power plant compartment.

The works on maintainability mainly include the analysis and determination of the maintenance of major equipment of system, the preventive maintenance work items and the requirements for maintenance guarantee as well as the maintainability analysis related to compartment arrangement. These works are used to improve the maintainability of system equipment and provide information for the maintenance guarantee.

The analysis and determination of the maintenance of major equipment of system, the work items of preventive maintenance and the maintenance guarantee requirements shall be implemented against the maintenance level of the nuclear power unit to provide input information for the maintenance guarantee of the nuclear power plant.

The maintainability analysis related to the compartment arrangement shall be implemented against the maintenance accessibility and maintenance operation space of the major equipment in compartment. The maintainability analysis of the compartment arrangement can be implemented by relying on the computer 3D simulation design software.

In the maintainability analysis of the compartment arrangement, the consistency of all the involved systems with the pipes, equipment and installation rules, the interaction of systems, the significant influence of maintenance shall be described to satisfy the maintainability requirements of the system and important equipment.

(3) Analysis of failure modes and effects and maintainability

The analysis of failure modes and effects and maintainability shall be implemented based on the failure modes and effects analysis. Each potential failure mode shall be analyzed to determine the methods of detecting the failures and the measures of repairing the faulty products.

(4) Maintenance strategies and measures

In the design of the nuclear power plant, the analysis of the maintenance strategies shall be implemented according to the system and equipment composition, life and tasks, the environmental conditions and the possible maintenance measures, to determine:

① The maintenance level and conditions of the nuclear power plant; ② the maintenance ways adopted by the nuclear power plant; ③ the frequency of the preventive maintenance of the nuclear power plant; ④ the maintenance tools and special maintenance measures required by the nuclear power plant.

11.4.3 Allocation and Prediction of Maintainability

(1) Maintainability allocation

The maintainability allocation refers to the allocation of the maintainability quantitative requirements of the products to various components according to the given criteria.

In the development of the nuclear power plant, the overall maintainability indexes at device level shall be allocated to various systems and equipment at each function level to make clear the maintainability indexes of the developed systems and equipment, to facilitate management and ensure that the specified maintenance requirements are finally met.

The most common maintainability allocation indexes include the mean time to repair, the mean preventive maintenance time, the failure isolation rate, the failure detection rate, etc.

There are many methods for the maintainability allocation, wherein the common ones include the allocation by equivalence, the allocation by failure rate, the allocation by relative complexity, the allocation by similar products, the allocation by weighting factor, etc.

(2) Maintainability prediction

The maintainability prediction refers to the works implemented for estimating the maintainability of the products under the given working conditions. Although the maintainability design is implemented in the development or improvement process of devices, the maintainability prediction shall be implemented to judge whether the specified requirements are met and whether further improvements are required. Based on the expected maintenance level of the device, we can know whether the specified indexes are met and timely find the defects in the maintainability design and guarantee to be used as the basis for changing the design.

The maintainability level of various relevant systems and major equipment of the nuclear power plant shall be predicted according to the design conditions, to verify whether the design results meet the specified maintainability indexes. The most common predication parameter is the mean time to repair, or the longest repair time, the man-hour rate or the preventive maintenance time can also be predicted according to the needs.

There are many methods for the maintainability prediction, wherein the common ones include the equivalent regression prediction, the unit comparison, the time accumulation, the sampling prediction. The functional level prediction method is usually adopted by the systems and equipment of the nuclear power plant, i.e. the overall mean time to repair of systems or equipment is calculated by counting the maintenance time and failure rate of each constituent unit or each maintenance task.

11.5 Tests and Evaluation of Reliability

The reliability tests are classified into the reliability engineering tests (including the environmental stress screening test and the reliability growth test) and the reliability statistical tests (including the reliability qualification test and the reliability acceptance test).

(1) Environmental stress screening test

The environmental stress screening tests on components shall be implemented in the product development process, to eliminate the initial failures caused by defective components, process defects and other reasons.

The schemes and main content of the environmental stress screening tests on electronic products shall be specified in the program of components.

(2) Reliability growth test

The reliability growth tests are a process of implementing tests in a planned way as well as analyzing and solving problems in the product development process. The failures are found through the tests under the actual environment, the simulation environment or the rapidly changing environment. The reliability of the products is improved through analysis and adoption of effective measures of improving design and process, etc., to realize a reliable growth.

In the development process of the nuclear power plant, to carry out the reliability growth tests on the key equipment, implement the idea of "test—analyze—improve" and improve the reliability of the test products by combining with the objects to be performed on.

The information of the reliability growth tests shall be included in the closed loop management according to the systematic requirements of failure reports, analysis and corrective measures, to gradually accumulate the reliability data of the products.

(3) Reliability qualification test

Used for the newly designed products and the significantly improved and modified products, the reliability qualification tests aim to verify whether the design of the products meets the reliability requirements or not.

(4) Reliability acceptance test

The reliability acceptance tests are used for verifying whether the reliability of the mass-produced products at delivery meets the specified requirements and checking whether the reliability of the designed products is maintained and realized in the production process.

(5) Reliability evaluation

For the products that are given certain reliability quantitative requirements and need verification, the reliability evaluation of the products can be implemented according to various test results and application experience of the low-level products (components) if no full-size reliability test on the products is given.

11.5.1 Environmental Stress Screening Test

The environmental stress screening tests are mainly applicable to electronic products electrical, electromechanical, photoelectric and electrochemical products. The standards also contain the environmental stress screening methods on electronic products as well as electrical, electromechanical, photoelectric and electrochemical products. When the environmental stress screening of the electronic products is actually implemented, the size and duration of the screening stress can be determined based on the relevant standards. For the non-electronic equipment, the environmental stress screening and the types and magnitude of stress can be determined by referring to the standards and combining with the structural characteristics of the products.

In the environmental stress screening tests, the following principles shall be focused on:

① The environmental stress screening tests shall be used for developing and implementing the schemes based on or by referring to certain standards;
② The environmental stress screening tests aim to eliminate the initial failures; therefore, it is not necessary to accurately simulate the actual application environmental conditions;
③ For the key products, 100% of the environmental stress screening tests at the component level, the circuit board level and the equipment level shall be implemented according to the provisions;
④ When the products are in the process of transferring from development to production, the schemes shall be adjusted according to the changes of manufacturing process, assembling technique, etc.

11.5.2 Reliability Growth Test

The core of the reliability growth test is the test, analysis and improvement process, which is actually a repeated implementation of the following activities:

① To induce failures by relying on the tests of simulating the actual application conditions and fully expose the problems and defects of the products;
② To locate the failures, analyze the failures and find out the failure mechanism;
③ To propose the corrective measures used for improving the design according to the analysis results of the failures;
④ To manufacture the newly designed hardware;
⑤ To test the products that are provided with new hardware continuously, verify the effectiveness of the improvement measures and expose the new problems.

The reliability growth tests shall be implemented according to the requirements of the specified standards.

11.5.3 Reliability Qualification Test and Reliability Acceptance Test

For the electronic products, the reliability qualification test and the reliability acceptance test shall be implemented according to the special specifications and standards. For the non-electronic products, as the time before failure is in accordance with multiple distribution forms, the cost, the test time and the practical significance of project shall be weighed in a comprehensive manner during the development of the test schemes.

The following principles should be followed during the reliability qualification test and the reliability acceptance test:

① The test schemes shall be developed and reviewed, and the preparations before the test and the works after the test also shall be reviewed;
② The test samples in the reliability qualification test shall be not less than 2 and shall be sampled randomly. The test samples shall be subjected to the functional test, the environmental test and the environmental stress screening and shall reach the final state;
③ The failure criteria of the products shall be in accordance with the provisions of the selected test schemes;
④ The test conditions of the reliability qualification test shall be as real as possible;
⑤ The deteriorated components that do not exceed the allowable tolerance shall not be replaced in the tests;
⑥ The test samples shall be adjusted after the tests and shall be delivered after being restored to the specified technical conditions and being subjected to the relevant acceptance procedures.

References

1. Fengge, Pang, and Peng Minjun. 2000. *Marine Nuclear Power Unit*. Harbin: Harbin Engineering University Press.
2. Ganyue, Xie. 2012. *Introduction to Reliability, Maintainability, Guarantee, Testability and Safety*. Beijing: National Defense Industry Press.
3. Qingxiang, Gong, Yu. Zhao, and Gu Changhong. 2007. *Model Reliability Engineering Manual*. Beijing: National Defense Industry Press.
4. Tingxiao, Lu, and Zheng Pengzhou. 1993. *Reliability Design and Analysis*. Beijing: National Defense Industry Press.

Chapter 12
Accident and Safety Analysis

12.1 Overview

In order to prevent the accidents, a series of safety principles shall be followed from design to operation, such as reasonable design, reliable equipment, well-established operating procedures and good safety culture of operating personnel. However, the accidents in marine nuclear power plant cannot be eliminated by all these efforts, that is, such efforts cannot ensure the success of accident prevention totally. Therefore, during the design, we shall also take into account severe accidents properly several engineered safety features are designed to prevent the accidents and mitigate the consequences. Each engineered safety feature is specific to the accidents for certain control. The deterministic analysis of control ability is performed to ensure that such features have sufficient design margin. An accident that requires the safety features to reach the maximum design ability is called design basis accident.

Accident analysis is an important part of the safety design of marine nuclear power plants. Accident analysis aims to study the behaviors of marine nuclear power plants under the failure conditions, and the analysis results are used as the basis of the design of safety features for marine nuclear power plant, and to assess the safety of nuclear reactors under the failure conditions, that is, to determine if the three safety functions of nuclear reactors are available, which are control of reactivity, removal of reactor residual heat and the containment of radioactive products.

Accident analysis shall include: a. performance of marine nuclear power plant under anticipated events. b. safety features under design basis accidents, and c. event sequences that may cause severe accidents. The analysis results can be used to evaluate the ability to prevent the postulated initiating events and accidents, and verify the effectiveness of safety systems and safety related items or systems, and prepare the emergency response procedure.

J. Yu, *Marine Nuclear Power Technology*,
https://doi.org/10.1007/978-981-15-2894-1_12

12.2 Accident Analysis Methods

Accident analysis methods are divided into deterministic accident analysis and probabilistic safety analysis (PSA). The deterministic method is a traditional measures which is used in the design of marine nuclear power plants for a long period, and it is based on the principle of defense in depth. The reactor is designed as safe and reliable as possible, and multiple engineered safety features are arranged to minimize the accident consequences during the most severe postulated accident. When the types, capacity and response time of safety features need to be determined, a reference postulated accident shall be used as the design basis and deemed as the maximum credible accident, that is, if the safety features arranged can prevent this accident, they can certainly prevent other kinds of accidents. The PSA method is different from the deterministic method which applies the probabilistic risk theory to evaluate the safety of marine nuclear power plants. It presumes that the accidents of marine nuclear power plants are random events, and there are many potential factors that can cause the equipment failure , and the safety shall be expressed by the mathematical expectations of all potential failure.

The deterministic method divides the accidents into *credible* accidents and *incredible* accidents based on past experience and socially acceptability. For the PWRs, the double-ended guillotine break of main coolant piping is deemed as the maximum credible accident, and shall be carefully considered in the design and strictly protected. Due to these safety features such as emergency core cooling system, severe consequences would unlikely occur even if such a severe initiating event occurs. The accidents with less severe consequences, such as small break LOCAs in primary pipes and transient operating conditions during the operation of marine nuclear power plants, are not studied in detail and given due attention in the operation management and personnel training. The main cause of the US Three Mile Island accident in 1979 is the lack of sufficient understanding of small break LOCAs. During the accident, incorrect operations were implemented due to the misjudgment of the operators and the less severe accident at the beginning became more and more severe and finally resulted in a severe core damage accident in the history of commercial nuclear power operation. The PSA presumes that there is no definite limit between *credible* accidents and *incredible* accidents and that only the probabilities of accidents vary. There are thousands of potential accidents for a marine nuclear power plants, and the social risk caused by each accident should be expressed by the mathematical expectation of all potential accident consequences. Such mathematical expectation is the risk. The risk study for the marine nuclear power plant indicates that the core meltdown is the main factor causing the release of radioactive substances into the environment, while the small break LOCAs are the main causes of core meltdown. Three Mile Island accident shows that the PSA can discover the weaknesses in the reactor design. Therefore, the PSA can be used in the design as an important supplement to the deterministic method to balance the design.

12.2.1 *Deterministic Accident Analysis*

There are four basic elements in the traditional deterministic accident analysis process,

① Determining design basis accidents,
② Selecting the single failure causing the most severe consequences to the safety system under a specific accident,
③ Verifying the models and all input parameters for the analysis are conservative,
④ Comparing the final results with the prescriptive acceptance criteria and confirming the design adequacy for safety systems.

The selection of design basis accidents is to meet the minimum safety design requirements specified in the laws and regulations and constrain design to test the design margin of safety system by design basis accidents. The selection of design basis accidents is mainly based on the top-level safety objectives of nuclear power plants. The classification of design basis accidents is mainly based on the frequency of occurrence and consequences of accidents.

During the reactor design, the corresponding design basis accidents shall be determined, and the analysis and computer simulation shall be performedto analyze and evaluate the capability of the reactor to prevent the accidents. The reactor design and the safety analysis assessment is an iterative process to reach a compromise between the reactor safety and the overall performance of nuclear power plant.

In 2003, the International Atomic Energy Agency (IAEA) published an important report summarizing the classification of reactor accident analysis methods, as shown in Table 12.1.

The completely conservative method was applied in the 1970s and covered all uncertainties of the knowledge at that time. However, this method may predict the

Table 12.1 Procedure and assumptions of safety analysis methods

Analysis method	Computer program	Available systems	Initial boundary conditions	Methods
Completely conservative method	Conservative	Conservative assumptions	Conservative input data	Deterministic
Realistic analysis method	Best estimate	Conservative assumptions	Conservative input data	Deterministic
Best estimate plus uncertainty analysis method	Best estimate	Conservative assumptions	Real input data with uncertainty	Deterministic
Risk-informed safety analysis method	Best estimate	Based on probabilistic risk assessment assumptions	Real input data with uncertainty	Deterministic + probabilistic

unreal behaviors, change the events sequence and obtain the misleading results. Therefore, this method is not used in most cases.

The realistic analysis method is currently the typical method for safety analysis. It selects the conservative input parameters that are determined by engineering experience, qualitative analysis and a large number of sensitivity calculations to determine the conservative inputs. It obtains the analysis results based on the conservative envelope. It is generally presumed that the value predicted by the program is worse than the actual value. For example, the calculated cladding temperature is higher than the actual temperature in the LOCAs.

The best estimate plus uncertainty analysis method is based on the statistical method or experiment results extrapolation. It considers the factors causing uncertainties in the program calculation, and quantitatively gives uncertainty zone of the results predicted by the program and the accuracy of the program predicted results is statistically significant (such as meeting the requirement of double 95% probability). This method, as recommended by IAEA, is the development trend of reactor safety analysis technology. This method can facilitate the reactor safety design and make the performance of reactors better. Therefore, it has been gradually applied to the reactor design. For the marine reactors, the cost for improving safety system capacity is far higher than that for land-based NPPs because of constraints of the space and weight. This method is particularly suitable for marine nuclear power plant, because it is very likely to ensure the safety of reactors and makes the performance of reactors better, thus reducing unnecessary safety restriction.

IAEA gives the concept of safety margin [1] and shows the difference between the results from the conservative analysis and the results from the best estimate analysis, as illustrated in Fig. 12.1. The real safety margin is unknown, and the safety margin usually refers to the difference between the calculated results and the acceptance criteria specified by the safety authorities. The analysis methods will affect the value of the safety margin. The best estimate method quantitatively gives the uncertainty zone of the results. The results obtained by this method meet certain probability requirements and are usually lower than the calculated results by the conservative envelope method. Therefore, the accident assessment is more real and provides the potentials for improving the performance of marine nuclear power plant.

Besides the traditional deterministic safety analysis methods, the probabilistic safety analysis methods are often used as an important complement during the design of nuclear power plant. The probabilistic safety assessment (PSA), also known as probabilistic risk analysis (PRA), is a system engineering method developed since the 1970s. It uses system reliability assessment techniques (i.e. fault tree and event tree analysis) and probabilistic risk analysis methods to comprehensively analyze the occurrence and development of various potential accidents in complex systems, and consider the probability of occurrence of such accidents and corresponding consequences. The Plant States published *Reactor Safety Research* (WASH-1400) in 1975. Since this milestone research, there has been substantial development in the methods, and the probabilistic safety assessment has become a standard tool for the safety assessment of marine nuclear power plant.

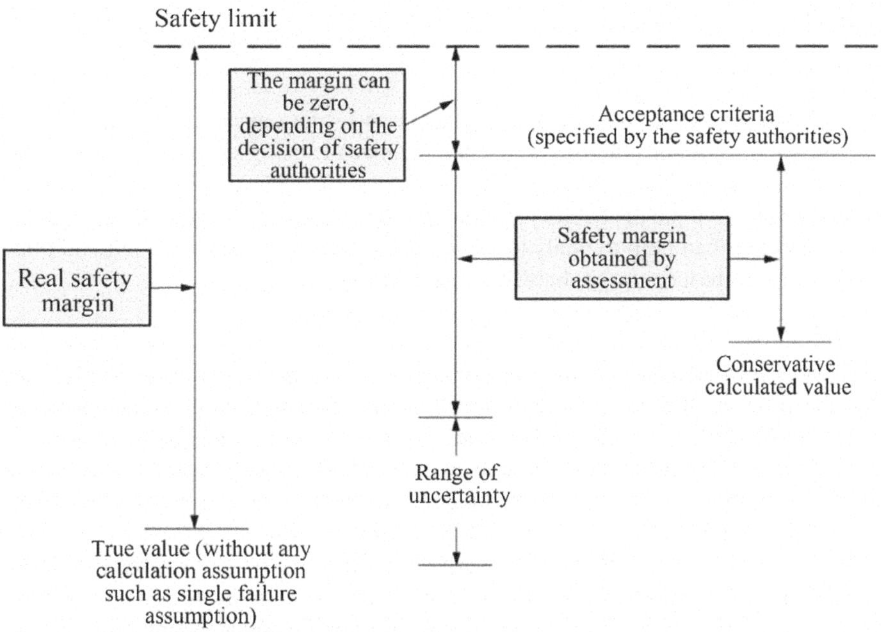

Fig. 12.1 Safety margin and assessment method

12.2.2 Probabilistic Safety Analysis

After the US Three Mile Island accident in 1979, it was discovered that the entire development process of this type of accident had been clearly forecast in WASH-1400. Since then, the probabilistic safety assessment has been widely recognized. It has been studied by the special teams successively in many advanced countries and some developing countries, and widely used in various aspects to analyze the weaknesses in design, improve the design, diagnose the failures, guide the operations and formulate the maintenance strategies. It has gradually become an important tool for safety assessment and safety decision-making.

The PSA of nuclear power plant is actually a comprehensive review and understanding of such plant, as well as a comprehensive analysis for the safety of complex process systems of marine nuclear power plant from various aspects. In the analysis, the comprehensive analysis is also performed for various aspects such as system correlation, personnel interaction, the uncertainty of results and *relative importance* of different accident sequences.

The event tree and fault tree analysis method used in the PSA can also be used for demonstration of design, safety review, searching for system weaknesses, assessment and establishment of accident management regulations, guidance on operation and maintenance and other aspects, and better effects have been obtained.

The PSA can be divided into three levels by its applications, as described below.

(1) Level I PSA

The reliability analysis for operating systems and safety systems of reactors shall be carried out to determine the accident sequences causing core damage, perform the quantitative analysis, calculate the frequency of occurrence of each accident sequence, and give the probability of core damage per operating year of the reactor.

The analysis can help to analyze the weaknesses in the reactor design and point out the ways to reduce the probability of core damage.

(2) Level II PSA

The level I PSA results plus the assessment of containment response. Analyze the physical process of core meltdown and the containment response characteristics, including the analysis of the loads on the containment under the core damage accidents, the failure modes of containment, and the release and migration of radioactive substances in the containment. Determine the frequency of radioactive release from the containment according to the level I PSA results.

The analysis is to investigate the severity of radioactive release caused by various core damage accident sequences, identify the weaknesses in the design, and provide specific suggestions for accident consequence mitigation and accident disposal after the core damage.

(3) Level III PSA

The level II PSA results plus the assessment of off-site consequences. Analyzing the migration of radioactive substances in the environment and calculating the change in radioactive concentration with time at different distances outside the reactor. Determining the off-site consequences caused by radioactivity accidents according to the level II analysis results and the concept of public risk.

Level III PSA is to investigate the relative importance of consequence mitigation measures and support the formulation of emergency response plans.

12.3 Classification and Analysis Requirements for Design Basis Accidents

12.3.1 Accident Classification and Limit Criteria

In 1970, based on the analysis of operating conditions of nuclear power plant, the American National Standards Institute classified the operating conditions of nuclear power plant into four (4) categories by the estimated probability of reactor accidents and the radioactive consequences of such accidents for the residents. This classification principle is mostly adopted in the world. According to the estimated frequency of occurrence of reactor accidents and the radioactive consequences for the staff and the environments, all operational states of marine nuclear power plant are divided into four (4) typical conditions, a. normal operating-Conditions I, b. incidents of moderate frequency-Conditions II, c. infrequent incidents-Conditions III, and d. limiting accidents-Conditions IV.

Condition I is defined as the events that often occur during the operation, maintenance, reloading and testing of marine nuclear power plant. Since such events may occur frequently, the possibility of accidents (Conditions II, Conditions III and Conditions IV) caused by such events must be considered. Therefore, the initial conditions selected for specific accident conditions are the normal operating conditions that can give the most severe accident consequences. As a result, the analysis is conservative. The events defined by Conditions I will lead to some physical parameters changes, but the changes do not reach the reactor shutdown protection setpoints, and the reactor shutdown system will not be triggered.

Condition II is defined as the accidents occurring at a moderate frequency during the operation and testing over the lifetime of marine nuclear power plants. Such conditions may trigger the reactor scram, and the low power operation or normal operation can be recovered after the corrective measures are taken or the malfunction is eliminated. Such accident conditions will not further develop into more severe accidents (Conditions III and Conditions IV). In the accidents defined in Conditions II, the integrity of the fuel element cladding s be maintained, the fluid pressure changes shall not jeopardize the reactor coolant system, and the radioactive product release limits shall meet the provisions in the laws and regulations.

Condition III is defined as the occasional accidents during the operation and testing over lifetime of marine nuclear power plant. Such accidents take place rarely with a frequency of occurrence of about $10^{-4} - 3\times10^{-2}$/(reactor·year), and few fuel elements are damaged in such accidents. During the disposal of such accidents, in order to prevent or limit the radiation hazards to the environment, the engineered safety features shall be put into operation, but the accidents shall not develop into limiting accidents with more severe consequences (Conditions IV). According to the definition of infrequent incidents (Condition III), in such accidents, the reactor core geometry shall not be severely damaged, the number of damaged fuel elements must be limited, and the core can still be properly cooled. In addition, the integrity of reactor coolant pressure boundary and reactor compartment must be maintained.

Condition IV is defined as accidents that are extremely unlikely to occur during the operation and testing over the lifetime of marine nuclear power plant. The frequency of occurrence of such accidents is 10^{-6} – 10^{-4}/(reactor·year), so such accidents are also called *postulated accidents*. Once such accidents occur, a large number of radioactive substances will be released, so such accidents must be considered in the design of nuclear power plant. Such accidents are the accidents with an extremely small probability but extremely severe consequences, and it may result in release of a large number of radioactive substances. Therefore, such accidents must be considered in the nuclear safety analysis, and shall be used to represent the limit design conditions. The engineered safety systems must function reliably under Conditions

IV to ensure the transition of reactor to cold shutdown in time. The release level of radioactive fission products shall neither severely endanger the health of personnel nor severely contaminate the environment inside and outside the ship.

The basic requirements for the safety design of marine nuclear power plant are as follows, in case of common failures, there shall be no or only extremely few radioactive hazards, and in case of limiting accidents, the engineered safety features shall ensure the structural integrity of reactor coolant pressure boundary, the safe shutdown of reactor, and the control of accident consequences.

In order to ensure the safety of marine nuclear power plant, it is stipulated that the safety analysis report shall give detailed analysis and calculation for the accidents under Condition II, Condition III and Condition IV, provide the quantitative results and assess whether its results meet the current specifications and standards.

The accidents to be analyzed for the marine PWRs are shown in Table 12.2. It can be seen that the reactor accident analysis involves a wide range of problems in reactor physics, thermal hydraulics, control, structure, shielding and dose protection.

Since the risk is equal to the consequences times the frequency of occurrence of the accident, the above-mentioned accident classification is mainly based on the frequency of occurrence and the consequences of the accident. The design must ensure that the consequences of accidents with high frequency are not severe, and that the frequency of accidents with severe radioactive consequences is very low,

Table 12.2 List of accidents to be considered in the analysis of typical accidents in marine nuclear power plant

Incidents of moderate frequency	Infrequent incidents	Limiting accidents
• Feedwater temperature decrease • Feedwater flow increase • Steam flow increase • Loss of condenser vacuum • Main steam system depressurization • Total loss of feedwater flow • Partial loss of reactor coolant flow • Uncontrolled lifting of a group of control rods • Spurious start-up of outage loop • Unexpected start-up of reactor coolant pump during natural-circulation operation • Spurious start-up of high-head safety injection system • Ship blackout	• Total loss of reactor coolant flow • Uncontrolled lifting of a cluster of control rods • Spurious opening of pressurizer valve • Steam generator tube rupture • Small break of reactor coolant piping • Small break of main steam line	• Large break of main steam line • Feedwater line break • Reactor coolant pump rotor seizure • Reactor coolant pump shaft break • Control rod ejection accident • Large break of reactor coolant piping • Ship blackout (loss of Class 1E power supply) • Leakage and diffusion of radioactive gas to other compartments due to the reactor compartment seal damage

so as to effectively control the risks related to marine nuclear power reactors. The purpose of accident analysis is to verify whether the design of the reactor and its engineered safety features can ensure the sufficient reactor safety.

As can be seen from Table 12.2, the accidents and incidents studied in the design and construction of marine nuclear power plant can be divided into two categories,

① Accidents without break, mainly refer to the normal transients, such as reactivity insertion accidents, loss-of-flow accidents and loss of heat sink accidents.
② Line break accidents characterized by loss of primary and secondary fluids, such as steam line break accidents, feedwater line break accidents and LOCAs.

This section will introduce the causes, phenomena and results analysis for all kinds of accidents based on the accident characteristics.

12.3.2 Reactivity Insertion Accidents

A reactivity insertion accident refers to an accident that leads to reactor power excursion because of the unexpected reactivity insertion in the reactor. If such an accident takes place during the start-up, the prompt criticality may take place and the reactor may be out of control. If such an accident takes place during the power operation, the reactor will be severely overheated and the integrity of primary loop system pressure boundary may be damaged. It should be noted that, because of various reactivity feedback effects in the reactor, all accidents in marine nuclear power plant, whether the thermal accidents or physical accidents, will eventually lead to changes of core reactivity. This section only discusses the reactivity insertion accidents directly caused by improper reactivity regulation measures.

For the PWR cores, the reactivity insertion accidents can be divided into the following two categories based on the potential factors.

① Uncontrolled control rod withdrawal accidents. Due to the malfunction of reactor control system or control rod drive mechanism, the control rods are withdrawn uncontrollably and continuously produce the positive reactivity, thus causing the continuous power increase. This phenomenon is called the uncontrolled control rod lifting accident (also called control rod withdrawal accident). Uncontrolled control rod withdrawal can be a Category II accident (such as uncontrolled lifting of a group of control rods) or a Category III accident (such as uncontrolled lifting of a cluster of control rods).
② Control rod ejection accidents: during the operation of a PWR, because of the rupture of the sealing housing of control rod drive mechanism, the control rod drive shaft causes the rapid ejection of control rod out of the core under the differential pressure of core. Such an accident is called rod ejection accident. The consequence of this mechanical failure is the loss of coolant and the insertion of the step change of reactivity in the reactor. The amount of step reactivity insertion

is the integral worth of reactivity of the part of ejected rod originally inserted in the reactor, and the effect of loss of coolant from the break is equivalent to a small break of primary pipe accident. Rod ejection accidents belong to Category IV accidents.

12.3.3 Loss-of-Flow Accidents

The reactor of a marine nuclear power plant is cooled by forced circulation by means of coolant transported by main coolant pump. For a reactor in power operation, if the main pump is shut down because of power supply failure or mechanical failure, the coolant flow will decrease or be totally lost, thus cause mismatch between the coolant flow and the reactor power and further leading to fuel cladding temperature increasing rapidly in core. This phenomenon is called a loss-of-flow accident.

The loss-of-flow accidents are divided into four (4) categories, i.e. partial loss of flow, complete loss of flow, main pump rotor seizure and main pump shaft break. Main pump rotor seizure and main pump shaft break are limiting accidents. The partial loss-of-flow accident refers to the condition that some main pumps idle due to blackout or failure, while the complete loss-of-flow accident refers to the condition that all main pumps idle due to blackout or failure.

12.3.4 Loss of Heat Sink Accidents

A loss of heat sink accident refers to an accident that finally results in the insufficient cooling ability or total loss of cooling ability of the core because secondary loop failure which cannot remove all heat generated from the core. For the primary loop system of a reactor, in order to remove the heat released by fuel fission at the rated power, there must be a heat sink, i.e. a secondary loop cooling system that functions normally. If there is a failure in the secondary loop or the third loop, the heat generated in the primary loop cannot be removed in time under normal conditions. As a result, the temperature at the core inlet of primary coolant will become excessively high, thus causing the overheating of core or even the damage to the fission product barriers. Such accidents are referred to the loss of heat sink accidents. For the PWRs, there are two kinds of initiating events for the loss of heat sink accidents,

① The feedwater is partially or totally interrupted. This is a typical loss of heat sink accident with high probability. The flow decrease or interruption of feedwater may be caused by the mechanical failure or loss of power supply of feedwater pump, the accidental closure of valve, the damage to feedwater heater, or even the damage to equipment or pipes such as condensate pump. Once the flow of feedwater decreased, the steam generator level will decrease, or even the steam generator will be fully filled with steam. As a result, the heat transfer coefficient will decrease significantly, and the primary side will be in an approximate adiabatic heating state. In this case, the reactor scram must be performed and the residual heat removal system must be started up to remove the decay heat to prevent the core damage.

② The turbine is shut down and the bypass valve is not open. Whether it is emergency shutdown, turbine generator failure or power grid failure, in order to protect the turbine, the steam inlet valve shall be closed immediately after the turbine is shut down and the bypass valve to the condenser must be opened immediately. Otherwise, a loss of heat sink accident will take place. If the bypass valve is not opened after a trip, the secondary loop will be filled with steam, and the energy stored in the primary loop of the core cannot be removed. In this case, the temperature and pressure of primary coolant will increase rapidly under an approximate adiabatic state.

12.3.5 Steam Generator Tube Ruptures

A steam generator tube rupture accident (SGTR) refers to an accident caused by the rupture of one or more tubes in the steam generator (also including the crack causing the minor continuous leakage). SGTR will cause the second barrier (i.e. the reactor coolant pressure boundary) of nuclear power plant to lose its integrity and lead to the connection between the primary loop and the secondary loop, thus causing the contamination of secondary loop by the radioactive primary coolant. In addition, it should be pointed out that the steam generator tube rupture accident will also make the radioactivity directly bypass the third barrier (i.e. the reactor compartment) of nuclear power plant enter other compartments or condenser.

The main causes of steam generator tube rupture include the mechanical and thermal stress of the tube and the corrosion caused by the feedwater, especially the local tube wall thinning above the tube sheet caused by the deposits at the tube sheet, the tube rupture and the corrosion caused by primary coolant.

12.3.6 Loss of Coolant Accidents

A loss of coolant accident (LOCA) refers to the break or rupture of reactor coolant pressure boundary that causes the leakage of coolant. The LOCA phenomenon is very important in the reactor safety research because of its complexity and severe consequences.

There are many kinds of LOCAs caused by the rupture of primary loop system of a PWR. The classification and possible consequences mainly depend on the fracture characteristics, i.e. the position and size of break.

The most severe LOCA is the catastrophic rupture of core vessel below the core water level, in which there is no cooling water in the core, so the core meltdown cannot be prevented and large radioactive substances will be released. In fact, the accurate calculation indicates that the probability of reactor vessel leakage (or break) is smaller than that of pipeline break by several orders of magnitude. Therefore, only the large size pipeline break is used as the limiting design basis accident.

According to the differences of break sizes and physical phenomena, the LOCAs are generally divided into large breaks, small and medium breaks, small breaks in steam chamber and steam generator tube ruptures. There are not absolute boundaries among the large, medium and small breaks. Generally, they are determined according to the capacity of safety injection system. The areas of large, medium and small breaks of different nuclear power plant are also different.

The blow down out process of LOCAs in PWRs is more complex and the emergency core cooling is more difficult. Therefore, in order to analyze the physical process and consequences of the LOCAs and ensure the safety of reactor, the more complex models must be established for the more detailed transient analysis and calculation, and a series of experiments shall be performed to verify the calculation results.

12.3.7 Ship Blackout Accidents

The failure of power system in nuclear power plant may cause the loss of power supply for the entire ship, thus causing the loss of normal function of electric equipment. The ship blackout accidents are generally divided into two levels, the first level is the ship blackout that standby power supply is available on board, and the second level is the loss of all power supplies including the standby power supply.

Since ship with reactors sails on the sea, it is impossible to obtain the electric power from the power supply system outside the ship when ship blackout accident occurs, the consequences of which will be very severe. Therefore, there are very high standards for the arrangement of standby power supply and the operations of the operators under various types of emergencies. In addition, a set of passive residual heat removal system shall be arranged to use the seawater as the ultimate heat sink, so the passive system can be used to ensure the core safety for a long period when the ship blackout accidents occur.

If the ship blackout accident occurred, the reactor scram will take place, the standby power supply on board will be automatically put into operation, and the scram signals will put the active residual heat removal system into operation to remove the residual heat of the core.

If ship blackout accident occurred while standby power supply is not available, the control rod can be automatically inserted under the loss of power supply and the reactor scram will take place. Because of the loss of power supply, all electric equipment including the main pump will stop working, and the residual heat of core can only be removed by the passive residual heat removal system.

12.3.8 Anticipated Transients Without Scram

The anticipated transient without scram (ATWS) refers to the transients without scram or plant trip. In these transients, although the parameters of the primary loop or secondary loop exceed the protection setpoints, the control rod assemblies are not inserted into the core. The initiating events for ATWS that are generally analyzed include loss of main feedwater, turbine shutdown, loss of AC power supply, loss of condenser vacuum, accidental withdrawal of control rod group and accidental opening of pressurizer relief valve.

After an ATWS accident, the negative feedback characteristics are mainly relied on. The primary loop temperature rise will cause the reactor power drop, so the corresponding heat removal means to remove the core heat shall be considered in the accident analysis to ensure the reactor safety for a time as long as possible, and the operators shall take all possible means for manual shutdown after the identification of accidents.

12.4 Accident Analysis Cases

As can be seen from the previous section, for the PWRs used in most marine nuclear power plant, the methods and content of accident analysis are similar to those of commercial NPPs. However, since there are great differences between the environment and safety system configuration of marine nuclear power plant and those of commercial NPPs, there are also great differences in the accident handling and mitigation means.

The ships with marine nuclear power plant sail on the sea for a long period, so they can only depend on their own corresponding systems and equipment to handle and mitigate the accidents without the support of other power supply, water supply or other supplies. This section will take the ship blackout accident as an example to introduce the methods and procedure of safety analysis for marine nuclear power plant.

12.4.1 Causes of Ship Blackout Accidents

The failure of power system will cause the loss of normal power supply. After an accident, the low speed signals of main pump trigger the reactor scram. At the early stage of the accident, the core coolant flow decreases rapidly. If the core is not cooled for a long time, the heat transfer on the surface of fuel element deteriorates sharply and the local departure from nucleate boiling may occur, thus causing the element burnout.

After the standby power supply is connected, the active residual heat removal system can be put into operation to remove the residual heat of core and ensure the reactor safety. Once the standby power supply (marine diesel generators and batteries) is lost, the active residual heat removal system cannot be put into operation because of the ship blackout. In this case, during the entire ship blackout (including the loss of standby power supply) accident, only the passive residual heat removal system can be used to remove the residual heat of core, thus ensuring the safety of nuclear power plant.

The ultimate heat sink of a marine reactor can be provided by the seawater with infinite capacity, and the passive means can be used for a rather long time to remove the residual heat of core in an accident. On the contrary, the ultimate heat sink of a land-based NPP can only be a large pool with limited capacity, and the large pool only has the capacity of removing the decay heat for a rather short time without the support of AC power supply since the power of land-based NPP is rather large. The advantage of using the seawater with infinite capacity is usually utilized effectively in the marine reactor design. In this case, the reactor elevation shall be below the waterline of ship to a certain height to provide a sufficient elevation difference for the establishment of passive means.

12.4.2 Frequency of Occurrence and Limiting Criteria of Ship Blackout Accidents

According to the classification of accidents mentioned above, the ship blackout accidents belong to moderate frequency events and Accident Conditions II. The following two acceptance criteria shall be complied with:

① The core will not have any departure from nucleate boiling after the accident, and

② The residual heat can be removed for a long time to ensure the reactor safety.

12.4.3 Analysis Methods and Assumptions of Ship Blackout Accidents

Generally, in the ship blackout accidents, the reactor scram system and the residual heat removal system can be timely put into operation to ensure the reactor safety, the negative feedback caused by the core overheating can also timely protect the core safety, and the short-term safety margin is rather large. Therefore, the conservative analysis methods are usually used to obtain sufficiently conservative calculation results. The principal calculation assumptions are as follows,

(1) Initial conditions

The initial power of reactor is equal to 100% of the rated power plus the maximum steady-state thermal measurement deviation. The initial average temperature of reactor coolant is the nominal value plus the maximum steady-state control band and measurement deviation. When the DNBR is calculated, the initial pressure of the pressurizer is the nominal value minus the maximum steady-state fluctuation and measurement deviation. When the system pressure need to be calculated, the initial pressure of pressurizer is the nominal value plus the maximum steady-state fluctuation and measurement deviation. The initial flow of reactor coolant is the thermal design flow.

(2) Initial events and functional assumptions

It is assumed that at the time of 0 s, the ship blackout takes place, the main pump start to idle, the main feedwater is cut off, and the low speed signals of main pump trigger the reactor scram. After the scram, the quick closing valves of primary equipment and auxiliary equipment are closed quickly.

(3) Assumptions related to core

The minimum values are taken for the fuel temperature reactivity coefficient and the moderator temperature reactivity coefficient.

(4) Control and protection system

The low speed signals of main pump trigger the reactor scram.One conservative assumption is that the pressurizer sprayers and the relief valves are not available.

12.4.4 Analysis Results of Ship Blackout Accidents

After a ship blackout accident, the main pump coast, the core flow decreases, and the average temperature of primary loop increases. The low speed signals of main pump trigger the reactor scram, the power decreases after the scram, and the temperature and pressure of primary loop decrease accordingly. The quick closing valves of primary equipment and auxiliary equipment are closed quickly after the shutdown, the secondary loop loses the heat sink, and the temperature and pressure of primary loop increase. After residual heat removal system is put into operation, it cannot immediately decrease the temperature and pressure since the core power level is still high. After a period of time, the temperature and pressure of the primary and secondary loops decrease, so the reactor safety is guaranteed.

If the standby power supply is lost during the ship blackout accident, the residual heat removal system cannot remove the residual heat. Meanwhile, the residual heat removal cannot be realized by the startup and shutdown feedwater system of the secondary coolant circuit. In addition, since the coolant-charging system cannot be put into operation to remove the residual heat of reactor, all active means for removing residual heat are lost. In this case, only the passive residual heat removal system can be used to remove the residual heat of reactor. The ship blackout accident analysis is done to study whether the passive residual heat removal system can remove the residual heat of reactor. After the passive system is put into operation, it shall ensure the removal of residual heat, thus ensuring the reactor safety.

During the design of marine reactors, it is necessary to perform the calculation and analysis to all accidents for nuclear power plant listed in Table 12.2, from the perspective of deterministic analysis, ensure that the nuclear power plant meet the corresponding limiting criteria after accidents, thus ensuring the safety.

Finally, the results of safety analysis need to be reviewed by the safety authorities. Only after the preliminary safety analysis report passed the construction can be started, and only after the final safety analysis report passed the commissioning and operation can be started. After the above steps are completed, the safety of marine nuclear power plant can be effectively guaranteed.

12.4.5 Severe Accidents

Severe accidents refer to a series of processes that begin with the large-area failure of fuel cladding in the core of marine nuclear power plant, which threatens or destroys the integrity of reactor pressure vessels or containments vessel, and cause the leakage of radioactive substances. In general, the severe accidents of marine nuclear power plant can be divided into two categories. The first category is the core meltdown accidents, and the second category is the core damage accidents. A core meltdown accident is a process of core exposure, temperature rise and meltdown caused by insufficient cooling of the core. The development of such accident is rather slow, with the time scale at the magnitude of hour. A core damage accident is a process of power surge and fuel fragmentation caused by the rapid insertion of huge reactivity. The development of such accident is rather rapid, with the time scale at the magnitude of second. Because of the inherent negative reactivity temperature feedback characteristics of PWRs and the arrangement of engineered safety features, the probability of core damage accidents in PWRs is minimal.

12.4.6 Major Phenomena and Processes of Severe Accidents

The core meltdown in the PWRs can be roughly divided into two categories, that is, high-pressure core meltdown and low-pressure core meltdown. At present, most designs of marine nuclear power plant include the depressurization system, so the probability of high-pressure core meltdown is low.

The initiating events of low-pressure core meltdown are the large and medium break LOCAs causing rapid depressurization. If the injection function or the recirculation function of emergency core cooling system cannot be performed, the core becomes exposed and melted down, and the zirconium alloy cladding reacts with steam to generate a large amount of hydrogen. The core level drops below the lower lattice plate, the reactor support fails and the molten core falls into the water in the lower plenum, generating a large amount of steam. There is a risk of melt through of vessel bottom head. Once the vessel bottom head is molten through, the molten core will fall into the reactor cavity, and the ablation of the reactor cavity wall occurs, which further cause the melt through of the hull.

12.4.7 Severe Accident Prevention and Mitigation

In the design of reactors, many design methods such as defense in depth, diversity design, and redundancy design are used to prevent severe accidents. However, for the PWRs, any safety system may fail, so the severe accidents of PWRs cannot be completely prevented. Instead, the probability of severe accidents can be continuously reduced. Once a severe accident occurs, the corresponding mitigation measures shall be designed to prevent the further expansion of accident consequences.

The basic goal of severe accident mitigation is to maintain the cooling of severe damaged core as much as possible, achieve a controllable final steady state and maintain the integrity of reactor pressure vessels and containment vessel as long as possible, thus gaining time for the implementation of emergency plan, minimizing the radioactive release to the environment and minimizing the long-term pollution of the seawater. Both the experiments and the analysis indicate that, after the core meltdown, there is a very obvious time effect on the settlement and retention of radioactive substances in the containment. Therefore, it is extremely meaningful to avoid the initial failures of containment vessel and delay the failure time as far as possible. The mitigation of severe accidents mainly includes the following aspects.

(1) Prevention of high-pressure core meltdown

From the perspective of accident mitigation, in order to prevent the high-pressure core meltdown (DCH) from jeopardizing the early integrity of the containment, the high-pressure core meltdown shall be converted into a low pressure process as soon as possible.

The studies indicate that converting the primary loop to a low pressure process can be achieved by operators' actions (opening the pressurizer safety valve at the right time for depressurization), automatic depressurization system (if any) or natural process (natural circulation cooling). Some countries have also specially designed the operating procedures involving the system depressurization.

During the primary loop depressurization, it should be noted that the pressurizer safety valve shall be opened at the right time. If the safety valve is opened too early, there will inevitably be more loss of primary coolant inventory, which makes the core heating more obvious at the early stage.

(2) Containment vessel heat removal and depressurization

The internal pressure of the containment vessel has a certain relationship with the heat accumulated in the containment vessel, and the containment vessel depressurization process is also the heat removal process. Spraying is an important means of containment vessel heat removal and depressurization. Spraying has two functions: one is to condense the steam in the containment vessel to maintain a rather low pressure, and the other is to decontaminate the radioactive iodine and aerosols by spraying and additives, thus minimizing the possible release of radioactivity. The spraying mechanism analysis shows it is better to use the intermittent operation mode with small flow. This mode can ensure that the water in the refueling water storage tank is saved under the premise that the containment vessel pressure is not higher than the design setpoint, thus generally prolonging the time of spraying effect, i.e., postponing the overpressure time of containment vessel. According to the results of the analysis of severe accident timing sequence, one of the effective mitigation measures is to ensure the availability of at least one circuit of spraying injection for 4–5 h after an accident.

In fact, the simple spraying injection does not remove the heat from the containment vessel. Instead, it only uses the cooler spraying liquid to absorb some heat released from the core, temporarily suppressing the temperature rise of the containment vessel. The removal of heat in the containment vessel further relies on safety injection and recirculation spraying. The rather hot main coolant and spraying liquid accumulated in the pit are pumped out, and the heat is transferred by the heat exchangers to the component cooling water and then released to the environment. The cooled main coolant is re-injected into the main system or sprayed into the containment vessel. Therefore, safe injection and recirculation spraying are important cooling means for the heat removal of the containment.

Spraying and recirculation spraying are effective heat removal and depressurization measures, but they also have relatively large negative impact. The alkali-containing spraying liquid will cause the corrosion of equipment, and the remedial work is very complex. If the spraying is performed late after the accident, then most zirconium will be oxidized, and other metals will slowly react with the steam to generate the hydrogen. The spraying will quickly condense the steam and may substantially increase the hydrogen partial pressure in the containment atmosphere and even achieve the concentration for hydrogen explosion. Therefore, the spraying at the late stage must be treated with prudence.

Another available means of containment vessel heat removal and depressurization is the containment vessel air cooling system. The air cooling systems of some marine nuclear power plant are designed as Class 1E systems, which can automatically switch to the emergency operation state under accident condition, thus reducing the fan speed, increasing the base flow of service water and dehumidifying the air flow before the air flow enters the activated carbon absorber. For this marine nuclear power plant of this type, the air cooling systems are prior to the spraying. In addition, the air cooling systems of many marine nuclear power plants are only used to remove the heat generated by the main system equipment during the normal operation. They do not belong to the Class 1E equipments and their design capacity is rather small. Therefore, their contribution is not considered in the accident analysis. At the accident mitigation stage, if their supporting systems (power supply and cooling water) can provide the support, they may be put into operation. They can at least remove a considerable number of decay heat after the shutdown, thus it helps reduce the requirements on other mitigation systems.

(3) Hydrogen elimination measures

In order to eliminate the threat of hydrogen explosion and combustion and solve the problems related to the operation of spraying at the late stage, a perfect containment hydrogen recombination system shall be considered. The marine nuclear power plants are generally equipped with the Class 1E containment vessel hydrogen recombination systems. Part of air in the containment vessel is evacuated by the system, and to contact the metal catalyst net with a temperature of about 800 °C, thus facilitating the chemical combination between hydrogen and oxygen to eliminate the hydrogen.

The hydrogen igniter is an important hydrogen elimination device, which is similar to the mine safety light. This small device is arranged in the proper compartment. The small electric sparks in the igniter can trigger the chemical combination between hydrogen and oxygen.

Apart from the hydrogen igniter, the other device for mitigating the hydrogen deflagration is recombiner. These two kinds of devices can be used separately or simultaneously according to the progress of the accident. The operating principle of recombiner is the catalysis of $2H_2 + O_2 \longrightarrow 2H_2O$ to react under rather low hydrogen concentration. This reaction is passive, that is, it is self-started and self-supplied. There are neither moving parts nor needs of external energy supplies. As long as the concentration of hydrogen in the containment vessel begins to increase, recombiner will be actuated spontaneously.

(4) Final guarantees of containment vessel functions

In case of spraying and air cooling failures, there are two possible ways for the final guarantee of containment vessel functions:

① Depressurization by filtered exhaust

When the containment vessel is expected to experience an overpressure failure, part of gas in the containment can be exhausted in a controllable manner to achieve the depressurization. This measure will destroy the seal integrity of the containment. How to reduce the radioactive release to the environment is the key of the problem. Therefore, the exhausted gas shall be appropriately filtered.

② Flooding of containment vessel and reactor pit

If the water source can be guaranteed and the accident has become extremely severe, injecting a large amount of cold water into the containment is another possible measure to delay the containment overpressure. After cold water is injected into the containment, the water temperature will increase after absorbing the sensible heat and decay heat. It is impossible for the containment to exceed the pressure before reaching the saturation temperature under design pressure.

Reference

1. Safety Margins of Operating Reactors: Analysis of Uncertainties and Implication for Decision Making [R]. IAEA-TECDOC-1332, IAEA, 2003.

Chapter 13
Operation and Operation Analysis

13.1 Overview

The operation of marine nuclear power plant is similar to that of NPPs [1], mainly including initial start-up, normal cold start-up, hot start-up, steady-power operation, variable condition operation, abnormal operating condition, accident operating condition, hot shutdown and cold shutdown. However, unlike the NPPs, the marine nuclear power plants have to frequently change their operating conditions to meet relevant mission requirements. Therefore, during the reactor design, the calculation and analysis shall be performed for a large number of operating conditions, and the results shall be used as an important basis for the formulation of operating procedures.

Design transient analysis, accident analysis and operation analysis shall be performed for the reactor during the design of marine nuclear power plant. The design transient analysis is to provide the design inputs for the design of some control parameters and the mechanical calculations. The accident analysis is mainly to grasp the safety performance of the reactor and ensure that the reactor design can guarantee the reactor core safety under accidents. The methods and assumptions used in such transient analysis are completely different from those in operation analysis studies. For example, the methods and assumptions used in accident analysis shall be conservative and envelope-based. This combination of the most adverse factors is actually a very conservative and envelope-based assumption. Generally, such state generally does not exist in the actual operation of marine nuclear power reactors.

The operation analysis uses the quasi-realistic theoretical modeling of nuclear power plant to reflect the actual operation of nuclear power plant. The results can rather realistically and objectively reflect the response of different parameters under operation transients (such as normal transition of operating conditions) and accident conditions, reflect the maneuverability and operability of nuclear power plant, and provide the operators with information that is closer to the actual operating conditions. The operation analysis can systematically evaluate the various measures for accident intervention and mitigation, optimize the operating procedures for the operators, and provide theoretical support for formulating more reasonable operating

J. Yu, *Marine Nuclear Power Technology*,
https://doi.org/10.1007/978-981-15-2894-1_13

procedures for operating the nuclear power plant. The operation analysis can also be used to enrich the operation databases and provide the basis for fault diagnosis techniques.

The rationality and effectiveness of the operation analysis results depend on the collection and processing of a large number of basic data, including the manufacturing data of systems and equipment, and the feedback on actual operating data during the operation of nuclear power plant. These data can be used as the basis for continuously revising and perfecting the inputs of operation analysis, thus improving the reliability of operation analysis results.

13.2 Operation

13.2.1 Initial Cold Start-Up

The initial cold start-up is the first start-up of a newly built or newly refueled reactor. For new reactors or newly reloaded reactors, the physical and thermal parameters must be calculated before the start-up and various tests must be performed at zero power to obtain certain information of physical and thermal performance. However, because of the differences between the assumptions used in the calculation and the real conditions, the data obtained from the initial start-up often have errors when compared with the theoretical calculation results. Therefore, a number of start-up physics tests shall be performed during the initial start-up, so as to check the calculated the zero-power test data, accurately grasp the reactor performance. Finally the reactor power is increased to the design rated level.

The cold start-up requires the support of the auxiliary systems for primary and secondary loops of the marine nuclear power plant, and mainly includes such processes as system inspection, system cleansing, hydrostatic test, comprehensive system commissioning, fuel loading and criticality supervision, initial criticality, tests and experiments.

The main processes of cold start-up include initial temperature rise and deaeration of systems, establishment of pressurizer steam chamber after temperature and pressure rise and stable operation at certain power.

13.2.2 Normal Cold Start-Up

The normal start-up of a marine nuclear power plant is the normal start-up of its reactor, also known as the routine start-up. It refers to the cold start-up of the reactor after a scheduled shutdown or accidental shutdown.

The primary loop system has two types of cold start-up. In the first type of start-up, an external heat source (instead of depending on the heat source generated by the neutron fission in the reactor, the electric heating elements of pressurizer are put into operation and the main coolant pump is operating at a high speed, thus realizing the heating) is used to heat the reactor and the primary loop system to a certain temperature (generally after the establishment of steam chamber), and the control rods are lifted for the reactor start-up and operation. This type of start-up is usually called the electrical heating start-up. If the start-up is performed at a dock, the power will come from the power supply system of the dock. If the start-up is performed at sea, the power will come from the storage battery and diesel generator set of the ship. In the second type of start-up, the reactor is put into operation at a low power from the beginning of reactor start-up, and the heat generated by the reactor fission is used to heat the primary loop. This type of start-up is usually called nuclear-electrical combined heating start-up.

These two methods have their own advantages and disadvantages. The electrical heating start-up takes a rather long time, but the control is simple and the core is relatively safe. The nuclear-electrical combined heating start-up takes a rather short time, but it may bring risks to the reactor utilizing steam for pressurization, mainly because of the possibility of prompt supercriticality and low-temperature overpressure accident in primary loop system. Therefore, it is generally prohibited to use the nuclear heating cold start-up or nuclear-electrical combined heating cold start-up for the land-based NPPs. However, for marine nuclear power plant, the mission sometimes requires the rapid departure, and the nuclear heating can shorten the start-up time and reduce the dependence on external power supply. The nuclear-electrical combined heating cold start-up imposes high requirements for operation and control and requires the effective accident prevention measures. It shall ensure that there will be no spurious actuation of control rods and that the rates at which the temperature and pressure of reactor coolant rise do not exceed the limits. Although the nuclear heating can increase the maneuverability and vitality of marine nuclear power plant, it is strictly forbidden to use the nuclear heating for marine reactors before the delivery or without the verified nuclear heating start-up procedures.

13.2.3 Steady-Power Operation

Steady-power operation refers to the operating conditions under which a marine nuclear power plant maintains a constant power level in accordance with the requirements for berthing or propulsion at a uniform speed. Generally, steady-power operation refers to the operation mode in which the power of a marine nuclear power plant does not change with time.

After the reactor start-up is completed, the reactor will be switched to the operating mode of automatic load-following of the secondary loop loads, the reactor coolant system will be put in the automatic control mode, the primary loop auxiliary systems will be put in the normal operation or standby state, the reactor protection system will be put into operation, the process measurement system will monitor the safe operation of systems and equipment to periodically inspect the parameters, and the nuclear power plant will enter the steady-power operation.

The systems that are put into continuous operation when the reactor is in normal operation include reactor coolant system, pressure safety system, coolant purification system, component cooling water system, primary shielding water system and sampling system. The auxiliary systems that are put into intermittent operation as required include coolant-charging system and waste treatment system. The auxiliary systems that are put into standby state include safety injection system, residual heat removal system and chemical shutdown system.

Theoretically, if the reactor power does not change with time, the reactor reactivity will not change with time. However, in fact, when the reactor is in steady-power operation, the reactor core will inevitably undergo a series of physical and thermal effects. These effects will cause the change in reactor reactivity, resulting in the change in reactor power. Therefore, the stability of steady-power operation is relative, while the imbalance in stability is common. This fact should be treated seriously. In fact, the reactor reactivity is determined by various factors. In addition, there are many causes of reactivity change, mainly coolant temperature, density, burnup and fission product changes. Although the reactivity change caused by these effects is rather slow or comes closer to balance after a certain time, if the timely compensation is not performed, the negative reactivity insertion will cause the power cutback, thus resulting in the reactor shutdown. If the positive reactivity is inserted, the power will increase, thus damaging normal operation of the reactor. Therefore, in order to maintain the steady operating conditions, the operating conditions of all systems and unit equipment must be monitored. The state parameters to be continuously monitored during the normal operational states include coolant temperature at the inlet and outlet of reactor; coolant flow; pressurizer water level, pressure and temperature; reactor coolant quality; and steam pressure and water level at secondary side of steam generator. Once the parameters deviate by a certain degree, they shall be automatically or manually restored to the normal operating parameters according to the operating procedures.

13.2.4 Variable Condition Operation

Variable condition operation refers to the transition process in which the reactor operating power changes with time. Variable condition operation is an important feature of marine nuclear power plant. The speeds of ship with marine nuclear power plant often changes on the voyage, and different speeds correspond to different reactor core powers of marine nuclear power plant. Variable condition operation is similar to the gear up or down for automobiles. In addition, the reactor power or ship speed is frequently changed when the ships dock or undock, so the nuclear power plant shall have good maneuverability.

For the PWRs, the variable conditions generally adjust the reactor power, whether by increasing or decreasing the power, to the power matching the power required by the secondary loop according to the requirements for the secondary loop load change. During the variable conditions, the reactor power changes with time, until it stabilizes at the required power, after which the PWRs enter the steady-power operation.

In order to prevent the reactor shutdown during the variable conditions, it is necessary to perform the operation analysis to grasp the change of parameters during the variable conditions of nuclear power plant and optimize the operating procedures, so as to ensure that the actuation of reactor protection system will not be triggered to shutdown the reactor during the condition transition and that the marine nuclear power plant can successfully reach the specified power and keep in operation.

13.2.5 Natural-Circulation Operation

The natural-circulation operation conditions are special operating conditions. Such conditions do not require the operation of reactor coolant pump, so they are of great significance for reducing vibration and noise of nuclear power plant. In addition, such conditions can ensure that the reactors have can output certain power output upon the loss of power supply to or failure of the main pumps, so they are significant to the vitality improvement of marine nuclear power plant.

The reactors are usually subject to the transition from forced-circulation power operation conditions to natural-circulation operation conditions, during which the reactor coolant flow is rather low, the temperature difference between the core inlet and outlet is rather large, and there are great differences between the operation parameters under such conditions and those under the power operation. There is a set of special control and protection signals determined by the transient analysis for corresponding of reactor operation monitoring and protection. When the reactor operating points have any deviations, the operating points shall be corrected or corresponding protection actions shall be triggered as soon as possible to ensure the reactor safety.

13.2.6 Cold Shutdown of Reactor System

The cold shutdown of reactor is usually a process of directly cooling a marine nuclear power plant from a hot state operating with certain power to a normal temperature state for equipment overhaul, refueling or long time repair.

The cold shutdown consists of the following processes: power reduction, temperature fall, depressurization, elimination of pressurizer steam space, further temperature fall to normal temperature, and further depressurization to normal pressure. During the cold shutdown, the coolant-charging system shall be put into intermittent or continuous operation. Before the elimination of pressurizer steam space, the minimum pressure for the operation of main pump shall be maintained, and the coolant shall have a certain subcooling. After the elimination of steam space, the coolant contraction cause by temperature fall shall be compensated, the pressurizer shall be fully filled with water, and the system pressure for main pump operation shall be maintained.

The principle for normal cold shutdown is to shut down the secondary loop first and then the primary loop. The power of primary equipment shall be reduced first. In this process, in order to ensure the safety of turboset, the load shedding rate shall be limited by the allowable temperature fall rate of the cylinder metal, and the power shall be maintained for a period of time after each load shedding so that the temperature of the turbine cylinder and rotor can decrease uniformly.

In order to prevent the main pump from cavitation, the system pressure shall not fall below a certain limit, and the system pressure can be reduced to the normal pressure only after the main pump is shut down. During the temperature fall, the temperature fall rate of the coolant in the pressurizer and that in the loops shall match each other; the water temperature in the pressurizer shall be higher than that in the loops but the difference shall not be too large; when the coolant temperature in the loops is rather low, the water temperature in the pressurizer and that in the loops can gradually approach each other and finally become the same. During the cold shutdown, before the main pump, the residual heat removal pump and the purification pump are shut down, the component cooling water system shall be always in operation to provide relevant components with cooling water. When the pressure of reactor coolant system drops to a certain value, the signal for automatic operation of low-pressure safety injection is removed so as to prevent the spurious operation of this the safety injection system. During the cold shutdown, the rate of temperature fall shall not be larger than the limit. When the temperature approaches the brittle transition temperature of structural material, the system pressure shall not be too high, and the temperature fall and the depressurization shall match each other.

13.2.7 Hot Shutdown and Hot Start-Up of Reactor System

During the hot shutdown, the reactor is shut down, the system temperature and pressure drop to the hot standby state, the pressurizer maintains the steam space, the auxiliary systems are put into standby or operation, and the electrical equipment is powered by reliable power supply or on-shore power supply.

The main process of hot shutdown is as follows: the power of primary equipment is reduced, the primary equipment is shut down, the automatic power regulation function is removed for the reactor, the power supply load is transferred, and the turbo-generator set is shut down; later, the control rods drop, the reactor is shut down, the operation of all electric heating elements is stopped, two main pumps run at a low speed, and the residual power in the reactor after the shutdown is used for the pressure and temperature maintenance of loops; however, as time goes on, the temperature of reactor circuit systems falls continuously because of external heat dissipation; in this case, the main pump shall be switched from low speed running to high speed running, and the electric heating elements shall be put into operation according to the situation, thus increasing the temperature of the circuit and the pressurizer and compensating for the heat loss of the circuits.

Under normal conditions, in the hot shutdown state, the reactor coolant temperature is maintained at about 200 °C and the pressure is maintained at a value above 3.0 MPa. The initial stage of hot shutdown is the same as that of cold shutdown. After the above conditions are reached, the temperature and pressure will not fall any more but be maintained in this state.

The special problem for hot shutdown is the iodine pit problem, which mainly affects the restart of reactor after the hot shutdown. The iodine well is related to the operating power of reactor before shutdown and the time of steady operation at this power. Because of the existence of iodine well, the reactivity change in the reactor upon shutdown becomes complex, so the start-up after hot shutdown faces special safety problems. Other steps are relatively simple, which is equivalent to the cold start-up for establishing the pressurizer steam chamber.

Most fission products generated by the nuclear fission of fuel during the operation of marine nuclear power plant have radioactivity. After a series of decay, such fission product forms many new isotopes. Some isotopes have very large thermal neutron absorption cross sections, including ^{135}Xe and ^{149}Sm. In addition, the precursors of above two isotopes have a rather large fraction.

Assuming that the reactor is in long-period operation, after a sudden shutdown, the neutron flux density decreases sharply, causing the sharp decrease of disappearance rate of ^{135}Xe, while ^{135}I still decays to ^{135}Xe, so the concentration of ^{135}Xe does not decrease, instead, it increases to a value much larger than the value in the balance state. This is the xenon poisoning. The xenon poisoning can cause the sharp fall of effective multiplication factor after the shutdown, causing the insufficient excess reactivity. The loss of reactivity at the maximum xenon concentration is called the maximum reactivity equivalent of iodine well.

The extent to which the excess reactivity of marine nuclear power unit drops to the minimum after shutdown is known as iodine well depth. The iodine well depth is related to the neutron flux density before the shutdown of marine nuclear power plant. The larger the thermal neutron flux density, the larger the iodine well depth. In addition, the iodine well depth after shutdown is also related to the shutdown mode. If the shutdown is performed by gradually reducing the power, instead of sudden shutdown, since some ^{135}Xe and ^{135}I are consumed via neutron absorption and decay during the shutdown, the iodine well depth after shutdown will be smaller than that after sudden shutdown.

Special attention shall be paid to the start-up of marine nuclear power plant with the iodine wells, because the concentration of ^{135}Xe in the reactor is rather large. Since the neutron flux density increases suddenly, ^{135}Xe will be consumed heavily after the neutron absorption, the reactor will rapidly release the positive reactivity to rapidly increase the excess reactivity in the reactor. In this case, although the automatic rods automatically follow the reactivity change and are inserted, the following range will be limited or the following will be lost sometimes, thus causing the accident [2].

At the last stage of the operation of marine nuclear power plant, if the excess reactivity after shutdown is smaller than the iodine well depth, the marine nuclear power plant shall be restarted as soon as possible after the shutdown, or restarted after maximum xenon poison reactivity disappears and the excess reactivity value is larger than zero. Therefore, at the last stage of the operation of marine nuclear power plant, the operating personnel must know the excess reactivity in the reactor and the amount of xenon poison, and determine the allowable operating power level and reasonable shutdown mode of the marine nuclear power plant, ensure that the marine nuclear power plant can be started up at any time. All the above results shall be considered in advance through operation analysis, and reasonable and effective operating procedures shall be formulated to ensure the vitality of marine nuclear power plant.

13.2.8 Reactor Operation Under Abnormal Conditions

Abnormal conditions are the non-normal conditions between normal conditions and accident conditions, mainly including loop flow asymmetry, loop temperature difference inequality, single-loop operation, local control rod abnormality, partial damage to fuel element cladding in the reactor and partial equipment failure. The operating capacity of a marine reactor under abnormal conditions is the embodiment of ship vitality. To ensure the nuclear safety, under certain measures and conditions, it is extremely important for a nuclear power plant to operate at a certain power level to fulfill its mission. Therefore, the operation under abnormal conditions is also an important part of the operation of marine nuclear power plant. This section will focus on several typical abnormal operating conditions.

When the reactor is operating under normal conditions, the main pumps in the loops can operate at a high speed or at a low speed. Let's take a two-loop marine nuclear power plant as an example. In the normal operation, the loop flow is symmetrical, but under certain special conditions, the main pump in one loop is operating at a high rotational speed, while the main pump in the other loop is operating at a low rotational speed. The difference in rotational speeds of the pumps in two loops will inevitably cause the flow inequality in these two loops, thus resulting in flow asymmetry of these two loops. In principle, this operation mode is allowable in the design, but it is disadvantageous in the long term. It can only be used under special conditions and it must be performed with reduced power.

Under normal conditions, the temperature difference between the inlet and outlet for the operating loops is almost the same. For some abnormal conditions, the temperature difference between the inlet and outlet of different loops may be different. Let's take a two-loop marine nuclear power plant as an example. When the reactor is operating at the same flow, the temperature difference between the inlet and outlet of the left loop is unequal to that of the right loop. In principle, the operation under these conditions are allowable, but when the temperature difference of left and right loops exceeds the design limit, the power must be reduced for operation. The operating experience indicates that the causes of temperature inequality in two loops mainly include the leakage of one loop or the failure of one feedwater pump in one loop.

Single-loop operation refers to the operating conditions under which one loop is shut down and the other loop is operating normally. Single-loop operation is only allowable under extraordinary circumstances. For example, when either loop stops operating because of main pump failure, steam generator damage, damage of the main pipe or auxiliary pipe other than the pipe between main gate valve and pressure vessel or equipment failure, but the marine nuclear power plant shall accomplish its mission, the single-loop operation is allowable. For the single-loop operation of a double-loop reactor, the reactor power shall be less than 50% of the rated power.

The reactor core contains thousands of fuel element rods, all of which are operating under the conditions of high temperature and high radiation. The fuel element cladding may suffer from erosion, mechanical vibration and thermal stress impact because of process defects, missing inspections or other causes, so some fuel elements will inevitably be damaged. The fission products in the damaged fuel will enter the coolant, thus increasing the radioactivity level of the primary loop and enhancing the dose level in the reactor compartment environment. Therefore, for the marine power reactors, the element damage conditions shall be continuously monitored during the navigation by measuring the core dose level and determining the nuclides of fission products in the coolant. Once the dose in the reactor compartment exceeds the standard and the nuclides such as cesium and iodine generated by ^{235}U fission are detected in the coolant, it can be judged that the damage to fuel elements have occurred. After it is confirmed that the fuel element damage accident has taken place, when it is necessary to continue the navigation, the reactor shall operate at a reduced power while strengthening the supervision of the dose to the personnel and watching for the leakages in steam generators and component cooling water to prevent more severe compartment contamination and dose increase.

13.3 Operation Analysis of Reactor Accident Conditions

13.3.1 Purpose

When an accident occurs in the reactor, the response principle is to judge the accident in a timely and accurate manner, deal with the accident decisively and correctly, minimize the accident hazards and ensure the reactor safety. Therefore, it is necessary to correctly understand the causes, process and consequences of the accident, grasp the special phenomena during the development of the accident, so as to provide important guidance basis for rapid accident diagnosis and intervention by the operators, and for the formulation of accident disposal measures. This is extremely important for the safe operation of marine nuclear power plant.

13.3.2 Methods

At present, the commonly used best estimate analysis codes for thermo-hydraulic transient analysis mainly include RELAP5, TRAC, RETRAN and CATHARE. Let's take RELAP5 as an example. This code is based on non-uniform and non-equilibrium hydraulic models, completely uses the two-phase models to solve the mass and energy conservation equations of the system, perform the simulation as accurately as possible for the basic physical processes, provide satisfactory description of various phenomena, and cancel many unnecessary conservative assumptions. The best estimate codes are rather suitable for the experimental evaluation and estimation of the safety margin of conservative evaluation modeling codes, and they are also used as the main tools for the operation analysis.

In addition, the initial values of all reactor parameters used in the analysis take the design values, and the steady-state operation deviations and measurement errors are not considered, so as to make the values closer to the true values of reactor parameters and obtain more real post-accident transient processes.

13.3.3 Content

The operation analysis shall include all steps in each reactor activity such as start-up, steady operation, operation during condition transition, operation under abnormal conditions, operation under accident conditions and reactor shutdown, thus developing the corresponding operating procedures.

For the operation analysis, the operation during condition transition and the operation under accident conditions are rather complex. In the above operation modes, the reactor parameters change sharply, making it rather difficult for the operators to control and operate the reactor. However, the intervention means used by the operators must be appropriate. Therefore, it is particularly necessary to formulate accurate operating and disposal procedures to ensure the safe operation of reactor.

In the following text, let's take the operation during condition transition and the operation under accident conditions as the examples to introduce the calculation methods, process and content of operation analysis.

13.4 Operation Analysis Cases

13.4.1 Analysis of Transition Between Forced Circulation and Natural Circulation

The condition transition of marine nuclear power plant generally includes transition from high rotational speed to low rotational speed and transition from low rotational speed to high rotational speed for main pump, transition from forced circulation to natural circulation, transition from natural circulation to forced circulation, rapid power ascension of primary equipment and rapid power reduction of primary equipment, etc. The condition transition between forced circulation and natural circulation, as an extremely special transition process for marine nuclear power plant, involves the coordinated operation and actions of a large number of systems and equipment.

The mutual transition between forced circulation and natural circulation belongs to Conditions I. In order to ensure that the marine nuclear power plant has good stability in the transition process, the following requirements shall be satisfied:

① In the transition process, no reactor protection actions shall be triggered;
② In the transition process, triggering the actuation of pressurizer relief valve is not allowed; and
③ In the transition process, the steam parameters shall meet the requirement of steam consumption.

To meet the above requirements, the control and regulation systems shall operate reasonably, and the main pump control system, the reactor power regulation system and the steam generator feed water regulation system shall be put into operation, so as to match the change in reactor power.

During the mutual transition between forced circulation and natural circulation, the main pump control system is used to start up or shut down the main pump at the right moment. Different power control methods are adopted for the reactor power regulation system according to different operating conditions. The regulating rod movement speed is obtained by comparing the reference values of temperatures at reactor inlet and outlet under the target conditions with the actual average temperatures at the core inlet and outlet, thus obtaining the required power according to reference to feedwater flow and comparing such power with the high auctioneer nuclear power signal (see Fig. 13.1).

In the above regulation, the feedwater flow shall be obtained by the steam generator feedwater regulation system. According to the operating characteristics of nuclear power plant, the steam generator feedwater regulation system has two control modes:

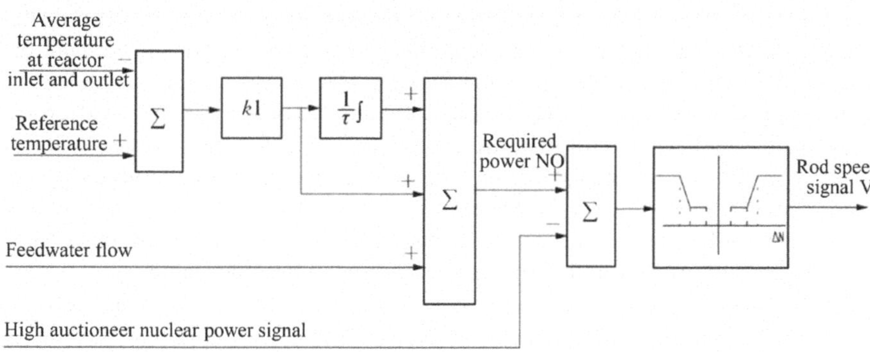

Fig. 13.1 Principle of reactor power regulation

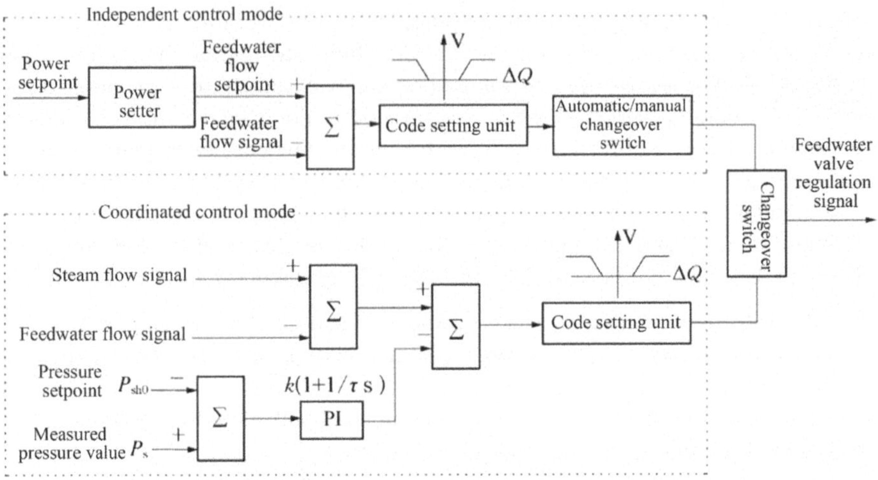

Fig. 13.2 Principle of steam generator feedwater regulation

independent control mode and coordinated control mode. These two modes can be mutually switched (see Fig. 13.2). In the independent control mode, the feedwater valve speed regulation signal can be obtained based on the deviation between the feedwater flow setpoint and the measured feedwater flow, or can be directly given manually to make the feedwater valve change its position at a fixed speed, thus changing the feedwater flow and the reactor power. This control mode may eventually fail to regulate the steam pressure to the setpoint during the transients. In the coordinated control mode, the measured value and the setpoint of steam pressure are compared to generate the primary regulation signal, and the steam flow signal and the feedwater flow signal are compared to generate the secondary regulation signal. According to the signal obtained by the superposition of the primary and secondary regulation signals, the feedwater valve speed regulation signal is obtained. This control mode can make it easier to regulate the steam pressure to the setpoint during the transients.

For the transition from forced circulation to natural circulation, as illustrated in Figs. 13.3, 13.4 and 13.5, at the initial stage of the transient, since the main pump has been shut down, the core flow decreases rapidly, which causes the coolant temperature increase in the core, thus resulting in the negative reactivity insertion that makes the reactor nuclear power decrease rapidly. Meanwhile, the primary loop flow decrease results in the rapid decrease of pressure and steam flow at the secondary side. Later, the water level at the secondary side and the inlet temperature at the primary side of steam generator increase, and the pressure and steam flow at the secondary side of steam generator begin to increase. Finally, under the function of regulation systems, the parameters of primary and secondary loops tend to become stable.

For the transition from natural circulation to forced circulation, as illustrated in Figs. 13.6, 13.7 and 13.8, at the initial stage of the transient, the power regulating system is put into operation, so the reactor power and the average temperature and

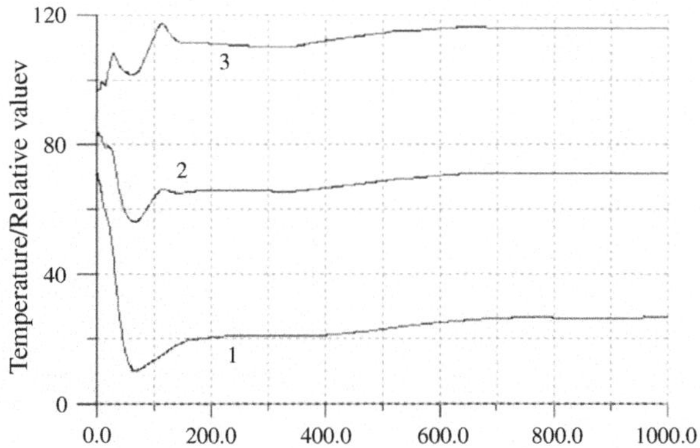

Fig. 13.3 Reactor coolant temperature change during transition from forced circulation to natural circulation. 1—Inlet temperature; 2—average temperature; 3—outlet temperature

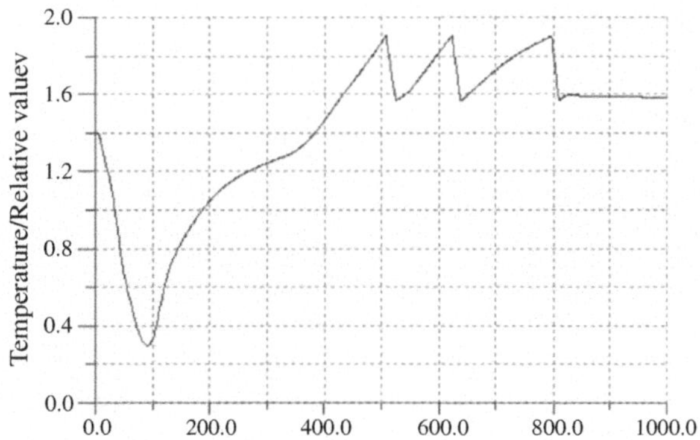

Fig. 13.4 Pressurizer pressure change during transition from forced circulation to natural circulation

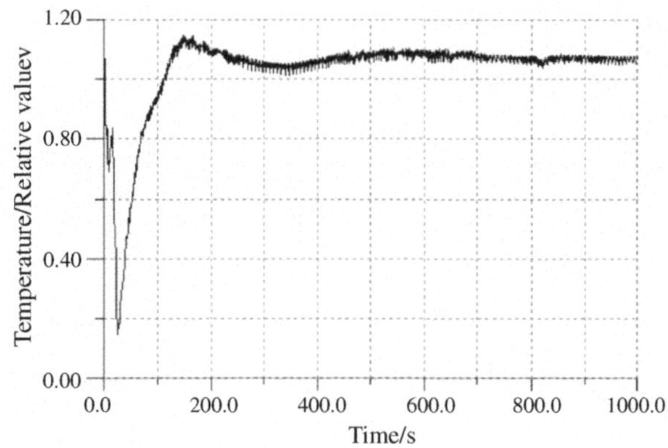

Fig. 13.5 Changing in pressure at secondary side of steam generator during transition from forced circulation to natural circulation

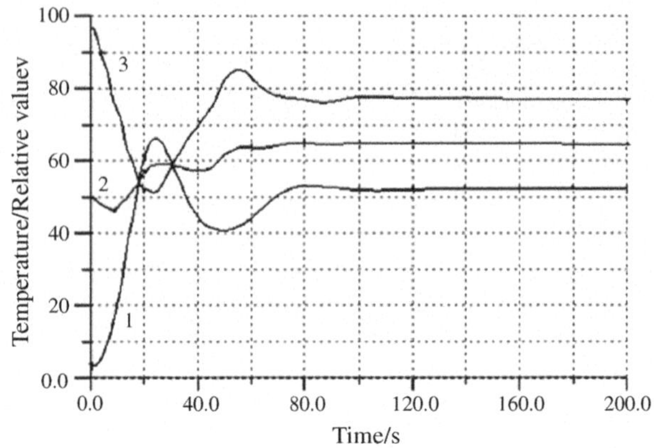

Fig. 13.6 Reactor coolant temperature change during transition from natural circulation to forced circulation. 1—Inlet temperature; 2—Average temperature; 3—Outlet temperature

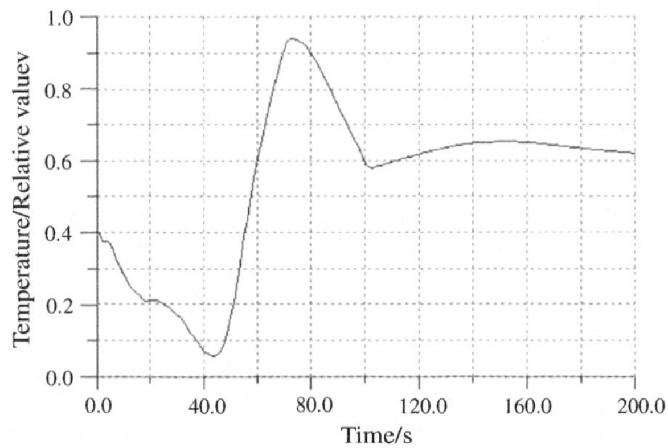

Fig. 13.7 Pressurizer pressure change during transition from natural circulation to forced circulation

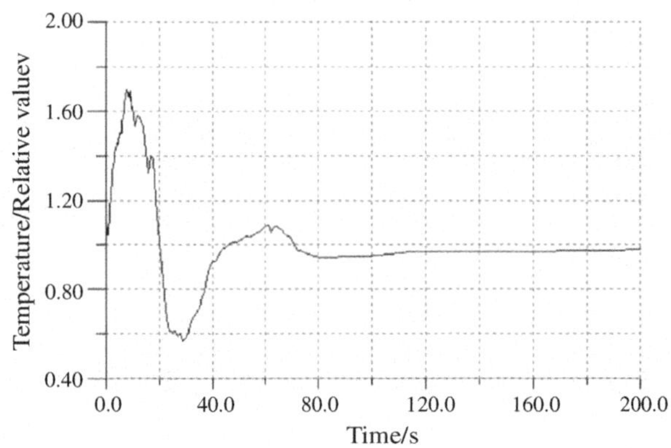

Fig. 13.8 Pressure change at secondary side of steam generator during transition from natural circulation to forced circulation

pressure of primary loop begin to decrease. After a certain time, the reactor coolant pump is put into operation, so the primary loop flow increases rapidly, the primary loop temperature and pressure decrease rapidly, the reactor power increases rapidly, and the steam temperature and pressure at the secondary side of steam generator begin to increase. Subsequently, under the regulation of the reactor power control system, the reactor power begins to decrease, and the temperature and pressure of the primary loop begin to increase. After the reactor power becomes stable, the temperature and pressure of the primary and secondary loops gradually become stable, and a new stable operating condition is finally reached.

13.4.2 Operation Analysis of LOCAs

A LOCA mainly refers to the loss of the primary coolant, such as a primary loop pipe break, a break in auxiliary pipe connected to the primary pipe, with the isolation valve jammed at opening position. LOCAs are generally classified into large, medium and small LOCAs according to the rate of loss of coolant.

Generally, in the reactor coolant inventory reduction accidents, the large break LOCAs are generally the most severe. In a large break LOCA, since the size of break is very large, the reactor core becomes exposed quickly, the pressure decreases rapidly, and the safety injection system is put into operation soon to inject the coolant for cooling the core. In a small break LOCA, the depressurization rate in the primary loop is low, and it is possible that only the high-head pressure safety injection system is put into operation during the accident (the medium and low-head pressure safety injection systems cannot be put into operation because of the high primary loop pressure). Hence the loss of coolant from the break cannot be compensated, causing the core exposure and the fuel element temperature increase and damage.

According to the positions of breaks, the small break LOCAs can be divided into three (3) types: cold leg break, hot leg break and steam chamber. Generally, the small break LOCAs related to cold leg break are more severe than hot leg break. The small break LOCAs of steam chamber refer to the LOCAs caused by small breaks in the piping system above the pressurizer steam space, and they shall attract special attention because the positions of such breaks may not be isolated.

The small break LOCAs require the operators to take rather many disposal actions and make judgement within a short time to mitigate the consequences of such LOCAs. Therefore, during the comprehensive special operation analysis of small break LOCAs, it is necessary to study the symptoms of such accidents, define the criteria for the operators to conduct accident diagnosis and intervention and study the disposal measures for such accidents, thus providing technical basis for formulating corresponding accident disposal procedures. For small break LOCAs,the effects of the coolant-charging system to compensate for the loss of coolant vary with the break sizes. Therefore, during the operation analysis, the calculation and analysis shall be performed respectively for the situations where the coolant-charging system is available, partially available or unavailable and there is no intervention of operators. The break sizes that cause the continuous loss of reactor coolant but cannot reduce the reactor pressure to a level triggering the actuation of low head safety injection system are the core of the operation analysis of small break LOCAs.

During the accident diagnosis research, the first step is to study the methods to judge the occurrence and type of accidents. The operators can make the following judgement based on the alarm signals: the operators can monitor the pressurizer parameters such as water level and pressure according to the high dose alarm signals for reactor compartment or the low water level alarm signals for pressurizer. If there are high dose alarms for the reactor compartment, and the pressurizer pressure drops or the makeup water pump is stared frequently, it can be judged that there is a LOCA. If the pressure and water level of the pressurizer decrease after the coolant-charging system is put into operation, but, for a long time, the low head safety injection system cannot be put into operation, it can be judged that there is a small break LOCA. If the pressure and water level of the pressurizer increase after the coolant-charging system is put into operation, it can be judged that there is a small break LOCA, and the diagnosis and disposal measures shall be determined according to the operation analysis results for the small break LOCAs. If the pressure and water level of the pressurizer decrease after the coolant-charging system is put into operation and the reactor coolant system pressure decreases to the pressure that can trigger the operation of low head safety injection system, it can be judged that there is a medium break LOCA, and the diagnosis and disposal measures shall be determined according to the operation analysis results for the medium break LOCAs. If there are high dose alarms for reactor compartment after an accident, and the pressurizer pressure decreases, but the pressurizer water level increases, it can be judged that there is a branch pipe break in the pressure safety system, and the break diagnosis and disposal measures shall be determined according to the operation analysis results for the non-isolable small break LOCAs.

As can be seen from the study results of the accident symptoms and the accident progress, if there are no disposal measures taken, the small break LOCAs will decrease the pressure and water level of the pressurizer, the reactor will be shut down, and the core may be exposed. Therefore, the study of disposal measures for small break LOCAs shall focus on identifying the loop with the breaks and recover the water inventory of reactor coolant system as quickly as possible, so as to prevent the core damage caused by core exposure, effectively remove the residual heat of core in a long time and ensure the reactor safety. At the same time, for marine nuclear power plant, on the premise of ensuring the safety, the accident shall be terminated as soon as possible to restore the power operation. If the breaks cannot be isolated, it is necessary to study the methods such as cooling, depressurization and water makeup to ensure that the core is sufficiently cooled as soon as possible.

After a small break LOCA, if the high head safety injection system cannot compensate for the loss of coolant and the medium and low head safety injection systems cannot be put into operation, the following disposal measures can be considered:

It is recommended that the operators take the following measures when it is confirmed that there is a small break LOCA in a nuclear power plant:

① Reduce the power and isolate the damaged loop. When the operators take isolation measures, they shall closely observe the pressurizer pressure. If the pressurizer pressure increases, then the break is in the isolated loop, and the break is successfully isolated. During the isolation of the loop, pay attention to switching the residual heat removal system and the coolant purification system to the non-isolated loop. If the break cannot be successfully isolated, refer to the disposal procedure for non-isolable small break LOCAs for the accident disposal measures.

② After the successful isolation of the break, if the reactor is not shut down:

A. At least one makeup water pump is available: isolate the corresponding secondary loop and maintain the power operation of reactor. B. Both makeup water pumps are unavailable: manually shut down the reactor, and switch the residual heat removal system or passive residual heat removal system to the non-isolated loop to remove the residual heat of core.

③ If the reactor has been shut down before the break isolation, open the relief valve to relieve the pressure, until the low head safety injection system is put into operation for a period. After the relief valve is closed, if the pressure of reactor coolant system increases to a value higher than the setpoint for putting the low head safety injection system into operation and it does not decrease for a long time, open the relief valve again to put the low head safety injection system into operation, and finally switch to the recirculation of low-head safety injection system.

As can been seen from the above analysis, when a small break LOCA occurs in the actual operation, the dose in the reactor compartment increases and the pressurizer pressure and water level decrease, triggering the high dose alarms for reactor compartment and the low water level alarms for pressurizer. After the accident, the high-head safety injection system cannot compensate for the loss of flow from the break, and, for a long time, the low-head safety injection system cannot be put into operation. If there is no timely intervention of the operators, the core will be exposed and the accident will further deteriorate. Therefore, the operators shall timely and accurately determine the type of the accident and intervene as soon as possible.

④ For non-isolable small break LOCAs, if the pressurizer pressure is already rather low but does not trigger the operation of low-head safety injection system when the operator intervenes, the core may already contain gas, so the operators can have to actively depressurize the primary loop or start up the passive residual heat removal system to decrease the temperature and pressure of the primary loop; if the pressurizer pressure is rather high upon the intervention and the core does not contain gas, the operators can start up the active residual heat removal system or the passive residual heat removal system to decrease the temperature and pressure of the primary loop system.

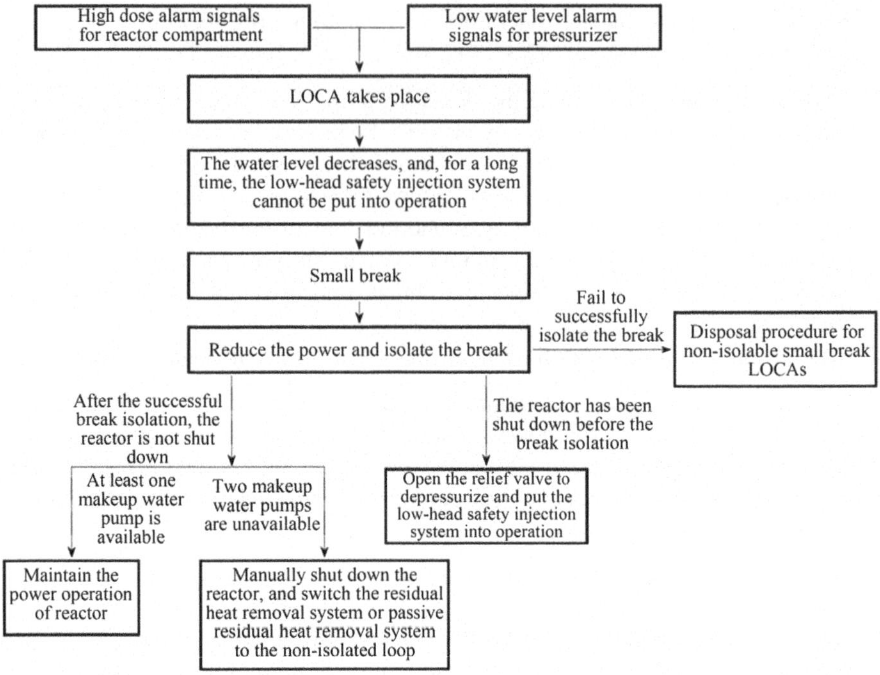

Fig. 13.9 Disposal of break LOCAs

Figure 13.9 illustrates the disposal flow obtained after the operation analysis of small break LOCAs.

References

1. Dafa, Zhang. 2009. *Operation and mManagement of Marine Nuclear Power Plants*. Harbin: Harbin Engineering University Press.
2. Xie, Zhongsheng. 2004. *Analysis of Nuclear Reactor Physics*. Xi'an: Xi'an Jiaotong University Press.

Chapter 14
Accident Management

14.1 Overview

In order to ensure the safety, safety systems are designed for each marine nuclear reactor to deal with various typical accidents, and the possibility of a failure in the safety systems is considered. Such accidents for which the single failure of the safety systems is considered and the targeted measures are taken called the design basis accidents. There is a very low possibility that the safety systems may have two or more failures, resulting in accidents beyond the design basis. Since there is a lack of targeted treatment measures, the core temperature may increase continuously. Such accidents are called the beyond design basis accidents. For the beyond design basis accidents of marine nuclear reactors, the prevention of accidents shall be considered. Once the preventive measures fail after the accident, the post-accident management shall be considered. In addition, if the accident further deteriorates, the emergency handling under extreme conditions shall be considered.

As an important part of the key steps for the reactor protection of marine nuclear power plant, the accident management means that the reactor operation and management personnel shall take measures during an accident to prevent the core damage, terminate the core damage process, contain the corium in the pressure vessel, maintain the integrity of the containment and minimize environmental release. It forms the defense in depth with other steps, providing a comprehensive countermeasure for safety measures and facilities of the marine nuclear power plant. The defense in depth mainly uses several barriers (fuel matrix, cladding, reactor coolant boundary and containment vessel) to prevent the release of radioactive substances and uses a series of measures to protect such barriers, including conservative design, mass and state monitoring, operational limits and functionally and spatially independent safety systems. If the barriers fail, the accident management will be the last barrier to reduce and terminate the environmental release.

The first priority in the nuclear safety is to prevent the accidents in nuclear reactors, including design basis accidents and beyond design basis accidents. A lot of work has been done in the discussion of the accident possibility reduction. However, only

J. Yu, *Marine Nuclear Power Technology*,
https://doi.org/10.1007/978-981-15-2894-1_14

a small amount of work focuses on the ways to mitigate the consequences of an accident. Currently, the greater emphasis is being placed on providing the measures for short-term or long-term accident control process and maximizing the prevention or mitigation of accident consequences. The accident management is the response when the preventive measures have become invalid, the core cooling has been lost or the core damage has become very critical. It includes pre-planned ad hoc responses and some recovery actions. It is beyond the design basis accidents considered in most emergency operating procedures (EOPs). The main purpose of the latter is to maintain proper core cooling to prevent the core damage. Although the accident management shall be independent from the emergency preparedness for the response to the release of radioactive substances to the environment, but it shall be coordinated with the latter.

Based on accident management programs and procedures developed by some countries, it is appropriate to incorporate the research results of beyond design basis accidents into the concept of accident management. Especially after the Three Mile Island accident, the necessity of beyond design basis accident management was highlighted for the first time. Most beyond design basis accidents anticipated in the PWRs may be much slower than the Chernobyl nuclear accident, so many additional measures can be used to delay the total exposure of core, terminate the further deterioration of core performance, restore core cooling capability and maintain the integrity of containment. For the NPPs, the use of engineered safety systems, external systems and related facilities in the plant and many normally operating auxiliary systems in nuclear power plant provides the operating personnel with many intervention opportunities, particularly in areas where the capabilities and shortcomings of equipment and personnel are well understood. For the marine nuclear power plant, there are not many intervention means available and the external support is limited, so it is necessary to consider the use of existing systems and equipment as reasonably as possible. In other words, during the accident management, the operators shall be able to use all systems and equipment, not just those required in current laws and regulations.

14.2 Objectives of the Accident Management

The overall objective of accident management for marine nuclear power plant is to effectively use the personnel and hardware resources to prevent or mitigate the design basis accidents and the beyond design basis accidents, thus further reducing the risks to the staff and the public. To meet this objective, the marine nuclear power plant management department must have the measures to develop the accident management programs. It is reasonable to include the basic requirements (such as the national radiation environment protection regulations) for mitigating the release of radioactive substances at the initial stage of the accident management programs. The purpose is to set in advance the scope of measures that must be taken to further reduce the risk of release. This guideline can be expressed separately by organizations of government departments, nuclear safety management departments, and marine nuclear power plant management departments.

According to the severity of accident progress and the feasibility of countermeasures, the accident management must consider the safety objectives to be achieved. Four (4) main safety objectives have been identified, and they are related to core, containment and release management:

① Preventing the core damage,
② Preventing the failure of reactor vessels and other reactor coolant system boundaries,
③ Preventing the containment vessel failure,
④ Reducing the release of fission products.

14.3 Accident Management Methods

The accident management programs are developed for the marine nuclear power plant to realize the accident management. The structured top-down methods are mainly used to develop the accident management guidelines. From the development of goals and strategies to the completion of the procedures or guidelines, both prevention and mitigation are involved. At the top level, the objectives of accident management are defined. In order to achieve these objectives, a variety of strategies shall be developed. Effective accident management measures shall be developed based on the strategies. The procedures or guidelines are used to develop the appropriate accident management guidelines for the personnel responsible for the implementation of accident management measures, so as to convert the policies and measures into the procedures used at the prevention stage and the guidelines at the mitigation stage.

The four main aspects of the accident management program are as follows, a. identification of the weaknesses of nuclear power plant: study how a particular accident threatens the critical safety functions, What extent will the core damage to if these safety functions are lost and cannot be timely recovered, and what threats the integrity of other barriers of fission products will be faced; b. identification of the capabilities of nuclear power plant: study all available capabilities of nuclear power plant to perform the safety functions, including mitigation capabilities of equipment and personnel; c. development of accident management strategies: develop appropriate accident management strategies for each threat or each weakness of nuclear power plant from the view of both prevention and mitigation; and d. development of procedures or guidelines: inducing objectives and strategies, start-up criteria, time windows for actions to be taken, possible duration of each action, necessary equipment and resources, actions to be taken, matters needing attention, exit and termination guidelines, and monitoring of the response of nuclear power plant.

The specific analysis methods of accident management mainly include deterministic analysis method and probabilistic analysis method.

The deterministic method is a method that has long been used for the nuclear power plant. The basic idea is to determine the response capacity and response speed of safety facilities and the overall safety performance of marine nuclear power plant on the basis of one or several postulated accidents. The deterministic safety assessment method is a mature assessment method that has been widely used up to now, and it is also a traditional safety assessment method approved by various countries. For the marine nuclear power plant, the shear ruptures at both ends of main coolant piping, which are deemed as the maximum credible accident, shall be carefully considered in the design and strict protection measures shall be taken. Since there is strict protection of safety features such as emergency core cooling system, there may not be severe consequences even if such a severe initiating event occurs. The accidents with minor consequences, such as small break LOCAs in primary pipes and transient operating conditions during the operation of marine nuclear power plant, are not intensively studied. In the deterministic safety assessment, the mechanistic code is used to study the physical process of nuclear power plant under the failure conditions. In the accident analysis, it is assumed that at most one single failure will occur in the safety systems, and the system will not lose its functions in such a failure. That is, the single failure criteria are met. The failure conditions considered in the deterministic accident analysis contain artificial assumptions , and neither the probability of failure occurrence is considered nor the consequences of operator intervention after the occurrence of the failure is analyzed. These assumptions are often based on many years of practical experience and some conservative assumptions, so many assumptions do not conform to the objective reality and the results of accident analysis are also conservative. Therefore, the countries gradually tend to use the probabilistic method as a supplement to the deterministic analysis method.

The main objective of probabilistic safety assessment (PSA) is to provide qualitative and quantitative information and insights on the design, performance and environmental impact of the systems, including the identification of key risk factors, the quantitative risk estimation and the comparison of risk reduction options. Compared with the traditional deterministic safety analysis, the PSA can not only determine the accident sequences caused by various kinds of initiating events, but also systematically and realistically determine the frequency of occurrence and the consequences of such accidents. In addition, the PSA can also give the quantitative analysis of some uncertainties in the safety assessment and quantitatively deal with opinions or judgments of the experts. For example, for marine nuclear power plant, the PSA can provide useful insights and inputs for the decisions regarding the following issues: a. design and change; b. operation; c. safety analysis; and d. safety management.

It is necessary to select the accident management strategies after a comprehensive review of the design basis accidents and the beyond design basis accidents of the nuclear power plant. The materials for the review can be from the following sources: a. accident research results from various institutes or laboratories; b. potential accident management strategies obtained from other channels; c. research results of accident management guidelines in the industry; and d. PSA study or test results of nuclear power plant. Based on the above materials, the strategies for nuclear power plant shall be selected from a variety of accident management methods and implemented. The effectiveness of each action in the strategies shall be demonstrated, and the pros and cons shall be checked. In addition, the uncertainties in the accidents shall also be considered.

14.4 Objects of Accident Management

Accident management is implemented as a part of the overall emergency plan, and this principle applies to all reactor types. The accident management involved in this book is mainly for the marine PWRs, but it can also be used as reference for other types of nuclear reactors, including existing nuclear reactors and nuclear reactors to be built. The accident management in this book is mainly for the power operation state, and also applies to other operating modes, including the shutdown state. Detailed accident management measures depend on the type of reactor. For example, the liquid metal cooled reactors and the light water reactors use different accident management methods.

14.5 Diagnostic Methods for Thermo-Hydraulic Phenomena in Typical Accidents

The basic purpose of accident management for marine nuclear power plant is to restore the core cooling, control the reactivity and maintain the integrity of reactor coolant system and containment vessel. Once this integrity cannot be maintained, the consequences of the release of radioactive substances can be minimized by delaying the failure of the reactor coolant system or the containment vessel boundary.

It is necessary to use specific procedures in the accidents. However, there are many failure combinations that may lead to a very large number of accident scenarios, so it is necessary to diagnose the thermo-hydraulic phenomena in typical accidents. That is, once the protection systems are triggered, the operators shall follow predetermined written procedures (such as EOP) to verify the automatic actuation of the safety systems, diagnose the reactor state according to the predetermined logical process to select the proper procedure, and take specific actions according to the selected procedure. The common practice is to provide an EOP that covers the single initiating events considered in the design. The diagnosis is to identify which events can explain the deviation of main parameters measured in the marine nuclear power plant. In this event-oriented method, after the identification of events, the event-oriented procedures for the corresponding specified necessary orderly actions are used. In recent years, the operating experience has shown the need to consider more complex situations in which it is difficult or even impossible to identify the initiating events and to take a corrective action to bring the marine nuclear power plant back to a controllable state. In order to cover a wider range of accidents and to consider diagnostic errors and inappropriate operator interventions, a more generic EOP has been developed in some countries.

According to the requirements for the PSA research program and the PSA work program, corresponding thermo-hydraulic analysis shall be performed to determine the related success criteria and accident consequences of Level I PSA event sequences. The specific objectives are as follows: a. determine the response process of the reactor coolant system and the auxiliary systems of the nuclear power plant in various identified postulated initiating events, and the behaviors and consequences of reactor core; b. determine the types and quantities of equipment important to safety that must keep operating or be put into operation to ensure the safety of reactor cores in various initiating events (i.e. the success criteria); and c. determine the allowable time for the operators to perform the interventions to mitigate the accidents. The thermo-hydraulic analysis is mainly aimed at the success criteria of each front-line system in the event sequence analysis of Level I PSA. The preliminary principles for determining the conditions of thermo-hydraulic analysis are as follows: a. the content of the calculation and analysis shall be less quantity but more quality(less but better), highlight the key points, and fully reflect the design features; b. if the assumptions of success criteria in safety analysis are considered to be too conservative and that lowering the success criteria will obviously affect the consequences of event sequences (core damage frequency, CDF) and the weight, the thermo-hydraulic analysis shall be performed to obtain rather realistic success criteria; c. the analysis results of similar sequences can be referred for the success criteria of non-dominant event sequences; and d. the relevant success criteria (including time windows for operator actions) cannot be obtained from safety analysis reports and system design specifications, and the engineering judgment cannot be made.

Referring to the Level 1 PSA research experience and the characteristics of accident safety criteria for the nuclear power plant, the criteria for determining the core damage are preliminarily considered as follows: a. the maximum temperature of fuel element cladding reaches 1,204 °C; b. the pressure of reactor coolant system is higher than 125% of the design pressure; and c. the temperature and pressure inside the containment vessel are higher than the design limits of the instruments and equipment. The radioactivity released to the environment after different initiating events is disposed by referring to the PSA for nuclear power plant and not specifically used as the judgment basis of core damage. However, for the event tree models for specific accidents, such as the steam generator tube rupture accident, the influence on the radioactive consequences shall be considered for the measures taken. In order to simplify the analysis, the success criteria of the relevant measures in the event tree models are based on existing radioactivity analysis results.

14.6 Emergency Response to Accidents

The emergency response refers to some immediate actions beyond the normal working procedures to avoid the nuclear accidents in the nuclear power plant or to mitigate the accident consequences. It is sometimes also called emergency state. It also refers to immediate actions beyond the normal working procedures. The emergency management refers to the work that shall be performed by the operating organizations to maintain and control all human, material and technical resources and levels to ensure the timely and effective emergency response under the emergency conditions. The administrative and logistical support shall be provided for the emergency management to meet appropriate and reasonable requirements for traffic, transportation, accommodation, material supply, procurement, finance, insurance and legal affairs. The emergency management of marine nuclear power plant shall take into account the appropriate available design and construction manpower for solving the engineering and technical problems arising out of the emergency response and supporting the medium and long-term actions for emergency response. The focus on emergency response in this book is to ensure adequate core cooling or effective off-site emergency response in case of an imminent core damage, rather than accident management after significant core damage.

In the emergency management, the following features of marine nuclear power plant when compared with the NPPs shall be considered:

① Movable location. Although a marine nuclear power plant also has a relatively fixed docking station, it may travel deep into the ocean or berth near the harbor district, so the accident sites are not fixed.

② Density of systems and equipment. Compared with a NPP, the marine nuclear power plant is located inside a limited space, so the equipment is highly dense and the redundancy is sometimes not considered for some compatible functions. The manipulation is difficult and the requirements are high. Because of the functional requirements, the marine nuclear power plant must be subject to start-up, shutdown and power level change at any time. There is a lack of adequate equipment and technical support in terms of supporting conditions. The reactor of a marine nuclear power plantdoes not have as sufficient primary water resource as that of an NPP and the amount of pure water it carries is limited. If the desalting plant fails, it may sometimes be shut down because of water shortage. The logistical and technical support for marine nuclear power plant is mainly available at the bases. Once the marine nuclear power plant leaves the wharfs, it will become more and more difficult for the marine nuclear power plant to get supply and support, and the existing support capabilities for open sea emergency response are not sufficient. The operating environment is harsh and complex. The marine nuclear power plant has no fixed sea areas for operation. They are affected by moisture, salt mist and marine corrosion, and also threatened by other factors such as typhoons, tsunamis and reefs. As a marine nuclear power plant is mobile, the probability of it colliding with other ships is greater than the probability of a NPP being hit by foreign objects. In order to meet the requirements of marine

environment, the marine nuclear power plant must adapt to large heels and trims.
f. The emergency rescue of marine nuclear accidents is difficult. The nuclear accidents of marine nuclear power plant at sea have a relatively minor hazard to the public and the environment, but they have great international influence. The current nuclear emergency preparedness still has certain deficiencies, and the equipment construction, the material reserves and the emergency measures are lacking.

Specifically, when considering the emergency management in extreme conditions, the attention shall be paid to the following aspects. First, the design of marine nuclear power plant shall draw lessons from past nuclear accidents and be improved from all aspects to deal with various extreme conditions. However, the occurrence of certain extreme conditions is often beyond the existing pre-judgment of people. Some unforeseeable conditions will make the nuclear power plant unable to meet the safe operation requirements in the design. Therefore, the simple and reliable emergency measures shall be studied. Corresponding facilities and equipment shall be prepared in the normal conditions to ensure the normal operation under extreme conditions. Such emergency measures are relatively easy to take under the extreme conditions (for example, it is possible to design the diesel generator system of Fukushima Nuclear Power Plant to have the capability of withstanding the extreme conditions in the tsunami). Second, the multiple extreme conditions shall be considered to improve the nuclear emergency systems of marine nuclear power plant. The effect of superimposed or additional catastrophic consequences under the extreme conditions on the safety of nuclear power plant shall be fully analyzed, the effect of superimposed or additional accidents on the successful implementation of various kinds of emergency measures shall be fully given, and the overall planning shall be performed for the specific implementation of emergency plans and emergency preparedness measures. For the offshore (berthing) conditions, under normal earthquake and tsunami conditions, all land-based support facilities can be used to provide the platform with all necessary support. The emergency plan shall be implemented inside the ship and across the entire area, the internal or external force shall be used to quickly restore the power, communication and other key facilities, the emergency organization shall be established, and corresponding disposal shall be performed. If the amount of nuclear substance leakage cannot be controlled within a short period of time, or if the related disposal work takes a rather long time, the corresponding unit may be placed in an enclosed facility for disposal. Under the extreme conditions, for example, when the hull is washed ashore by a tsunami and loses the self-rescue measures, it will totally rely on the land-based emergency measures. The mobile power supply system, water supply system, residual heat removal system and other facilities shall be used to take various supporting measures. In this case, the range of the emergency response area shall be expanded and the nuclear substance diffusion shall be controlled under the limited conditions. For the open sea conditions, because of the complicated conditions, the extreme conditions for marine nuclear power plant may be abnormal impact of the external force, and different impact locations and magnitudes of acting forces may result in different consequences. However, in general, the accident probability of ships is larger than that of reactors. Therefore, the emergency preparedness shall be considered at the beginning of ship design and building.

For the operating organizations of marine nuclear power plant, the basic principles and requirements to be followed in the emergency preparedness for nuclear accidents mainly include the following aspects:

(1) Establishing the emergency organization

① The operating organization of a marine nuclear power plant must establish an emergency command organization to uniformly command the emergency response under emergency conditions. The main responsibilities of the emergency command organization are as follows: a. implement the guidelines, policies and instructions regarding the nuclear accident emergency response given by the superiors; b. specify, revise and organize the implementation of the emergency response plan for the operating organization, and organize the implementation of the emergency preparedness; c. give the order to initiate or terminate the emergency standby or in-ship emergency state; d. organize forces within the harbor district to carry out engineering rescue and medical rescue; e. report to the superiors according to the emergency plan, and maintain close contact during the accident; f. record and save emergency various events, data and evaluation results at various stages in an emergency; g. organize the evacuation of personnel in the harbor district as required, and provide suggestions on whether to take emergency actions outside the harbor district; h. conduct radiation monitoring in and around the harbor district; and i. conduct the prediction and evaluation of accident consequences.

② The emergency organization must adhere to the principle of versatile compatibility and maintain a certain number of professionals involved in the decision-making of emergency operations.

③ The members of the emergency organization must be properly trained.

④ Various emergency response teams shall be established under the emergency command organization. The scope of responsibility of the emergency response teams shall include at least the following aspects: a. operation of nuclear power plant and dock support facilities; b. communication support; c. radiation monitoring and dose assessment and control inside and outside the harbor districts; d. marking of controlled channels and areas; e. damage control; f. engineering rescue; g. nuclear engineering disposal; h. medical rescue; i. decontamination; j. order management inside the harbor districts (including the management of personnel movement and material removal and the security work); k. remedial work; l. personnel verification (i.e. calling the roll); and m. keeping records.

⑤ The scope of responsibilities and work of each emergency response team must be expressly specified.

⑥ In the assignment of emergency tasks, the emergency command organization must consider the effectiveness of each shift. The personnel responsible for the operation and maintenance of nuclear power plant (mainly those for safety shutdown and power guarantee) shall not undertake any other emergency tasks.

(2) Preparing the emergency plan

① The operating organization of a marine nuclear power plant must have a thorough emergency plan in order to be fully prepared for the organization and technology.
② For the size, consequences and scope of each type of possible accident, the clear judgment criteria and emergency measures to be taken must be clearly defined.
③ When preparing the formulation of emergency plans, the operating organizations shall consider not only expected operating conditions and accident conditions, but also accidents with very small possibilities but severer consequences, including beyond design basis accidents with environmental consequences more serious than those of design basis accidents.
④ The factors to be considered when preparing the emergency plans are as follows: a. types of emergency conditions that may occur in the marine nuclear power plant; b. types and quantities of radioactive substances that may be released into the environment; c. environmental characteristics (including meteorology, hydrology, physiognomy, population, traffic and water source); d. exposure pathways (such as smoke plume and ingestion); e. characteristics of the potential release and exposure change with the time; f. areas requiring the emergency protection measures; g. actions to be taken to control the emergency conditions; h. organization and coordination in an emergency; i. emergency procedures, and steps to implement each emergency measure and time for each step; j. manpower and equipment required for evaluation, remedial measures, protective measures, communications, medical care and record retention; k. habitability and accessibility of specific areas in a ship, especially the central control room and the compartment that also serves as the emergency command center; l. rally points and evacuation routes for personnel in the harbor district; m. training, drills and inspections; and n. records and reports that must be prepared and kept.
⑤ The emergency plans must define the alternative sequence of the Emergency Commander in Chief. In addition, the name and function of the superior emergency organization, the responsibilities related to emergency response and emergency preparedness, and the contact method with the superior emergency organization shall also be described.
⑥ Warning signals used for various emergency situations must be clearly defined in the emergency plan. These signals must be easy memorize and recognize, so that the emergency personnel can respond quickly and promptly.
⑦ The emergency plan shall include the restoration measures after the termination of the emergency, mainly including: a. formulation of relevant regulations for the release of area control; b. formulation of pollutant treatment and disposal plan; c. continuous measurement of surface and harbor

radiation levels and radioactivity content in environmental samples such as soil, plant and water, and estimation of the exposure dose to the public; and d. reassessment of reactor safety.

(3) Determining the emergency state

The emergency states of marine nuclear power plant are divided into emergency standby, in-ship emergency and harbor emergency.

① When certain unusual events may lead to the unacceptable unusual release of radioactive substances in the marine nuclear power plant, an emergency standby state shall be declared. An emergency standby alarm may be initiated by any of the following events: a. failures of engineering safety facilities that are related to the safety of marine nuclear power plant and may result in unacceptable release of radioactivity; b. forecasts and notifications of serious natural events (such as earthquakes, strong typhoons and tsunamis) near the ship harbors; c. alarms of ship damage (such as flooding and fire); d. release of toxic or harmful gases around or inside the ship that may endanger the lives of the operating personnel; e. man-made sabotage; f. accidents in other nuclear facilities near and inside the ships. In an emergency standby state, the field personnel shall take active measures to limit the development of the accident as much as possible.

② If the assessment indicates that the radiation consequences of the accident may be limited to a specific area inside the ship, or that there will be no significant impact on the terrestrial environment even if the radiation consequences spread outside the ship, an in-ship emergency shall be declared. It can be considered that the impact is only inside the ship in any of the following circumstances: a. the area monitoring, effluent monitoring and other special monitoring indicate that the radiation dose outside the ship is not unacceptable; b. there is no significant failure in primary circuit system pressure boundary or containment; or c. the ship is navigating in the sea area far away from the harbor district.

③ When the dose rate in the harbor district reaches the specified limit, a harbor district emergency state must be declared. The emergency state classification system used in the emergency plan must be standard. Each type of accident must be associated with an emergency state, so that the corresponding actions can be taken immediately when a specific state occurs. Different categories of emergency states shall require different levels of response.

If the accident occurs in a sea area that may have a significant impact on the terrestrial environment, corresponding emergency actions can be taken according to the development of the accident.

(4) Emergency facilities

The emergency facilities mainly include emergency response centers, emergency command centers, technical evaluation centers, communication systems, evaluation facilities and radiation protection facilities. The emergency facilities shall be arranged to comply with the principle of concurrent use. If the existing facilities cannot guarantee the minimum demand for emergency response, some appropriate additional facilities shall be provided.

(5) Emergency response

When there is an emergency accident, the emergency command organization shall promptly communicate the order of entering the emergency state to the relevant organizations and personnel. The following communication requirements shall be met: a. the communication shall be strictly performed according to specified procedure and terminology; b. the notified initial information must be simple and clear, and include the name and title of person giving the alarm, the time of entering the emergency state, the emergency level, the scope of emergency and the meteorological conditions; and c. all types of emergency organizations shall immediately repeat the original information upon receipt of it to ensure its reliability.

The evaluation in an emergency response shall include the following: a. monitor the operating state of the reactor devices by the instruments in the central control room (or additional dial plates); b. collect and master the evolution of the accident, the source items, the meteorological conditions of the harbor district where the ship, and other required data; c. conduct the radiation monitoring inside the harbor district (including inside the ship) or outside the harbor district, and conduct the monitoring of radioactive effluents of the ship; d. summarize and analyze the collected data to predict radiation dose under the accident conditions; and e. according to the evaluation conclusions, propose the confirmation or modification of emergency state level and the corresponding.

The specific emergency protection measures shall be generally as follows: a. according to the results of radiation monitoring, determine the contaminated areas and provide corresponding signs or warnings; b. monitor the harbor district, including the personnel in ships and the vehicles and materials leaving the harbor district, and, if necessary, perform the decontamination; c. strictly control the access and passages in the ship and the harbor district, and restrict the personnel from entering the heavily polluted areas; d. provide the places with good shielding, sealing, ventilation and filtration conditions as the shelters, or warn the personnel to close doors and windows and not to go out; e. distribute iodine tablets and personal protective clothing; f. when the contamination level exceeds the standard, the supply of food and beverages for the personnel shall be provided under the supervision; and g. determine the evacuation route and time required for evacuation in advance.

(6) Emergency termination

When it is confirmed that the accident has been controlled and that the amount of the radioactive effluents of the marine nuclear power plant and the contamination level to the environment have fallen below the acceptable limits, it may be considered to terminate the emergency state. The emergency command organization of the operating organization may determine the termination of emergency standby and in-ship emergency state according to the above principles. The superior emergency organization has the right to issue the order to terminate the emergency state in the harbor district.

(7) Emergency training and drills

The regular universal education of and training in the emergency knowledge must be provided for all emergency personnel, making them to become familiar with and grasp the basic content of the emergency plan and the basic knowledge and skills to complete specific emergency tasks. Before the ship is put into service, the operating organization must train and assess the following personnel: members of emergency headquarters, operators on duty in the control room, communication personnel, safety analysis personnel, evaluation personnel, engineering rescue personnel, chemical defense personnel and medical aid personnel.

The main content of emergency training is as follows: a. the basic content of the emergency plan and the basic knowledge and skills to complete the emergency tasks; b. emergency response procedure under the emergency conditions; and c. responsibilities of emergency personnel in an emergency state.

During the operation lifetime of a nuclear power plant, the operating organization must conduct the re-training for all emergency personnel once a year.

In order to test the effectiveness of emergency plan, the completeness of emergency preparedness, the adaptability of emergency response capabilities and the coordination of emergency response personnel, the operating organization must regularly conduct the emergency drills. The frequency and scale of drills shall be reasonably determined according to factors such as the actual needs of the emergency work and the capabilities of the operating organization.

Before the ship is formally put into service, the operating organization shall conduct a comprehensive emergency drill under the unified command of the superior emergency organization. After each drill, the effect of the drill, the experience obtained and the existing problems shall be evaluated, and the suggestions on the modification of emergency plan shall be provided.

Chapter 15
Ageing Management

15.1 Concept of Ageing Management

15.1.1 Concept of Ageing and Its Management

The internal environment of a marine nuclear power plant is normally a hydrochemical environment with high temperature and the primary loop system and equipments are exposed to fast neutrons, while the external is the marine environment. Meanwhile, the operating conditions of marine nuclear power plants change frequently, and swing and seawater impact may occur during the operation of marine nuclear power plant. The physical and chemical effects of the environment such as irradiation, temperature, medium, vibration, and impact would lead to performance degradation of the nuclear power plant. Ageing refers to the process that the structures, systems and components of a marine nuclear power plant gradually deteriorate over time or service. The ageing has various forms, which are closely related to the material type of structures, systems and components and the environmental characteristics. The ageing forms of different structures, systems and components vary from each other. In general, the ageing of nuclear power plant mainly includes degradation in ductility and toughness, crack propagation, thickness reduction and pitting, etc. The ageing of structures, systems and components would lead to the decline of original design functions or even loss of function, to jeopardize the operational safety and service life of nuclear power plants.

Ageing management refers to the engineering, operation and maintenance actions taken to control the ageing-induced degradation of above-mentioned structures, systems and components within acceptable limits. Examples of engineering actions include design, qualification and failure analysis; examples of operational actions include monitoring, implementation of operating procedures within specified limits and environmental measurements; examples of maintenance actions include preventative maintenance, corrective maintenance, and replacement. An all-around understanding of the equipment and its performance degradation caused by ageing and the effects of the degradation on the functioning of the components is the foundation

J. Yu, *Marine Nuclear Power Technology*,
https://doi.org/10.1007/978-981-15-2894-1_15

and prerequisite for effective ageing management, which is realized by integrating existing programs and activities related to ageing management.

15.1.2 Method for Systematic Ageing Management

The ageing management of nuclear power plants requires the coordination of all relevant programs and activities with the systematic ageing management method, including cognizing, controlling, monitoring, and mitigating the ageing effects of nuclear power plants. The general flow of this method is shown in Fig. 15.1, which is a closed Deming cycle of plan-do-check-act (abbreviated as PDCA) applied in the quality management system. This closed cycle is based on related operational experience feedback, R&D results, results of self-assessment and external comprehensive evaluations on the ageing management, thus it can enable the continual improvement of the ageing management program.

As the ageing management of nuclear power plants is based on the equipment and components, the equipment-level ageing management program shall also be developed to specifically stipulate the ageing management requirements for equipment and components in addition to the general ageing management program that specifies the overall management requirements. These ageing management programs of two levels form the systematic and grading ageing management of nuclear power plants

15.1.3 Relationship Between Ageing Management and Current Operation Management

Commissioning programs, maintenance programs of different levels, irradiation surveillance programs, water chemistry surveillance procedures, periodic test programs and in-service inspection programs shall be formulated during the design and before the operation of nuclear power plants. These programs and procedures provide reference and foundation for ageing management, and the ageing management program shall be compatible with these programs and procedures. Modifications or supplements to these existing programs and procedures may be needed depending on the requirement of ageing management. Therefore, the ageing management program together with other programs and procedures form the operation and maintenance requirements for nuclear power plants.

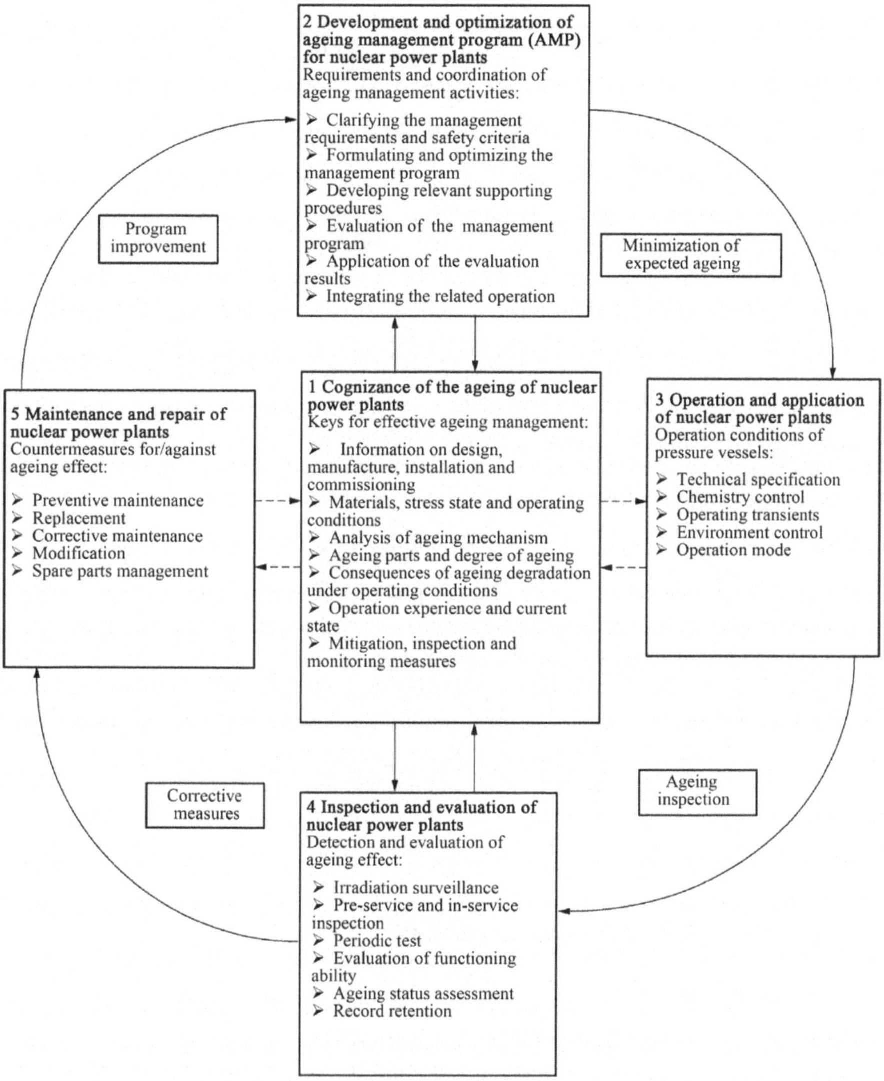

Fig. 15.1 Methods for systematic ageing management

15.1.4 Purposes and Significance of Ageing Management of Marine Nuclear Power Plants

(1) To ensure the reliability of marine nuclear power plants

Reliability is a crucial technical indicator for marine nuclear power plants, and ageing management is one of the most important means to ensure the reliability of nuclear power plants. Ageing management of the nuclear power plant helps in understanding of the ageing mechanism and the performance of its key equipment, so that maintenance and repair can be conducted in time to ensure that the equipment is available and can operate safely and reliably when required.

(2) To ensure the economical efficiency of nuclear power plants

Nuclear power plants are expensive and their operating and maintenance costs are very high, so the economical efficiency of nuclear power plants has to be considered. On the other hand, the design of the nuclear power plants is conservative, but the worldwide operation experiences show that the prolongation of the operating lifetime of the nuclear power plant is possible, which increases the economical efficiency of nuclear power plants. The ageing management is one of the most important means to improve the economical efficiency of the nuclear power plant, while ensuring its safe operation. When the ageing management is conducted for the nuclear power plant, the performance and condition of the key equipment related to the safe operation of the plants shall be evaluated based on the collected data on ageing management, and the feasibility forthe prolonged operation shall be analyzed.

(3) To ensure improvement in design technique for nuclear power plants

The long-term operation of nuclear power plants has accumulated a large number of valuable operational data, and the results from the comprehensive analysis of these data can be used to evaluate the advancements and deficiencies of the nuclear power plant design. The database established in the process of ageing management of nuclear power plants can help in the accumulation of operational data, and experience and lessons learned can be fed back to the design of new nuclear power plants, thus promoting the improvement of design techniques and operational technologies of nuclear power plants.

15.2 Status of Ageing Management

Since the 1990s, the ageing management of nuclear power plants has become the hot topic for the owners and nuclear safety regulatory authorities, with more and more nuclear power plants approaching the end of their lifetime. In particular, the ageing management of nuclear power plants has received unprecedented attention after the accident of Fukushima Nuclear Power Plant. Ageing management refers to the engineering, operation and maintenance activities to control the ageing process of structures, systems and components (hereinafter referred to as SSCs) of nuclear power plants within an acceptable limit. Ageing management is an extremely important task for ensuring the safety and economical efficiency of nuclear power plants. Some research institutes have carried out a lot of researches from the perspectives of the ageing mechanism of equipment in the nuclear power plants, ageing management methods, collection and preservation of data required for the ageing management, and equipment life assessment. International Atomic Energy Agency (IAEA) has issued a series of guides on the ageing management of nuclear power plants based on the existing research results, which mainly include:

① IAEA Safety Guide No. NS-G-2.12 Ageing Management for Nuclear Power Plants;
② IAEA Safety Series No.15 Implementation and Review of A Nuclear Plant Ageing Management Programme;
③ IAEA Safety Series No. 50-SG-012 Periodic Safety Review of Operational Nuclear Power Plant;
④ IAEA Safety Series No. 50-P-3 Data Collection and Record Keeping for the Management of NPP Ageing;
⑤ IAEA Technical Reports Series No. 338 Methodology for the Management of Ageing of NPP Components Important to Safety;
⑥ IAEA Safety Reports Series No. 57 Safe Long Term Operation of Nuclear Power Plants.

In the United States, a series of federal regulations, NRC regulatory guidelines, industrial guidelines and summary reports related to the lifetime extension of nuclear power plants have been successively issued in recent years, in order to increase enhance the economical efficiency. The key part of these requirements of lifetime extension is ageing management. For example, when updating the license of NPPs, in addition to the final safety analysis report and environmental assessment report, the integrity analysis report of NPPs and the ageing analysis report are required, which the key content of which is to validate that the ageing of NPPs is under effective control and certify that the ageing state at the end of lifetime extension still meets the requirements of safe operation. The relevant requirements of ageing management mainly include:

① 10CFR54 Requirements for Renewal of Operating Licenses for Nuclear Power Plants;
② R.G 1.188-2001 Standard Format and Content for Applications to Renew Nuclear Power Plant Operating Licenses;
③ NUREG 1800 Standard Review Plan for the Review of License Renewal Applications for Nuclear Power Plants;
④ NEI-95-10 Industry Guideline For Implementing The Requirements of 10CFR54-The License Renewal Rule.

The National Nuclear Safety Administration of China has successively issued some requirements on ageing management of NPPs, and now it is formulating technical policies for the extension of NPP licenses. The National Energy Administration of China has developed the ageing management guidelines for key SSCs of NPPs with the participation of related organizations in China. The most important part of these related requirements is the *Periodic Safety Review of Nuclear Power Plants* (HAD103/11) and the *Ageing Management for NPPs* (HAD103-12), which put forward general requirements for the ageing management of NPPs in China. Other relevant management requirements mainly include:

① *Methodology for the Control of Ageing of NPP Components Important to Safety* (HAFJ0068);
② *Guidelines for Formulating the Ageing Management Programme of Mechanical Equipment in Nuclear Power Plants* (GB/T28536);
③ *Management Guidelines on Ageing of Instrumentation and Control Cables Important to Safety for Nuclear Power Plants* (NB/T20087);
④ *Assessing, Monitoring and Mitigating of Ageing Effects on Class 1E Equipment Used in Nuclear Power Plants (N*B/T20086);
⑤ *Guideline for Preparing Ageing Management Programme of Pressurized Water Reactor Nuclear Power Plants* (NB/T20151);
⑥ *Guideline of Ageing Management for PWR Reactor Pressure Vessels* (NB/T20154);
⑦ *Ageing Management Guideline of Prestressed Concrete Containment in Nuclear Power Plants* (NB/T20153);
⑧ *Ageing Management Guideline of Piping in Nuclear Power Plants* (NB/T20152);
⑨ *Ageing Management on Class 1E Equipment Used in Nuclear Power Plants* (NB/T20155).

15.3 Strategies for Ageing Management

15.3.1 Overview

Ageing management strategy shall be implemented on marine nuclear power plants. Although the management at the operation phase is the most important part of the ageing management, the ageing management of marine nuclear power plants shall cover the entire lifetime of nuclear power plants, including the whole process from design, construction, commissioning, operation to decommissioning. Only when the issues of ageing of nuclear power plants such as prediction of ageing mechanisms, effective measures for ageing mitigation and the detection of ageing effects are considered fully during the design, manufacture, construction and commissioning, the feasibility and effectiveness of ageing management can be guaranteed in the operation phase.

15.3.2 Design

The ageing management of the marine nuclear power plant begins from the design phase, and the ageing issues considered in the design phase is directly related to the service lifetime of the nuclear power plants. The possible ageing mechanisms of systems and components of marine nuclear power plants in the future operating environment shall be thoroughly assessed and evaluated in the design phase, such as radiation embrittlement, fatigue, corrosion, wear, and reasonable design of structures, materials, operating environments and operating conditions shall be provided in combination with experience feedback. Primary methods for ageing management include:

① Make full use of worldwide research results and experience feedback, and select materials, process and manufacturing methods with better anti-ageing performance in the design;
② Take full consideration of the future operating conditions of nuclear power plants, and analyze the effects of operating conditions on the ageing of nuclear power plants and optimize the structural design and condition design;
③ Fully consider and optimize the equipment layout and structure to improve the accessibility during equipment repair, inspection, testing, monitoring, maintenance and replacement and minimize the occupational exposure during such activities;
④ Take a full account of the methods for monitoring the ageing effects, and try to provide pre-warning information on ageing status of nuclear power plants and provide data required for condition evaluation.

15.3.3 Fabrication and Construction

The ageing management methods applied in the fabrication and construction phase mainly include:

① Optimizing the fabrication process, reducing non-conformances and avoiding the consumption of designed margins for ageing analysis in this phase while meeting the design requirements;
② Re-evaluating the effects of technical changes or non-conformances on the ageing margins by the designer;
③ Installing specimens or special instruments for monitoring, detection and supervision of ageing according to the design requirements;
④ Keeping complete data recording of equipment fabrication and forming basic documents for future ageing management and status evaluation.

15.3.4 Commissioning

Methods for ageing management in the commissioning phase mainly include:

① Incorporating the deviations of operating conditions during the commissioning into the analysis of ageing effects of nuclear power plants;

② Evaluating the effects of temporary measures taken for commissioning on the ageing effect of nuclear power plants;
③ Conducting sufficient verification of the accessibility for the future plant repair, maintenance and inspection;
④ Keeping complete data recording in the commissioning and forming basic documents for ageing management and status evaluation in the future.

15.3.5 Operation

The ageing management in the operation phase is the core of the ageing management of nuclear power plants, and this section gives a brief description of the ageing management methods for the operation phase and more details are provided in Sect. 15.4. The ageing management methods applied in the operation phase mainly include:

① Establishing the ageing management organization and determining the relationship between ageing management and current operation management;
② Screening the ageing-sensitive systems and equipment of nuclear power plants and determining the scope of ageing management;
③ Formulating the ageing management program of marine nuclear power plants based on the organizational structure and management scope;
④ Analyzing the ageing mechanism of ageing-sensitive equipment in detail;
⑤ Formulating the ageing management program of equipment based on mechanism analysis and operation management system;
⑥ Collecting and preserving the data required for design, manufacture, commissioning and operation, to establish the ageing management database;
⑦ Establishing the evaluation methods for actual condition of equipment and carrying out the evaluation of actual condition of ageing-sensitive equipment;
⑧ Performing the ageing management review, and continuously improving the ageing management program.

15.3.6 Decommissioning

Primary ageing management methods applied in the decommissioning phase include:

① Sampling plan shall be considered in the decommissioning plan in addition to irradiation surveillance specimens, especially the performance test sampling which is helpful for the verification of the ageing mechanism analysis and ageing status evaluation;
② Comprehensive analysis and evaluation of the performance and condition of nuclear power plants shall be performed, with the help of irradiation surveillance and sampling performance test as well as the ageing management results in the

design, construction and operation phases, to provide basic data for subsequent design.

15.4 Ageing Management During Operation

15.4.1 Screening of Ageing-Sensitive Systems and Equipment

(1) The purposes and principles of the screening

Because of the huge number and various types as well as different sensitivity to ageing effects of systems and equipments of nuclear power plants, it is impossible and dispensable to carry out qualitative and quantitative analyses to the ageing degree of all the systems and equipment. Systematic methods shall be employed to focus on the systems and equipment that have an adverse effect on the safe operation of nuclear power plants and are sensitive to ageing effect, to save resources.

Screening shall follow the safety principles, and the safety shall be understood as any condition that has effect on one of the following condition: ① maintaining the integrity of the reactor coolant pressure boundary; ② the capacity to shut down the reactor and maintain the reactor in the safe shutdown state; ③ the capacity to prevent accidents or mitigate the accident consequences of the large release of radioactivity to the environment. The ageing management shall focus on the SSCs that have higher safety level and impose severe impact on the safe operation of plants due to ageing effect, such as reactor pressure vessels and steam generators. Meanwhile, close attention shall be given to the systems and equipment that do not fulfill any safety functions but belong to safety related SSCs and their failures would prevent other system and equipment from performing the safety functions, such as reactor internals.

Screening shall also comply with the economy principle. Some key components with larger dimensions but of low safety level would not be included in the scope of ageing management according to the economy safety principle. Nevertheless, they shall be taken into account in the ageing management if their high manufacturing and replacement costs seriously affect the economical efficiency of the nuclear power plant.

(2) Technical preparation before the screening

The Following information on nuclear power plants shall be collected before screening:

① Lists of systems and equipments;
② Drawings, including system flow diagrams, wiring diagrams and I&C logic diagrams;
③ Design documents, including system design manuals, equipment specifications and design specifications;

④ Operating procedures, including emergency operating procedures and general operation procedures.

(3) Screening process

The screening of ageing-sensitive systems and equipments in marine nuclear power plants is performed according to safety-related principles on the basis of requirements described in *Ageing Management for NPPs* (HAD103/12) and good practice experience worldwide. The objects of ageing management are divided into two levels for screening: system level and equipment level. Screening for marine nuclear power plants can be conducted with reference to the process adopted by land NPPs (see Fig. 15.2).

(4) Method for equipment classification

Systems and equipment to be considered in the ageing management are classified and ranked according to the importance to safety, the possibility of ageing and the degree of difficulty, and the probabilistic or deterministic analysis approach is adopted for the risk analysis when necessary. The scope of ageing management is graded with reference to the ageing management experience feedback from similar plants, with

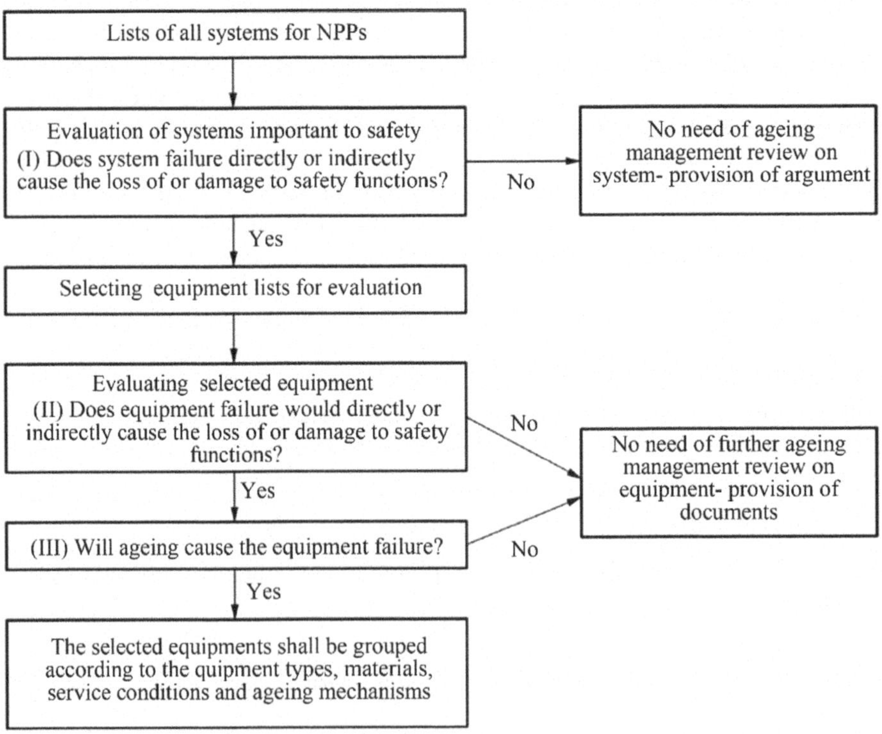

Fig. 15.2 Screening process of ageing management objects

full consideration of characteristics of mechanical components and components of electrical instruments. The mechanical equipment and electrical instrument equipment can be classified as that shown in Figs. 15.3 and 15.4 respectively. The scoring method is that the possible maximum scores for each indicator of each component in the equipment are ranked according to the product of the three indicators. Those with a product within range of 18–27 is defined as AM-1 (key), 6–17 as AM-2 (importance) and 1–5 as AM-3 (common). Such method has considered the replaceability of equipment components, i.e., non-replaceable equipment and components are given priority in ageing management.

In the process of ageing management, priority shall be given to AM-1 equipment. The ageing mechanism shall be clarified, and the target-oriented ageing management program and measures shall be developed. Methods for monitoring ageing parameters

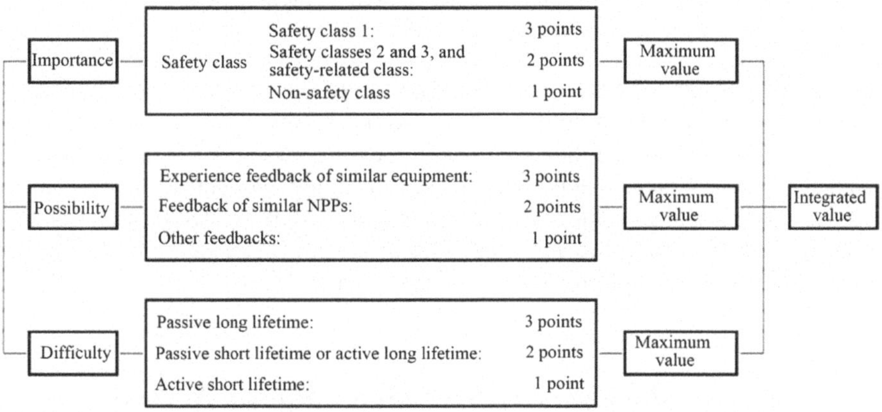

Fig. 15.3 Method for classifying mechanical equipment

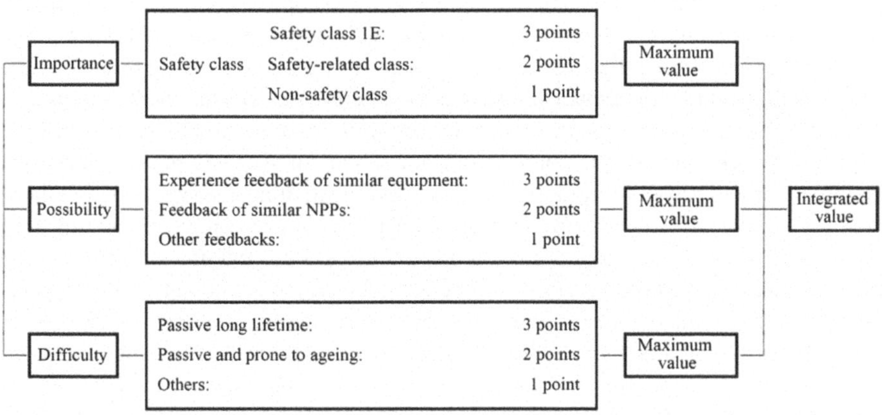

Fig. 15.4 Method for grading electrical instrument equipment

and mitigating ageing effects shall be provided, and the method for evaluating the ageing status shall be established, to periodically conduct the ageing management review and ageing status evaluation. For AM-2 equipment, the ageing mechanism shall be clarified, the ageing management review shall be performed periodically, and the ageing status shall be reviewed according to requirements. For AM-3 equipment, the ageing mechanism shall be clarified and the ageing management review shall be carried out when necessary, but such equipment is not the focus of the ageing management.

(5) Equipment classification

Systems and equipment to be considered in the ageing management shall be grouped based on the overall consideration of equipment types, materials, operating conditions, ageing mechanisms. Vessels, valves and pumps can be classified according to the aforesaid principle, to simplify the ageing management.

(6) Screening results

The screening results shall be provided in the form of lists of AM-1, AM-2 and AM-3 equipment, and similar equipment shall also be classified. Key equipment in the ageing management are as follows: ① reactor pressure vessel; ② reactor internals; ③ steam generators; ④ pressurizers; ⑤ reactor coolant pumps; ⑥ control rod drive mechanisms; ⑦ nuclear-class cables (as groups); ⑧ other safety-related heat exchangers (as groups); ⑨ other safety-related vessels (as groups); ⑩ other safety-related valves (as groups); ⑪ other safety-related pumps (as groups).

15.4.2 Ageing Management Program for Marine Nuclear Power Plants

After the determination of the ageing management scope of nuclear power plants, the ageing management program of the marine nuclear power plant shall be prepared with the consideration of the organizational structure of the organizations responsible for the operation of the nuclear power plants. The ageing management program of the marine nuclear power plant is a general program involving many parties. All the responsible departments shall share a common understanding and provide sufficient management and resource supports to ensure that the technical assessment and management strategies of the ageing management program are exhaustive and of high quality, and the actions specified in the program can be implemented. The ageing management program shall feature comprehensiveness, prospectiveness and systematicness, and it shall be used to guide the systematic ageing management activities of nuclear power plants, to coordinate management activities such as the operation, repair, in-service inspection and supervision, to ensure the effective management of the ageing of SSCs of nuclear power plants and the reliability, safety and economical efficiency of nuclear power plant, and to provide technical supports for the life extension and decommissioning.

Based on the experience of relevant NPPs, the ageing management program shall at least include purposes, policies, assignment of responsibility, scope, development of equipment ageing management programs, review and optimization of equipment ageing management program, data collection and record retention. Elements of the ageing management program for marine nuclear power plants are shown in Table 15.1.

Table 15.1 Basic elements in ageing management program for marine nuclear power plants

S/N	Basic elements	Description
1	Purposes	• Establishing a sound system for the ageing management of marine nuclear power plants • Planning the ageing management and ensuring the long-term safe and stable operation of marine nuclear power plants
2	Policies	• Clearly defining the role of ageing management in the operation of the marine nuclear power plants
3	Organizational labor division	• Clearly defining the role of different departments in the ageing management program • Clearly defining the organizational structure for implementation of the ageing management program and the assignment of responsibilities
4	Scope	• Determining the equipment to be considered in the ageing management program
5	• Development of equipment ageing management program	• Clearly defining the requirements of development planning and schedule of equipment ageing management program based on the operation and management conditions of the plant, and guaranteeing the integrity and capacity to fulfill the functions of concerned equipment • Tracking and monitoring the whole process for program development and ensuring the quality
6	Review and optimization of equipment ageing management program	• Clearly defining the methods and plans for program review and optimization • Reviewing and optimizing the program and ensuring the continuous effectiveness of the program
7	Data collection and record retention	• Clearly defining the purposes and functions of data collection and record retention • Clearly defining the type and range of data related to ageing and the data collection and retention methods

The key contents of the ageing management program of nuclear power plants shall be supported by corresponding management procedures and plans, and. normally, the corresponding ageing management procedures shall be developed, such as the long-term plan and the annual work plans for the ageing management of nuclear power plants, the assignment of the responsibilities in the ageing management, the technical requirements for the equipment ageing management program, the review process for the ageing management program and technical requirements of the ageing management database. These work plans, technical requirements and working procedures shall stipulate every detail of the key contents of the ageing management program so as to ensure the effective operation of the ageing management program of nuclear power plants. The ageing management program shall be reviewed according to the review process procedure developed, the applicability and effectiveness of which shall be analyzed. See Sect. 15.4.7 for the review details of the review.

15.4.3 Aging Mechanism Analysis for Aging-Sensitive Equipment

(1) Purposes and methods for aging mechanism analysis

Clear understanding of ageing mechanism is the basis and key part for the implementation of the ageing management. All the activities in the equipment ageing management are based on the ageing mechanisms. Therefore, the in-depth and comprehensive analysis of the equipment ageing mechanisms is the most important and critical step.

The ageing mechanism analysis of the ageing-sensitive equipment is carried out with reference to the results from worldwide scientific researches, IAEA guidelines and technical reports on the ageing management of NPP equipment, GALL reports on the operation experience for the ageing degradation of NPPs by NRC and relevant operation experience. The possible ageing mechanism shall be analyzed and evaluated. The existing ageing mechanisms and their occurrence positions, failure modes, impact effect on functions, major monitoring and mitigation measures shall be identified based on the analysis of the design characteristics, structures, materials, manufacturing and operation conditions of the equipment. This process is not only to identify the potential ageing mechanisms, but also to analyze and evaluate the current status and the sufficiency of inspection, monitoring, control and mitigation measures for each ageing mechanism, thus to provide suggestions on the data type and resource required by the ageing mechanism evaluation. The preliminary evaluation of the applicability and the adequacy of relevant procedures and programs shall be conducted, to be well prepared for the following development of the equipment ageing management program.

During the mechanism analysis, the equipment shall be subdivided into different parts and components based on the complexity of the structure. For each component, issues concerning the possible ageing mechanism, the possibility of component failure arising from ageing and the effects of component failure on the fulfillment of design functions shall be analyzed in the form-based structure; reasons for managing the ageing of the component and methods for monitoring its ageing shall be provided and indicators for ageing status shall be given if conditions permit.

(2) Data requirements for ageing mechanism analysis of ageing-sensitive equipment

The following data shall be collected and processed for the ageing mechanism analysis of ageing-sensitive equipment:

① Design information

i Design basis, design standards, design parameters, design lifetime, technical conditions of water chemistry, design transients, failure criteria; ii Material types and characteristics of critical components, and structure parameters and characteristics.

② Manufacturing (construction) information

i Composition and performance data of structural materials and welding materials; ii data of primary manufacturing techniques (smelting, forging, welding, thermal treatment, machining and surface treatment); iii results of non-destructive testing; iv results of important performance examination and tests; ⑤ data of non-conformance management during the manufacturing.

③ Information on transportation, installation and commissioning

Records of activities during the transportation, installation and commissioning, especially incidents, abnormal conditions and their treatment measures during the activities, such as monitoring and measurement results of water chemistry in the primary loop system and treatment results of the abnormal conditions.

④ Information on experience feedback

i TECDOC (one of the IAEA publications) regularly released by IAEA, covering reactor pressure vessels, steam generators, reactor internals, instrument & control equipment, in-core instrumentation and control cables, pipelines; ii American NRC GALL reports and relevant research results on ageing from the Electric Power Research Institute (EPRI for short); iii latest ageing failure cases of NPPs worldwide; iv latest research results in this field worldwide.

(3) Analysis steps for ageing mechanisms

The analysis of ageing mechanisms is generally conducted according to the following steps:

① To analyze the ageing mechanisms based on the actual situation of marine nuclear power plants; study the technical conditions concerning the design, materials, manufacturing and operation, focusing on the analysis of design characteristics; compare the differences in the design of the equipment in marine nuclear power plants with that in the land PWR power plants; and find out the improvement measures taken to avoid the ageing mechanisms appeared previously.

② To analyze the potential ageing mechanisms of equipment and components, including the theory theoretical conditions of the occurrence of ageing mechanisms, the possible parts prone to ageing, and the affecting factors to effects on ageing mechanisms. This step is the foundation for the ageing mechanism analysis, and the results from worldwide research on material ageing shall be fully considered.
③ To collect the worldwide experience on the operation of similar equipment. Such experience helps the determination of the occurrence positions, the possibility and the consequences of ageing degradation;
④ To analyze and judge the possibility and the sensitive positions of the occurrence of the ageing mechanisms under actual operating conditions, based on the afore-collected data and information;
⑤ To analyze the ongoing inspection, monitoring, control and mitigation measures, which includes whether these measures can be used to inspect, monitor and mitigate the possible ageing mechanisms timely and effectively, whether the current means are sufficient and adequate, and whether it is necessary to supplement or introduce new inspection, monitoring and mitigation measures;
⑥ To provide the methods, contents, input parameters and frequency of parameter acquisition for equipment ageing evaluation;
⑦ To provide comprehensive analysis for component ageing mechanisms based on above analysis results.

(4) Classification of ageing mechanisms

Different inspection, monitoring, detection and evaluation measures are implemented based on the possibility of ageing mechanism and the criticality, ensuring the comprehensiveness and priorities of the management of equipment ageing mechanisms. Based on the aforesaid analysis of ageing mechanisms, the equipment ageing mechanisms can be divided into three categories according to the possibility of ageing mechanisms and severity of consequences caused by ageing.

① Class I ageing mechanisms refer to those that are very likely to occur during the lifetime or the consequences of which seriously affect the equipment functions or safety margins, or those that have occurred in similar equipment as shown in the experience feedback. Such ageing mechanisms shall be strictly controlled to ensure that corresponding inspection and supervision measures shall be promptly and fully implemented to monitor the severity of the consequences caused by the ageing mechanisms; corresponding operation, maintenance and repair measures shall be taken to delay the development of such ageing mechanisms; and their severity and consequences shall be evaluated periodically.

② Class II ageing mechanisms refer to those that are likely to occur during the lifetime. Adequate attention shall be given to such ageing mechanisms. In addition, the development and status of such ageing mechanism shall be checked on a regular basis; active operation, maintenance and repair measures shall be implemented to control or delay the development of them; and their severity and consequences shall be evaluated periodically.

③ Class III ageing mechanisms refer to those that are less likely to occur during the lifetime. Sufficient attention shall be given to such ageing mechanisms. Worldwide reports on such mechanisms of similar equipment shall be collected periodically. The priority of such ageing mechanism is determined according to the results of experience feedback. The possibility of occurrence of such ageing mechanism on equipment shall be evaluated periodically (the evaluation period may be extended appropriately, compared with Classes I and II ageing mechanisms).

(5) Examples for ageing mechanisms and their classification (Table 15.2).

15.4.4 Equipment Ageing Management Program

Equipment ageing management program shall be formulated for the key equipment chosen based on the screening results. The equipment ageing mechanism analysis is the basis for development of the equipment ageing program. Analysis and determination of the primary equipment ageing mechanisms shall be implemented according to the requirements in Sect. 15.4.3, with reference to the worldwide research results and operation experience feedback, considering the factors such as material selection, water chemistry, weakness in the design and important non-conformances in manufacture. Then the equipment ageing management program can be formulated. The equipment ageing management program includes nine basic elements (see Table 15.3).

The equipment ageing management program focuses on the program development and optimization, operation supervision and monitoring, inspection and evaluation, maintenance and repair, and these four aspects together with the analyses of ageing mechanisms form the closed cycle process for continuous improvement of the equipment ageing management.

(1) Program development and optimization—clearly defining the relationship between the ageing program and existing commissioning program, repair program, irradiation surveillance program, operation program or procedures, to realize the objective of integrating different equipment management activities;
(2) Operation supervision and monitoring—clearly defining the specific supervision and monitoring contents required by ageing management, and providing detection and monitoring methods for different important ageing mechanisms analyzed;

Table 15.2 Major ageing mechanisms of reactor pressure vessels

Classification of ageing	S/N	Degraded components	Degraded parts	Ageing mechanism
I	1	Core shell	Shell and weld	Irradiation embrittlement
	2	Driving mechanism/temperature tube base	Penetration and weld	Stress corrosion (SCC)
	3	Driving mechanism/temperature tube base	Penetration of tube base	Fatigue
	4	Neutron measuring tube base	Penetration of tube base	Fatigue
	5	King bolt	Threaded portion	Fatigue
	6	Vessel flange—nozzle belt	Sealing surface	Pitting corrosion
	7	Vessel flange—nozzle belt	Sealing surface	Crevice corrosion
II	1	King bolt	King bolt	SCC
	2	King bolt	Threaded portion	Wear
	3	Head cover and shell	Stainless steel cladding	Pitting corrosion
	4	Inlet/outlet nozzles and safe end	Nozzle (region of high stress)	Fatigue
	5	Inlet/outlet nozzles and safe end	Welds between nozzle and safe end	Fatigue
	6	Vessel flange	Ligament area of king bolt hole	Fatigue
III	1	Core shell	Shell and weld	Fatigue
	2	Vessel	Stainless steel build-up welding	Intergranular corrosion
	3	Nozzle shell	Shell	Irradiation embrittlement
	4	Head cover	Forgings of head cover	Fatigue
	5	Transition ring of bottom/lower head	Forgings	Irradiation embrittlement
	6	Vessel	Vessel forging	Thermal ageing
	7	Head cover	Forgings of head cover	Boric acid corrosion
	8	Radial guide	Weld	SCC

Table 15.3 Basic elements of equipment ageing management program

S/N	Basic elements	Description
1	Management objects	(1) Understanding the ageing phenomenon; (2) Determining the list of parts considered in the ageing management and important ageing problems
2	Preventive measures	(1) Clarifying preventive measures, including condition monitoring or performance monitoring measures; (2) Validating parameters to be monitored or checked; (3) Clearly defining good practices conductive to the improvement of ageing effect
3	Detection of ageing effect	Analyzing and evaluating the techniques that can effectively detect the ageing effect of equipment, to determine the optimal methods and technologies of detecting ageing effect
4	Monitoring and trend prediction methods	(1) Clearly defining the status indicators and parameters for monitoring the ageing degree of nuclear power plants; (2) Incorporating requirements on the collection of monitoring data for relevant equipment; (3) Incorporating ageing-related evaluating technologies and methods, including data analyses, trend analyses and life evaluation
5	Mitigation measures	(1) Analyzing the feasibility of mitigation measures for equipment ageing effect; (2) Including the mitigation measures such as changing the operation conditions, maintenance, repair and replacement; (3) Formulating the corresponding mitigation measures
6	Acceptable criteria	Selecting the appropriate ageing evaluation methods and technologies, and clearly defining the acceptable criteria for status evaluation and those suitable for continuous service
7	Corrective measures	Providing the corrective measure program for equipment that fails to satisfy the acceptable criteria after evaluation, including temporary corrective measures and long-term corrective measures

(continued)

Table 15.3 (continued)

S/N	Basic elements	Description
8	Experience feedback	(1) Collecting, preserving and analyzing the operational experience feedback of equipment considered in the ageing management (2) Establishing the operational experience feedback mechanism, to ensure that the internal and external operational experience feedback of the plant is understood in time (3) Establishing R&D achievement feedback mechanism, to ensure the R&D achievements are fed back to the ageing management program timely
9	Quality management	(1) Including the administrative control process implemented under the ageing management program (2) Including indicators to easily evaluate and continuously improvethe ageing management program (3) Including the validation procedures to verify the sufficiency and rationality of the preventive measures and the completeness and effectiveness of the corrective measures (4) Including the requirements of data collection and record retention

(3) Inspection and evaluation—clearly defining the inspection items necessary for ageing management and the required contents and methods for analysis and evaluation, and establishing evaluation methods and life-prolongation criteria for important ageing mechanisms, so as to obtain the remaining service lifetime of equipment from the evaluation;

(4) Maintenance and repair—clearly defining the enforceable maintenance and repair strategies and measures required by ageing management, so as to prevent or mitigate the important ageing mechanisms of the equipment.

15.4.5 Data Collection and Retention for Ageing Management

To realize the effective ageing management of marine nuclear power plants, corresponding ageing management database shall be established and the information on operation, inspection/monitoring, repair/maintenance/replacement shall be collected in addition to design data, fabrication data, construction data, installation data, transportation data and commissioning data collected during the analysis of ageing mechanisms (see Sect. 15.4.3 for details), including: ① transient conditions during the operation of systems and equipment; ② operating temperature, pressure and other parameters of reactors; ③ operating data of water chemistry in the primary and secondary loops; ④ in-service inspection reports; ⑤ other monitoring records; ⑥ activity records and test results regarding irradiation surveillance specimen; ⑦ Repair and maintenance records; ⑧ records of replacement parts and pre-service inspection reports; ⑨ relevant data associated with equipment ageing analysis, including review and summary of equipment ageing-related information, equipment ageing mechanisms, ageing positions, analytical/empirical models for predicating equipment ageing, data inventory required by equipment ageing evaluation; ⑩ test methods for equipment ageing monitoring, periodic inspection methods, online monitoring methods and evaluation methods; ⑪ documents about equipment ageing effectiveness, including validity evaluation of functional parameters and condition indicators used in monitoring equipment ageing, existing monitoring technical capacity and practicability assessment used to detect significant ageing of equipment, current state evaluation of equipment ageing and functional performance and existing method evaluation used to predict future performance; ⑫ methods for mitigating equipment ageing, including validity evaluation of repair, replacement, applicable conditions and operation practices and design and materials; ⑬ equipment ageing evaluation reports, including special information on equipment, ageing analysis, ageing monitoring, ageing evaluation and ageing mitigation.

Relevant responsible departments shall establish the databases with collected data and sort themby equipment. The responsible department is accountable for the establishment, updating and maintenance of databases related to ageing management; interfacing and exchanging data with other departments; providing necessary assistance and information for persons using the databases. The databases shall be updated periodically. After each inspection or repair, the department responsible for database maintenance shall collect all the operation, inspection and maintenance data and relevant experience feedback, and then sort and store them in the ageing management database. As the ageing management database provides important basis for ageing status evaluation and ageing management review, the integrity, accuracy and promptness of the database shall be fully guaranteed.

15.4.6 Actual Status Evaluation of Ageing-Sensitive Equipment

The actual status evaluation of ageing-sensitive equipment is to analyze and evaluate important ageing mechanisms of the ageing-sensitive equipment, so as to obtain the quantitative indicators of various ageing mechanisms that affect the equipment lifetime and the remaining lifetime of the equipment, and to ensure the safe operation of ageing-sensitive equipment in the next operating cycle. The actual status evaluation shall be based on the worldwide research results, the operational experience feedback in combination with the design parameters of equipment, design material selection, design of water chemistry, weakness in the design analysis report, and the data collected during manufacturing, transportation, commissioning and operation. And actual ageing degree for each mechanism is obtained by analyzing each important mechanism acquired according to the aforementioned analysis of ageing mechanism. In the process of evaluating the actual status of reactor pressure vessels, for example, the mechanism analysis shall be firstly performed to determine that irradiation embrittlement, fatigue and corrosion are the most important ageing mechanisms, and then such three mechanisms shall be analyzed respectively with collected data. The analysis shall be carried out in accordance with the methods specified in the ageing management program for reactor pressure vessels. The actual ageing state indicators of the three major ageing mechanisms are obtained, so is the remaining lifetime of the reactor pressure vessel. The safe operation of next cycle can be determined based on this remaining lifetime.

15.4.7 Ageing Management Review

The ageing management of marine nuclear power plants shall be reviewed according to the requirements of continuous improvement of ageing management. The review is mainly based on the actual conditions of ageing management activities for nuclear power plants. The effectiveness of the operation of ageing management system and ageing management program is evaluated on the basis of operating conditions in all aspects, including objectives, organization, activities, results and continuous improvement, and suggestions on modification is made when necessary. Meanwhile, the consistency of other ageing-related programs and management procedures with the ageing management program shall be evaluated, and the documents to be evaluated mainly include: ① ageing management program for nuclear power plants; ② ageing management program for key equipment; ③ irradiation surveillance program; ④ water chemistry supervision procedures; ⑤ in-service inspection program; ⑥ periodic test procedures; ⑦ preventive maintenance program; ⑧ transient statistical procedures; ⑨ spare parts management regulations.

The review of the above-mentioned management documents shall focus on: ① the maintenance frequency and the rationality of the maintenance measures; ② the adequacy of monitoring and testing data; ③ the rationality of the frequency of qualification and periodic test; ④ the rationality of the frequency and scope of in-service inspections; ⑤ the sufficiency of spare parts; ⑥ the integrity and timeliness of the ageing management database; ⑦ the rationality of the sampling plan for irradiation surveillance.

Commonly applied ageing management evaluation methods include self-assessment, peer review and comprehensive review. A review shall be performed based on the above-mentioned documents and contents to complete the applicability evaluation of the existing programs and management procedures, and then modification of existing programs and management procedures is determined based on the evaluation results.

Special attention shall be paid to the use of experience feedback in the ageing management review, because the experience feedback is an important link of the long-term safe operation of nuclear power plants and can prevent or delay the reoccurrence recurrence of similar accidents. Experience feedback is one of the important means to carry out ageing management on nuclear power plants and is also one of the bases for ageing management review. The ageing management experience feedback of nuclear power plants mainly includes: ① operation experience of NPPs with similar reactor type; ② operating experience of same type of nuclear power plants; ③ experiences learned from tests or other activities.

15.5 Application of Ageing Management in Lifetime Extension

15.5.1 Application of Ageing Management Results in the Demonstration of Lifetime Extension

The lifetime extension practices carried out on NPPs in some countries have proven the possibility of the operation of a nuclear power plant during its extended lifetime, but related integrity demonstration analysis shall be completed according to the new lifetime requirement and the safety of the nuclear power plant at the end of extended lifetime needs to be proved. The United States has established a complete set of regulations, guidelines and guides and completed the analyses of lifetime extension for more than 70 units. IAEA is also releasing relevant safety guidelines.

The life extension demonstration of US nuclear power plants mainly involves the re-evaluation of environmental impacts, assessment report of NPP integrity and deadline analysis report. The latter two reports are the core of the life extension demonstration, and they aim to prove the ageing of NPP systems and equipment are managed in accordance with the established ageing management program, and such systems and equipment can operate safely at the end of the extended lifetime. Therefore, the results of the ageing management in preliminary design, construction, and operation phases can be applied to demonstrate the lifetime extension of nuclear power plants.

According to the lifetime extension practice of nuclear power plants in the United States, the ageing management program, supporting procedures for ageing management, ageing monitoring and inspection procedures, and ageing management review results formulated in the early stage of the ageing management of the marine nuclear power plants can be used for the integrity evaluation, analysis and demonstration of marine nuclear power plants, so as to prove that the ageing of the marine nuclear power plant is effectively managed in accordance with the established ageing management methods. The evaluation methods and results of the ageing status of nuclear power plants can be used to prove that the marine nuclear power plant can operate safely at the end of its extended lifetime.

15.5.2 Requirements of Ageing Management in the Extended Lifetime

The ageing management for marine nuclear power plants during the extended lifetime can still use the ageing management methods applied in the design lifetime, but the equipment ageing management program shall be continuously improved based on the results of the demonstration for lifetime extension. Especially with the ageing of equipment, the classification of ageing mechanisms, ageing evaluation methods, frequencies of ageing monitoring and inspection and irradiation surveillance plans need to be adjusted. Specific requirements are as follows.

(1) Adjustment of classification of ageing mechanisms

With the development of the equipment ageing and the continuous feedback of operating experience, the original classification of ageing mechanisms needs to be reassessed, and the current situation and margins of various ageing mechanisms shall be recomputed to provide the classification of ageing mechanisms for the extended lifetime. If the consumption number of design transients is much lower than the design parameter and there is a large margin for fatigue analysis due to good operating state, the classification of the fatigue mechanisms can be readjusted based on the new fatigue analysis results.

(2) Modification of ageing evaluation methods

With the development of technologies, the original ageing evaluation methods may no longer be applicable and need to be modified according to latest research results. Taking the irradiation embrittlement evaluation method as an example, as the prediction formula is continuously upgraded and new methods (e.g., the master curve method) are under development, such method needs to be adjusted based on these research results.

(3) Adjustment of methods and frequencies for ageing detection and monitoring

With the increasing of ageing degree and the development of detection methods and modifications of relevant laws and regulations, the original frequencies and methods of the ageing detection and monitoring may no longer meet the requirements for ageing management in the extended lifetime and shall be adjusted accordingly.

(4) Adjustment of irradiation surveillance plans

With the lifetime extension of nuclear power plants, new irradiation surveillance plans shall be formulated according to new lifetime requirements as the original irradiation surveillance plans do not cover the radiation conditions at the end of the extended lifetime. If possible, new irradiation surveillance or ex-core neutron monitoring methods shall be adopted to conduct the irradiation surveillance during the extended lifetime.

Chapter 16
Test Verification

16.1 Overview

A test of a piece of equipment or a system usually refers to testing and verifying the design performance or natural characteristics of it through the actual modeling of its working condition and operation status by means of full-scale or scaled simulation. In other cases, the test refers to verifying the actual performance of the real object or part of it through the commissioning or operation test of it in the actual environment and under real condition. For the R&D and design of marine nuclear power plant, on the one hand, tests are conducted to study the nuclear, thermal, mechanical, and its loop and system properties of the nuclear power plant; on the other hand, tests are conducted to verify the design assumptions or the results of calculation and analysis in the conceptual or preliminary design of nuclear power system so as to support the design. And the mooring tests and sea trials after construction are mainly conducted to verify the performance of individual system and the overall performance of the marine nuclear power plant.

According to the different purposes, the tests in the development of marine nuclear power plant are roughly classified into individual test, comprehensive test, reactor physical start-up test, mooring test and sea trials and special test. From the aspect of test methods, the tests are usually conducted with test beds, experimental reactor, construction of real nuclear power ship or engineering demonstration reactor for test.

With the development of marine nuclear power technology, the general requirements for marine nuclear power tests are getting stricter and stricter, and the test technology is under great development and tends to develop towards digital modeling and virtual simulation.

This chapter explains the brief classification of tests, outlines the comprehensive tests closely related to the marine nuclear power engineering development and construction, such as comprehensive verification tests of systems, physical start-up tests, mooring tests and sea trials, gives a brief review of the engineering demonstration

J. Yu, *Marine Nuclear Power Technology*,
https://doi.org/10.1007/978-981-15-2894-1_16

reactor that is an important test platform in the engineering development of nuclear power plant, and, provides a general introduction to the digital reactor test technology based on the development of test technology.

16.2 Classification of Tests for Marine Nuclear Power Plant

At the research and design stage of marine nuclear power plant, the tests involve many research fields. In terms of the disciplines, the tests are mainly classified into reactor physics and shielding tests, fluid mechanics tests, power equipment tests, thermal safety tests, water chemistry tests and structural mechanical tests. In terms of the radioactivity and nuclear power technology development stages, the tests are classified into non-radioactive individual tests, non-radioactive comprehensive tests, radioactive individual tests and radioactive comprehensive tests.

(1) Non-radioactive individual tests

The non-radioactive individual tests are mainly to test and verify the key parameters and functions of the components or equipment of the nuclear reactor system, and they are the technical basis for the R&D of marine nuclear power technology. The non-radioactive individual tests mainly include fluid mechanics tests, power equipment tests, thermal safety tests, water chemistry tests and structural mechanical tests. The non-radioactive individual tests can generally be conducted on general test facilities.

(2) Non-radioactive comprehensive tests

The non-radioactive comprehensive tests are to research and verify the comprehensive performance of the key technologies of the integrated power systems and equipment. For non-radioactive comprehensive tests, it is generally necessary to design the specific modeling test facilities for nuclear power plant.

The non-radioactive comprehensive tests mainly include flow-induced vibration tests for internals by the reactor integral hydraulic mock-up, comprehensive performance tests of steam generators, verification tests of engineered safety systems, verification tests for system operation characteristics under maritime conditions and verification tests for steady-state operation characteristics of systems. Among such tests, the comprehensive verification tests of systems to assess the overall performance of the marine nuclear power plant are comparatively important.

(3) Radioactive individual tests

The radioactive individual tests are mainly the experimental study of nuclear physical properties, irradiation, shielding and radioactivity of reactor materials, fuel assemblies and nuclear instruments and equipment.

The radioactive individual tests mainly include in-pile tests of nuclear instruments and equipment, nuclear physical characteristics tests of fuel elements, criticality physics tests, material irradiation tests, fuel element examination tests, burn-up measurement tests, radiochemical tests, shielding material performance tests, verification tests of radiation monitoring system and equipment, and radionuclide measurement tests.

(4) Radioactive comprehensive tests

The radioactive comprehensive tests are mainly the experimental study of reactor physics, shielding, fault diagnosis and other aspects of reactors, and the purpose is to verify theoretical calculations to ensure the operational safety of the reactor.

The radioactive comprehensive tests mainly include reactor zero-power physical tests and reactor radiation shielding and irradiation effect tests.

At the construction and commissioning stages of a marine nuclear power plant, the tests mainly include stand-alone recovery tests, mooring tests, reactor physical start-up tests, sea trials and other possible special tests.

16.3 Comprehensive Verification Tests of Systems

16.3.1 Function of Comprehensive Verification Tests of Systems

The comprehensive verification tests of systems are to use the dedicated non-nuclear comprehensive test facilities to simulate the operation of nuclear power plant, evaluate the overall performance of the nuclear power plant, reduce the risks of tests after loading nuclear fuel for nuclear power plant, and comprehensively resolve the non-nuclear technical problems of such plant.

The main contents of comprehensive verification tests of systems include verification tests of overall operation performance of nuclear power plant such as maneuverability, operation transients and accident transients, comprehensive performance verification tests of main equipment and coordination and matching tests of safety systems.

As an important means for the research and design of marine nuclear power plant, in general, the comprehensive verification tests of systems are required only when rather many new technologies are used in the design of nuclear power plant, the operating conditions and operating environment of such plant are rather complex, and there are major uncertainties in theoretical analysis and local tests.

16.3.2 Content of Comprehensive Verification Tests of Systems

(1) Comprehensive performance verification tests

The comprehensive performance verification tests mainly include power plant maneuverability tests, adjustment tests of power, pressure and water level, coordination and matching tests of primary and secondary loops, operation transient modeling tests and accident transient modeling tests. Such tests are to verify the rationality of system and equipment operation and the reliability and effectiveness of systems.

(2) Comprehensive verification tests of main equipment

The comprehensive verification tests of main equipment mainly include start-up, shutdown and switching tests related to rotary equipment, operation tests of faulty pumps under the reverse flow effect and performance verification tests of active equipment control systems and equipment prototypes. The thermo-hydraulic performance tests of pressurizers and steam generators are to verify the dynamic and static characteristics of the equipment.

(3) Tests of safety systems and main auxiliary systems

The tests of safety systems and main auxiliary systems are the relevant verification tests performed according to the system configuration.

Let's take the verification tests of passive safety systems as the example. The relevant tests mainly include tests of the passive heat removal system, tests of the passive cooling water injection and discharge system and tests for passive control of atmospheric environment in the compartment. The purposes of the tests are mainly to verify the system design process, assess the impact of system parameters and key equipment on the system operation sensitivity, check the reliability and time response of the control logic action, verify the results of thermo-hydraulic calculation and analysis and the rules for distribution and migration of relevant media/impurities, improve the system-related analysis and calculation methods and calculation models according to the test results, and develop the relevant calculation software and databases suitable for system analysis.

This chapter only describes the content of typical comprehensive verification tests of systems. Reasonable and feasible test items and verification scope shall be determined according to characteristics, technical maturity and integration technology risks of the simulated marine nuclear power plant.

16.4 Reactor Physical Start-Up Tests

16.4.1 Definition of Physical Start-Up Tests

The physical start-up tests refer to, after the reactor fuel loading, all the activities including operations and necessary tests in the entire process from water filling of reactor or lifting of control rods to the criticality of rated zero-power thermal state parameters. For the new reactors without reference reactors, such tests are generally performed after the zero-power test.

The zero-power test is conducted on a zero-power experimental facility using real fuel elements for a new reactor, so as to measure the critical fuel loading of the new reactor and the control rod worth, provide the technical basis for the physical start-up of the new reactor, and ensure the nuclear safety of the new reactor.

The purpose of the reactor physical start-up tests is to measure various physical parameters of the real reactor, so as to verify the nuclear design and provide technical support for the reactor start-up and operation. The physical parameters to be measured include criticality rod position in cold and hot states, total worth of control rods and individual worth of each rod, differential worth and integral worth of control rods and reactivity coefficient [1].

The physical start-up can be classified into hot physical start-up and cold physical start-up. The hot physical start-up refers to the physical start-up test performed after the increase of the temperature and pressure in reactor coolant system and the establishment of the steam chamber in pressurizer. The cold physical start-up refers to the physical start-up test performed when the primary coolant is in a relatively cold state (such as at the room temperature) and the pressurizer is fully filled with water and does not establish the steam chamber.

16.4.2 Stages of Physical Start-Up Test

(1) Division of stages of physical start-up test

The physical start-up activities can be implemented by stages and generally classified into the water filling and subcritical test stage ("Stage A" for short), the initial criticality and zero-power physical parameter measurement test stage ("Stage B" for short), the low power test stage ("Stage C" for short) and the power test stage ("Stage D" for short).

(2) Applicability requirements for various stages by different reactors

During the first physical start-up of a new type of reactor, the tests at Stage A, Stage B, Stage C and Stage D shall be completed. During the first physical start-up of a newly built reactor, the tests at Stage A, Stage B and Stage C shall be completed. During the physical start-up of a refueled reactor or a reactor in special conditions, the tests in Stage B, or the tests at Stage B and Stage C, as required, shall be completed.

16.4.3 Brief Introduction of Physical Start-Up Test

(1) Tests at Stage A. At this stage, the tasks are mainly to complete the first water filling and criticality safety supervision of the reactor, conduct the installation and commissioning of control rod drive mechanism, and perform the cold trial transmission of the control rods. During the water filling of the reactor, the amount of water filled in the pressure vessel and the water level of the pressurizer shall be continuously recorded. Extrapolation shall be performed on the basis of the pressure vessel water level and the neutron counting rate to determine whether the reactor is in a normal subcritical state.

(2) Tests at Stage B. At this stage, the tasks are mainly to complete the test items such as the first cold-state criticality test for the reactor, the measurement of "stuck rod" subcriticality and the "stuck rod" criteria verification, the cold-state integral worth measurement of control rod cluster, the measurement of cold-state shutdown depth, the symmetry test of radial neutron flux density distribution in the core, the measurement of differential worth and integral worth of control rod assembly, the measurement of reactivity temperature coefficient, the measurement of reactivity pressure coefficient, and the measurement of hot-state critical rod position.
(3) Tests at Stage C. At this stage, the tasks are mainly to complete the test items such as the test of operation characteristics of reactor and its relevant systems, the monitoring of neutron and γ radiation field at specific positions, and the verification of the compliance of the neutron detectors with the design requirements in the source range and intermediate range.

    After the reactor subject to the hot-state physical start-up test reaches the criticality, the reactor power shall be increased to the power level required for the tests at this stage according to the operating procedures specified in the design. It is necessary to limit the power increase rate and ensure the reactor safety.
(4) Tests at Stage D. At this stage, the tasks are mainly to conduct the test of operation characteristics at high power, the monitoring of neutron and γ radiation field at specific positions at different power levels, the verification of shielding performance, the reactor start-up and shutdown tests at different power levels, and the power calibration and measurement of reactor power coefficient.

    The tests at this stage shall be performed at different power levels from low power to rated power. The tests shall be maintained for sufficient time at each power level specified in the test procedure to meet the supervision and measurement needs.

16.5 Mooring Tests and Sea Trials

16.5.1 Overview

(1) Brief introduction of mooring tests and sea trials

After the ship building, the relevant tests must be performed according to the requirements as stated in the technical specifications to confirm that the ship building status meets the requirements in the design specifications or the development assignment before the official delivery for operation is made. Such tests are generally classified into mooring tests and sea trials. The tests of marine nuclear power plant are an important part of the mooring tests and sea trials of ship.

The tests under various working operating conditions of the nuclear power plant performed when the ship is moored at a special dock or wharf and kept in a static state are called the mooring tests of marine nuclear power plant. The tests under relevant operating working conditions of the nuclear power plant when the ship is navigating in the specified sea area are called the sea trials of marine nuclear power plant. The mooring tests and sea trials of nuclear power plant are sequential. Generally, only after the mooring tests are completed can the next-stage sea trials be carried out [2].

(2) Purpose and requirements for mooring tests and sea trials of nuclear power plant

The mooring tests of nuclear power plant are further classified into cold tests and hot tests. The mooring tests are to inspect the correctness, coordination and reliability related to the manufacture and installation quality and the operation of various types of equipment, devices and systems on the ship, and check whether the technical performance can meet the requirements in technical specifications, relevant norms, standards and other technical documents. If the requirements are not met, the causes shall be identified and relevant corrective measures shall be taken to meet such requirements.

On the one hand, the sea trials of nuclear power plant are to, at different power states, verify whether the systems and equipment can meet the design requirements during the navigation of ship, check the operation correctness and coordination, confirm the states of such systems and equipment during the operation of the entire nuclear power plant, and eliminate various kinds of faults. On the other hand, the sea trials of nuclear power plant can support various kinds of tests of other systems and the comprehensive performance tests of the ship as a whole.

It should be noted that only after the tests at a specific stage are completed according to the content of and requirements in the test program and confirmed to be acceptable can the tests at the next stage be performed.

16.5.2 Mooring Tests

(1) Division of mooring test stages

The mooring tests of nuclear power plant are divided into cold tests, hot tests, reactor fuel loading, criticality tests and joint commissioning of primary and secondary loops after reactor start-up. Among them, the reactor fuel loading, criticality tests and physical start-up are introduced in the relevant chapters of this book.

(2) Conditions of and preparations for mooring tests

Before the mooring tests are performed on a marine nuclear power plant, the following test conditions must be met:

① The installation of the ship has been completed, and the test items at the building stage have been completed and accepted.

② The pressurization and cleansing have been completed for the lubricating oil system that is used to support the nuclear power plant tests, and the water (gas or steam) filling, pressurization, cleansing have been completed for other systems, and the level calibration has been completed for evaporator, water tank and oil tank.
③ The land-based electricity, steam, gas, water and oil required for mooring tests can be provided according to the required quality indicators.
④ The measuring instruments and tools required for the tests have been calibrated and are within the validity period.

(3) Cold tests.

The cold mooring tests for marine nuclear power plant generally contain the following main tests:

① Hydrostatic tests for systems and equipment: The purpose is to inspect the leak tightness of connections and sealing surfaces of equipment, pipes and valves, and to assess the strength.
② Inspection of valve isolation performance: With the subsystems depressurized one by one, the valves for each system are opened successively according to their distance to the pressure boundary of the reactor coolant system and their design pressures, to inspect their isolation sealing performance.
③ System cleansing according to the requirements, and water sampling for testing the water quality: The indicators for the water shall meet the specified values. The deionized water can be used for the cleansing. The filling-discharging operation of the device or the inching of the pump is performed to flush the loop system, so as to achieve the purpose of cleaning.
④ Adjustment of pipe supports and hangers, dampers, equipment supports and bellows is performed in the cold state to reach the cold state values specified in the design.
⑤ Pipe vibration tests in the cold state: After the rotating equipment is started, the instruments are used to measure the pipe vibration conditions. The pipes with excessive vibration must be analyzed, and the supports and hangers or the supporting and hanging methods shall be improved, or even the piping shall be redesigned to eliminate the non-conformance.
⑥ Inspection of insulation resistance and wiring correctness of all electrical equipment.
⑦ Inspection of the actuation performance of electrically operated valves and solenoid valves, and setting and verification of safety valves. The safety valve tests can be performed on a specific test rig.
⑧ Actuation tests of various types of pumps.
⑨ Switching tests for equipment and components that have function switching and standby requirements.

⑩ Tests for indication, alarm, control and interlocking of process parameters such as pressure, water level, temperature and flow.
⑪ Control or automatic control tests for the automatic control equipment or components.

(4) Hot tests

The hot mooring tests for marine nuclear power plant generally contain the following main tests:

① Adjustment of pipe supports and hangers, dampers, equipment supports and bellows in the hot state of the system.
② Measurement of the thermal stress and thermal displacement at typical positions of the pipes.
③ Measurement of temperature at the outer surfaces of thermal insulation layers of equipment and pipes in the hot state, and inspection of thermal insulation effect.
④ Inspection of the actuation performance of electrically operated valves and solenoid valves in the hot state, and tests of safety valves.
⑤ Inspection of insulation resistance of all electrical equipment.
⑥ Actuation tests of various types of pumps.
⑦ Switching tests for equipment and components that have switching features or may be kept standby.
⑧ Tests for indication, alarm, control and interlocking of process parameters such as pressure, water level, temperature and flow.
⑨ Control or automatic control tests for the automatic control equipment or components.
⑩ Inspection tightness of sealing surfaces.

(5) Joint commissioning of primary and secondary loops

The following commissioning tasks shall be completed at the allowable low reactor power of nuclear power plant in the mooring state:

① Joint commissioning of the reactor with the measurement, protection and control systems.
② Joint commissioning of individual system with the I&C system.
③ Joint commissioning of rotating equipment in individual system with the power supply and electrical control system.
④ Verification of the correctness of the operation modes and operating procedures of systems and equipment.
⑤ Verification of the steady and transient characteristics of reactor core, coolant and reactivity control system.
⑥ Inspection of the radiation shielding performance of the reactor compartment.
⑦ Operating condition transition tests at low power.

16.5.3 Sea Trials

(1) Conditions of and preparations for sea trials

Prior to the sea trials of a nuclear-powered ship, the following preparations shall be made:

① Before the trials, the wharf shall be able to provide sufficient water, power, gas and chemicals as required.
② All tests as required in the mooring test program have been completed for the reactor and primary loop system, and they have been confirmed to meet the design requirements through reviews.
③ The instruments, meters, spare parts and special tools required for the trails have been fully prepared and meet the test requirements.

(2) Brief introduction of sea trials

During the sea trials of marine nuclear power plants, it is generally necessary to perform the reactor and primary loop system tests and the secondary coolant circuit tests. The reactor and primary loop system tests include reactor self-stabilization and self-regulation performance test, reactor and primary loop system stability test, verification test for steady-state operation characteristics of reactor and primary loop system and primary loop system maneuverability test. The secondary coolant circuit tests mainly include main steam turboset test and steam dump system test.

Taking the primary loop system maneuverability test as an example, the maneuverability test is mainly to check whether the changes of the parameters related to the primary loop caused by load change and load rejection of the nuclear plant are within the design limits and confirm whether the changes meet the required design analysis.

The load change test is to, after increasing (or decreasing) the power of the nuclear power plant within the time and load change amplitude specified in the design, monitor the temperature, pressure and flow of reactor coolant in the reactor and primary loop system, the feedwater temperature and flow and the steam pressure and flow of steam generator, the pressurizer water level and other process parameters. The purpose is to verify that the peaks of the change of such process parameters will not exceed the operational limits specified in the design during the transient processes of increasing or decreasing the power.

The load rejection test is to, after the load shedding of the nuclear power plant with the amplitude specified in the design, monitor the temperature, pressure and flow of reactor coolant in the reactor and primary loop system, the feedwater temperature and flow and the steam pressure and flow of steam generator, the pressurizer water level and other process parameters. The purpose is to verify that the peaks of the change of such process parameters will not exceed the operational limits specified in the design during the transient process of load shedding.

16.6 Engineering Assessment Tests of Prototype Reactors

16.6.1 Significance and Role of Prototype Reactors

During the advance research of the marine nuclear power engineering models, the prototype reactor, as the large-scale integrated test platform related to nuclear power, simulates the construction of a nuclear power plant equipped in a real ship, and such nuclear power plant can be close to the actual engineering model and truly and adequately assess the design performance and operation characteristics of the nuclear power plant. The prototype reactors have an important significance and play an important role throughout the development of marine nuclear power technology.

During the development of marine nuclear power technology in the world, the significance of the construction of prototype reactors is reflected in the following two aspects:

First, the prototype reactors have played a significant role in promoting the development of marine nuclear power technology. The prototype reactors are used to perform the experimental verification of reactor physics, thermo-hydraulics, operation characteristics, safety and reliability, test the design and manufacturing processes, verify the operation performance of systems and equipment, accumulate the engineering experience, and train the operators.

Second, the prototype reactors closely combine the scientific research with the engineering verification. The adequate tests on land-based prototype reactor are to verify the technology, expose the problems, accumulate the data, improve the design, enhance the technical maturity of marine nuclear power plants, and fully restrain the engineering risks.

The prototype reactors provide the platform for conducting the tests under the simulated actual operating conditions that cannot be duplicated in the laboratory. The important technical data and experience gained from these tests can make it possible to discover potential problems and solve such problems before such problems actually occur in the marine reactors in operation.

According to the construction conditions and experience of worldwide prototype marine nuclear power reactors, the main functions of prototype reactors are to

① Verify the overall performance of nuclear power plants;
② Verify the physical and thermo-hydraulic performance of the reactor cores;
③ Verify the key technologies of nuclear power plants and the overall performance after technology integration, and assess the reliability of main equipment;
④ Accumulate relevant experience in design, installation, commissioning and operation for guiding the construction and operation of marine nuclear power plants;
⑤ Provide professional operation and training for the operators of marine nuclear power plants;

⑥ Verify new technologies, new materials, new equipment and new processes to be used in nuclear power plants; and
⑦ Verify and develop the computer software related to marine nuclear power plants.

16.6.2 Content of Engineering Assessment Tests of Prototype Reactors

The prototype reactor is a comprehensive test platform for nuclear power plants loaded with nuclear fuel. It can be used to comprehensively assess the design specifications, performance, operation characteristics and construction techniques of nuclear power plants. The tests include the test items and content of comprehensive tests of systems, fuel loading and physical start-up, mooring tests and sea trials. The systems and equipment covered by such tests include reactor, primary loop system and conventional power units in secondary loop. In particular, the prototype reactors can also simulate the operational profile and conditions of the actual ships, and keep in operation for long to assess the overall performance of nuclear power plants and the performance changes of nuclear fuel, equipment materials, structures, shielding and equipment reliability. According to the needs, the detailed special test items and test contents can be prepared.

16.6.3 Development of Prototype Reactors

The nuclear power prototype reactor was first developed in the United States. In 1948, the United States began researching and constructing a land-based prototype reactor for naval reactors. The United States enjoys the earliest start, the most advanced technologies and the most perfect system in this field. It has designed and constructed nine land-based prototype reactors. Westinghouse constructed the land-based prototype reactors S1W and A1W, General Electric constructed the land-based prototype reactors S1G, S3G, S5G, S7G, S8G and D1G, and Combustion Engineering Inc. constructed the land-based prototype reactor S1G. Currently seven land-based prototype reactors at three sites have been shut down. The two land-based prototype reactors currently in operation are located at the Kesselring site in the State of New York.

France introduced the technology from the United States at the beginning, and gradually developed its own marine reactors. The four generations of PWRs developed by France have corresponding land-based prototype reactors respectively.

France constructed the PAT-type separately land-based prototype reactors with separate layout and equipped the "Le Redoutable"-class and "Unyielding"-class nuclear-powered ballistic missile submarines; constructed the CAP prototype reactors with integrated layout and developed the CAS-48 reactors to equip the "Rubis"-class nuclear-powered attack submarines; constructed the integrated RNG land-based prototype reactors, in which the reactor for submarines is K15, to equip the "to Triomphant"-class nuclear-powered ballistic missile submarines and the "Charles de Gaulle"-class nuclear-powered aircraft carriers; and completed the construction of the RES land-based prototype reactors in 2010 to equip the future "Barracuda"-class nuclear-powered attack submarines.

The former Soviet Union/Russia submarine reactors were developed along two directions: the PWRs and the liquid metal cooled reactors. Four (4) generations of PWRs have been developed and three land-based prototype reactors, i.e. 27/VM, OK-650BK and KB-2, have been constructed. Two generations of liquid metal cooled reactors have been developed and two land-based prototype reactors, i.e. 27/VT and KM-1, have been constructed.

The United Kingdom has built two types of land-based prototype reactors, i.e. DSMP and STF-2, and two generations of nuclear submarines were developed based on such reactors.

16.6.4 Development Trend of Prototype Reactors

The prototype reactors, most of which were built in the early stage of the development of marine nuclear power, were used to comprehensively master and verify the technologies for design, construction and operation of nuclear power plants, and to reduce the technical risks for engineering applications. The prototype reactors are closely related to the marine nuclear power engineering construction. The construction of prototype reactors can be used to fully verify the comprehensive performance of nuclear power plants before such nuclear power plants are installed in the ships, solve various problems that may be encountered in design, manufacture and engineering as far as possible, accumulate the engineering experience, and train nuclear power operation personnel to effectively ensure the successful development of marine nuclear power plants. For example, for S1W prototype reactor, as the land-based prototype reactor for the reactor S2W of "Nautilus" nuclear-powered submarine, the 48-hour continuous operation test and the 96-hour simulated underwater sea trial of the Atlantic were performed, the test items such as reactor performance under dramatic change of the operating conditions, reactor coolant leakage and shielding performance were conducted, and the feasibility and reliability of nuclear submarine operation were verified, so as to provide the experience for actual operation of nuclear-powered submarines and reduce the technical risks in actual operation. A1W prototype reactor, as the land-based prototype reactor for the reactor A2W of the first nuclear-powered aircraft carrier "Enterprise" in the world, provided personnel training for the surface nuclear-powered ships of the United States Navy.

The construction of early prototype reactors was also used for the selection of technical routes. Theprototype reactor was constructed respectively for two technical routes and used to identify the advantages and disadvantages of each technical route. For example, the United States built the PWRs and, in the 1950s, the S1C prototype reactor with sodium as the coolant, and the former Soviet Union built the 27/BT and KM-1 prototype reactors with lead bismuth alloy as the coolant from the 1950s to the 1970s. Due to limitations of technical conditions, industrial level, operating conditions of submarines and application of land-based prototype reactors, various countries have finally chosen the practical PWRs as the marine nuclear power plants for the Navy.

From the perspective of the subsequent development of prototype reactors, it is more likely that the existing prototype reactors are used as the general basis for the development of new engineering models of ships through the research and development of new technologies. Therefore, the prototype reactors mainly play a role of verifying the new technologies. For example, the land-based prototype reactor S8G of the United States, as the land-based prototype reactor for the reactors of "Ohio"-class nuclear-powered submarines, was successively equipped with the cores S6W and S9G, which were used to develop the "Seawolf"-class and "Virginia"-class nuclear-powered submarines. The data obtained were also used for the development of aircraft carrier reactor A1B and next generation of submarine reactors. Meanwhile, this prototype reactor also provides an important training platform for the reactor operators. In addition, a few prototype reactors are even only used for engineering application research and technical verification, rather than direct application in the construction of ship engineering model, such as the land-based prototype reactor S7G of the United States.

Based on the development trend of the prototype reactors, with the development of computer technology, the verification of technologies and the release of the technical risks in engineering construction previously performed by the prototype reactors can now be solved by the computer technologies such as 3D simulation elaborate software modeling. In addition, with the deepening of the understanding of the overall performance of nuclear power plants, it becomes unnecessary to explore and master the performance by the prototype reactors. The general prototype reactor platform is mainly used to grasp the technologies that cannot be accurately analyzed by theoretical calculations, to discover and verify various uncertain influencing factors caused by technology integration, and to examine the new technologies, equipment, materials and processes. The role of the engineering prototype reactors in the assessment tests of the engineering design and construction services of the marine nuclear power plants will become smaller and smaller.

16.7 Virtual Tests and Digital Reactor System Simulation Verification

16.7.1 Virtual Tests

With the development of test technologies, in addition to the physical modeling tests such as those on the constructed test beds, the virtual tests using the computers to simulate the physical prototypes, operating conditions and operating environment are also developing rapidly.

A virtual test generally refers to the research process of using the mathematical models of a prototype in the test environment virtually constructed by a computer to establish the digital models to replace the physical prototype (such as equipment, devices and test beds) and simulate relevant tests. Compared with the physical tests, the virtual tests based on the virtual test platform have various advantages such as low test cost, short test cycle, wide range of operating conditions and environment and good universality. Virtual test technologies mainly involve various fields such as numerical calculation models, numerical calculation methods, high-performance computer technologies, simulation and verification technologies and virtual reality technologies.

In general, virtual tests can be classified into computer simulation tests and computer modeling tests.

The computer simulation tests are to collect the parameters of relevant components, equipment, systems and devices at the actual prototype operating site, complete the simulation models of relevant components, equipment, systems and devices, and perform relevant tests to verify and improve the design theories and methods. The core of the simulation tests is the acquisition and reproduction of relevant parameters and the accuracy of simulation models. The computer modeling tests are to integrate and analyze the test data based on the physical tests and the simulation tests, integrate the established mathematical models into the computer by means of theoretical analysis and modeling, and utilize the computer to perform wider modeling tests for the virtual objects in the design. The credibility of computer modeling tests depends on data accumulation and model establishment during the physical tests and simulation tests.

The virtual test technologies are applied rather widely in the field of nuclear power development, such as the simulators for training in some NPPs and the elaborate modeling design assessment for reactor cores. The application of currently rather advanced virtual test idea is the simulation test verification of reactor systems by the digital reactors. The simulation test verification of reactor systems by the digital reactor technology is the manifestation of the advanced application of virtual test theory in the nuclear power development test technology. Therefore, it is necessary to present a separate introduction to the digital reactors.

16.7.2 Definition and Role of Digital Reactors

A digital reactor is a kind of virtual reactor that integrates numerical modeling to simulate various characteristics of a reactor in the whole lifetime by using calculation software and various databases in a super-large scale high-performance calculating system.

The digital reactors can be used to perform multi-disciplinary high-precision calculation and analysis for physics, thermal, fuels and structures on the integrated R&D platform, realize the multi-disciplinary and multi-scale coupling and comprehensive modeling, and predict the performances, parameters, behaviors and other characteristics during various stages of the real reactors such as design, manufacture, operation and decommissioning.

The digital reactors can be used to fully and comprehensively evaluate or optimize various design, test, manufacture and operation support schemes at the early stage of development, thus enhancing the overall performance, timely discovering the defects, fully guaranteeing the safety and reducing the development costs. The simulation test verification by using digital reactors is only one of the important applications of the digital reactor technology.

A digital reactor can be the comprehensive embodiment of the development of the reactor system theory, computer technology and test technology when developing to certain levels. The simulation verification tests for the nuclear power plants by using digital reactors can assess the performances, technical indicators and behaviors under various operating conditions, will become the main development direction of future nuclear power plants, to some extent can replace the verification by test beds, engineering prototype reactors and other physical prototypes, and can develop and verify the new types of reactor systems more quickly, more efficiently and more optimally.

16.7.3 Overview of Digital Reactor Research

A large amount of work has been carried out on the application of advanced digital technologies, especially in the fields of aviation and aerospace. For example, the Boeing 777 aircraft developed by the United States is the world's first aircraft developed fully based on the digital prototype technology. This technology plays an extremely important role in the design, manufacture, assembly and evaluation of the aircraft, and ensures the successful production at the first try. In the nuclear industry, the most representative projects related to digital reactors are the CASL/NEAMS Project in the United States and the NURESIM/NURISP Project in Europe.

(1) NURESIM platform in Europe

The NURESIM Project in Europe was initiated by the European Union in February 2005. It consisted of several multi-physical field sub-projects that were aimed at realizing the further development of dynamic coupling of neutron programs with thermo-hydraulic calculation programs required for light water reactor (LWR) simulation.

The NURESIM platform integrates the latest developments in core physics, two-phase flow thermo-hydraulics and fuel models to accurately represent the physical phenomena. The coupled core physical and thermo-hydraulic models for reactor safety in the platform have multiple scales and multiple physical fields, and different software are simply coupled by common data structures and class functions.

The platform also provides the data environment for testing and comparing different software. This function requires the standardization of the connection among software. Therefore, this standardization will be gradually established with the development of the platform. In order to make it more convenient to use the software, the platform implements pre-processing, post-processing and supervision functions by the open source SALOME software.

(2) NURISP platform in Europe

The NURISP Project is based on the NURESIM Project in Europe. The NURISP Project lasted from 2008 to November 2012, and totally 22 scientific research organizations from 14 countries participated in this project. The overall goal of the NURISP Project is to further develop and propose new ideas for the European reference simulation platform that can be applied to existing PWRs, boiling water reactors (BWRs) and future reactor types. The development route of simulation platforms will be part of the strategic development agenda of future sustainable nuclear energy technology platforms.

The new software will be connected to the platform. At the same time, software integration, model development, coupling, sensitivity and uncertainty analyses and result verification will be further developed and applied more widely. During the NURISP Project, the focus was placed on existing and future PWRs, VVERs and BWRs. More general way was used for the study to facilitate the expansion and application for the fourth generation reactors in the future.

The NURISP Project was completed in December2012. The research of the NURESAFE Project, as the follow-up of the NURISP Project, was started in January2013 and planned to be completed in 2015. It aimed to further expand the platform, perform the safety analysis application and expansion verification of the second and third generations of LWRs, and expand the scope of application of the platform.

(3) NEAMS platform of the United States Department of Energy

The Nuclear Energy Advanced Modeling and Simulation (NEAMS) project is developing towards the understanding of introducing novel and scientific new nuclear energy systems that can meet the energy demands of the United States. The deepening of this understanding can a. improve the safety; b. enhance the economic efficiency by improving operating efficiency and reducing unnecessary design margin; and c. accelerate the design, approval and construction of nuclear energy systems.

The development of NEAMS is not only important for the nuclear energy utilization, but also helpful in facilitating the fulfillment of various forms of development commitments for energy utilization. This is very important for the future energy development in the United States.

NEAMS consists of a series of sub-projects, including reactor simulation, fuel simulation, safety protection, waste treatment, computing technology, basic methods and models, verification and function transformation.

(4) CASL simulation platform VERA in the United States

In early 2010, the Nuclear Energy Authority (NE) under the United States Department of Energy (DOE) organized an open competition for selecting a team to operate a nuclear energy modeling and simulation hub. The results of this competition were announced in May 2010. The team that was selected to operate the NE Hub was the Consortium for Advanced Simulation of Light Water Reactors (CASL). CASL is led by and headquartered at the Oak Ridge National Laboratory (ORNL). The leaders of CASL and most scientists and engineers from multi-disciplinary and multi-research institutions gather here.

In terms of nuclear energy, CASL mainly focuses on three (3) key issues: investment, nuclear fuel consumption and safety. The urgent goal of CASL is to find a way for enhancing the reliable and safe power output of nuclear reactor fuels under normal operating conditions. The main tasks are to focus on a series of challenging issues that limit the performance of PWRs, and to develop and apply modeling and simulation (M&S) technology to address the performance of three (3) key areas in NPPs:

① To reduce the investment and operation costs per unit of energy by the power increase and life extension of the NPPs;
② To reduce the volume of nuclear waste by increasing the fuel burnup;
③ To improve the nuclear safety by high-precision prediction of component performance failures.

The CASL applies the advanced M&S capabilities to develop the high-performance and predictable LWR simulation environment. This simulation environment is called the Virtual Environment for Reactor Applications (VERA). VERA will shorten the cycle required to build a reactor.

16.7.4 Technical Route of the Digital Reactor Development

According to the current worldwide investigation and analysis, there are currently three (3) technical routes in the field of digital technology. The first is the structure design and virtual reality simulation based on 3D digital geometric models; the second is the overall calculation, analysis and multi-disciplinary collaborative R&D based on the unified data models; and the third is the accurate calculation of complex product performance and the visual simulation of the integration with calculation results. The application of digital technology in the advanced industries in the world has been relatively mature, but the research on digital reactor technology is still at the initial stage in the world.

The technical features of the technical route of virtual reality technology for 3D digital models include fully digital 3D geometric models, whole life cycle data management and engineering scheme verification, evaluation and optimization. The purposes are to identify the engineering structural design errors as early as possible to reduce the rework costs, scientifically and rationally arrange the engineering manufacture schedule to optimize the installation and maintenance schemes, and track the equipment information in the whole life cycle to strengthen the management. The technological approaches are 3D digital modeling and integration of equipment structures within the entire scope; dynamic simulation of construction and maintenance processes; integration of whole life cycle data by 3D structural models; and visual display by virtual reality technology. Typical technical solutions include the digital development in the cooperation between French AREVA and Dassault, and the digital NPPs of the ITER Project.

The technical features of the technical route of the real-time condition modeling are the construction of the full-fluid system response model of a reactor to simulate the normal and accident operating conditions of the reactor. The purposes are to timely discover the fluid system design defects and correct them in order to improve the design, verify the I&C equipment response performance to improve the reliability and interactivity, and conduct the personnel training and emergency response scheme exercises. The technical approaches are to establish a full-range fluid model, simulate and verify the semi-physical simulation platform, and simplify the core model to ensure the real-time performance. Typical technical solutions include the simulators for Longmen Nuclear Power Plant and the analog simulators for traveling wave reactor, with the simulators developed by the A D Technology Inc.

The technical features of elaborate calculation and analysis technical route are accurate analysis and calculation of key characteristics of the reactor core and 3D visual display of calculation results. The purposes are to deepen the burnup, improve the fuel utilization, accurately calculate the safety margins, improve the conservative designs, and accurately predict the material performance for reactor life extension. The technical approaches are generally super-large scale high-performance calculation, elaborate physical, thermal and fuel theoretical models, multi-disciplinary coupling, unified platform and high fidelity visual display. Typical technical solutions include the CASL Project led by DOE of the United States, and the NURESIM/NURESIP Project of the European Union.

With the rapid development of reactor technology and computer technology, the digital reactor technology will become increasingly mature.

References

1. Dafa, Zhang. 1996. *Operation and Management of Marine Nuclear Reactor*. Beijing: Atomic Energy Press.
2. Shao Kaiwen, Ma Yunyi. 2014. *Introduction to Technology and Design of Ship (Rev. 2)*. Beijing: National Defence Industry Press.

Chapter 17
Reactor Loading and Unloading

17.1 Overview

As an important process supporting facility for nuclear power plants, the reactor fuel handling and storage system is essential to the safe and reliable refueling so as to ensure the continuous operation of nuclear power plants.

17.1.1 System Functions

The fuel handling and storage system is utilized to handle and store fuel assemblies and the associated assemblies inside the nuclear power plants, and to perform the operations of the involved special equipment and tools at various stages from receiving fresh fuel assemblies to shipping spent fuel assemblies out of the plant. Such operations include the acceptance, inspection, storage, transfer and loading of the fresh fuel assemblies, and the unloading, handling, storage, inspection, canning of the spent fuel assemblies (including the associated assemblies) as well as the inspection, maintenance, storage, canning and other operations of failure fuel assemblies, after the arrival of the containers of the fresh fuel assemblies at the nuclear power plant.

17.1.2 System Composition

The fuel handling and storage system involves a number of facilities, equipment and structures, including equipment, tools, systems and structures used during the whole process from accepting fresh fuel assemblies, through loading, in-core operation, unloading, transferring, and on-site storage, to loading into the containers to leave from the plant.

J. Yu, *Marine Nuclear Power Technology*,
https://doi.org/10.1007/978-981-15-2894-1_17

The fuel handling and storage system can be divided into 4 subsystems based on different functions: the fresh fuel storage system, the fuel handling system, the handling and storage system for other irradiated components and the spent fuel storage system.

The fresh fuel storage system consists of devices used for inspecting, operating, storing and transferring fresh fuel assemblies and associated assemblies, temporary storeroom for fresh fuels and such systems, facilities and equipment as related to power supply, instrument, control, communication and drainage.

The fuel handling system consists of the equipment, the pool and its auxiliary structures, and sluice gates used for in-pile loading and unloading, guiding and positioning, underwater inspecting and recording, and receiving the fuel and associated assemblies in the pool.

The handling and storage system of other irradiated components consists of the equipment, the pool and its auxiliary structures, sluice gates, and the water level and temperature measuring systems used for in-pile loading, unloading, guiding and positioning the associated assemblies, for loading, unloading, and guiding the reactor internals, and for storing and underwater observing relevant internals.

The spent fuel storage system consists of pool cranes, auxiliary working bridges, equipment used for the storage, inspection and operation of spent fuel and failed fuel, pool, and cooling & purification and coolant-charging systems for the pool, water level and temperature measurement system, and the systems, facilities and equipment related to power supply, instrumentation, control and communication.

Some nuclear facilities that may not be equipped with special facilities and equipment must be provided with related facilities and equipment having corresponding functions.

The primary dedicated equipment belonging to which the fuel handling and storage system belongs will be described in details in following sections.

17.1.3 Main Process Flow

The functions of the fuel handling and storage system show that it is mainly used to complete operations related to loading, unloading and storage of reactor fuel assemblies and the associated assemblies of the nuclear power plant. The main activities include the receipt, inspection, storage and transfer of fresh fuel assemblies after their packaging containers have been transported to the site; loading & unloading and transfer of the in-pile irradiated fuel assemblies; inspection, storage and canning of the irradiated fuel assemblies, including the inspection, repair, storage and canning of the failed fuel assemblies; loading the fresh fuel assemblies into the reactor core.

The main processes for fuel handling and storage include: reception, inspection, storage of the fresh fuel assemblies; unloading and transfer of the irradiated fuel assemblies; loading of fuel assemblies; inspection and storage of the spent fuel assemblies; and canning the spent fuel assemblies into the storage casks.

17.1.4 Design Principles

The design of the fuel handling and storage system for nuclear power plants shall:

① Comply with the provisions of relevant codes and standards;
② Ensure no criticality will occur to fuels under the anticipated conditions;
③ Implement the principle of controlling the radiation exposure to workers under all conditions to a level as low as reasonably achievable, and meet the requirements of relevant codes and standards;
④ Provide sufficient cooling for the handling, transfer and storage of the irradiated fuels to prevent the fuels from becoming overheated;
⑤ Follow the principle that the design of fuel handling equipment shall prevent mechanical damage or dropping of fuel assemblies and avoid unacceptable loading and unloading stress applied to fuel assemblies;
⑥ Ensure the discharge to off-site unrestricted area is within the defined limit under various conditions, and detect the amount of radioactive materials and exposure doses of operators and maintain them within acceptable limits;
⑦ Be designed together with other relevant systems to ensure appropriate internal and external interfaces;
⑧ Take into account the design lifetime, maintainability and decontaminability of equipment and corrosion resistance of materials and parts and components;
⑨ Ensure that the structures, systems and components can bear the principal loads (e.g., floor loads, seismic and dynamic loads) that may be applied to them;
⑩ Satisfy the fire protection requirements of relevant regulations and standards.

17.2 Reactor Fuel Loading

17.2.1 Reactor Fuel Loading Technology

In order to ensure the safety during reactor fuel loading, the design of reactor fuel loading process shall comply with the provisions of related codes and standards after the influences of nuclear criticality safety, radiation protection safety and other factors are taken into full consideration. The equipment necessary for each process and its function requirements and primary performance requirements are then determined according to the process needs.

The reactor fuel loading of nuclear power plants (NPPs) is carried out in the loading and unloading area for reactor fuel assemblies by joint operation of the equipment of the fuel handling and storage system. The loading and unloading area for reactor fuel assemblies is located in the reactor building, crossing the reactor pool and the refueling pool. There is a gate between the two pools, which can separate the two pools when necessary. Both pools are equipped with water filling and draining pipes. The draining pipe remains fully open during operation to make the water flow down to the sump for reuse in case of accident spraying. The fuel manipulator crane grabs the fuel assemblies to be loaded from the fuel assembly transfer container and delivers them to the position above the reactor and then places them into the reactor according to the designated coordinates and position.

The fuel manipulator crane is the core equipment in performing reactor fuel loading (shuffling) of NPPs. The positioning accuracy in the process of fresh fuels entering the reactor is ensured by the movement precision and positioning accuracy of the fuel manipulator crane on the track. The combined error of the movement accuracy and positioning accuracy of the fuel manipulator crane is $\pm$ 3 mm. In the process of fuel loading, the fresh fuel assemblies are transferred from the fuel building to the designated location in the reactor building through the underwater transmission equipment, and meanwhile fresh fuel assemblies are captured by the fuel manipulator crane from the underwater transmission equipment and then placed into the designated position in the core. The fresh fuel assembly is loaded into the reactor core through the guiding of the fuel manipulator crane itself and the auxiliary guide device in the core.

As the marine fuels will be used for a long time, the safety requirements for refueling technology are higher; due to the influence of fluctuations in the water surface, the alignment and positioning of reactor refueling are required to be more accurate; because the fuel transmission must pass through the atmosphere, the requirements for shielding are higher. Therefore, compared with the fuel loading in NPPs, marine reactor refueling adopts targeted special technologies. Compared with worldwide similar technologies, the subsequent reactor loading technology of nuclear power plants has higher safety and better equipment maintenance accessibility, and the fuel loading equipment has higher positioning accuracy, more shielding performance designs and meets higher design requirements.

17.2.2 Reactor Fuel Loading Process

The process of nuclear reactor fuel loading technology must follow principles as below at the present stage:

① Adapt to the characteristics of the on-site environment and the subsequent reactor structure of nuclear power plants, and comply with relevant technical requirements;
② Comprehensively consider the feasibility and safety of the fuel loading operations in the fuel loading process;
③ Determine the fuel loading process route according to the site layout and related technical requirements before new fuels are put into the reactor.

17.2.3 Main Equipment for Reactor Loading

The equipment and tooling required during reactor loading to ensure the integrity and safety of reactor fuels mainly include the special equipment used for operating, guiding and positioning, transfer, inspection and storage of fresh fuels.

17.3 Reactor Fuel Unloading

17.3.1 Reactor Fuel Unloading Technology

In order to ensure the safety during reactor unloading, reactor unloading process shall be designed in accordance with the provisions of relevant codes and standards after the influences of nuclear criticality safety, radiation protection safety and other factors are taken into full consideration. The equipment necessary for each process and its function requirements and primary performance requirements are then determined according to the process needs.

The reactor unloading of NPPs is carried out in the loading and unloading area for reactor fuel assemblies by joint operation of equipment in the fuel handling and storage system. The fuel manipulator crane picks up the fuel assemblies to be unloaded from the reactor and transports them to the fuel assembly transfer container, which is transported to the fuel building by the transport trolley. The reactor unloading in the nuclear power plants is performed by unloading the in-core fuel assemblies in the pool filled with borated water group by group. All operations are conducted in the borated water, to ensure the criticality safety and radiation protection safety during fuel unloading.

The current marine nuclear power unloading technology is an advanced and safe technology dedicated to reactor fuel unloading. In order to ensure that the reactor unloading process meets the requirements of relevant codes and standards, special equipment is used to safely unload the irradiated assemblies and the associated assemblies from the reactor core and transport them to the storage pool for storage. The subcriticality and fuel thermal safety of both the already unloaded fuel assemblies and the fuel assemblies to be unloaded can be ensured in this process. The proper structural design of the unloading equipment and remote underwater operation can prevent excessive irradiation and ensure that operators and maintainer can obtain reasonable shielding protection, thus satisfying the radiation protection safety requirements.

Attention must be paid to the following aspects during the fuel unloading of marine nuclear power plants:

① Fuel assemblies shall be maintained in subcritical state during reactor unloading.

② The shielding design of structures shall ensure the radiation exposure of unloading operators is effectively controlled.

③ The unloading process shall meet the requirement of fully removing the heat released from the fuel assemblies.
④ Lifting, observation, commanding and maintenance shall be convenient in the whole process of fuel unloading to reduce the mental stress and labor intensity of on-site workers and improve the operational safety.

17.3.2 Reactor Fuel Unloading Process

The theoretical clearance between adjacent fuel assemblies inside the reactor core is very small and the radiation from the core must be shielded during unloading and subsequent transfer.

The refueling processes and unloading routes for containers are determined according to the layout of facilities such as the pools in the refueling hall and cranes, and the installation locations of special unloading equipment like receiving devices and spent fuel storage rack. The main processes include: installation of shielding devices for fuel unloading, unloading of fuel assemblies from the reactor core, transfer of fuel assemblies to pools for storage, inspection of spent fuels, loading the spent fuel into the storage casks, etc.

17.3.3 Main Reactor Fuel Unloading Equipment

(1) Spent fuel storage rack

The spent fuel storage rack is installed inside the spent fuel storage pool in the refueling hall. Because a large number of irradiated fuel assemblies with high radioactivity and residual heat are stored in such rack for a long period, the issues concerning subcriticality, thermal hydraulics and earthquake resistance arising therefrom shall be taken into full account and such rack shall be designed according to relevant standards. The centers of storage units of the spent fuel storage racks are arranged at equal intervals (larger than the safe distance) and all the storage units can store enough fuel assemblies with several storage units as the standby. The pool is equipped with sufficient safety features to ensure the nuclear safety.

(2) Shielding device

The shielding device is the essential equipment for refueling. As required by the reactor refueling process, when refueling, the shielding device shall be used to conduct coordinate positioning for the fuel assemblies to be handled inside the pressure vessel and guide the fuel assemblies to enter and leave the reactor in the handling process and meanwhile to shield the high radioactivity in the reactor core so as to protect the refueling operators working on the top of the reactor from radiation.

(3) Refueling machine

The refueling machine is the essential equipment for opening the reactor head for refueling. According to the requirements of the reactor refueling process, when removing the reactor head for refueling, the refueling machine shall be used to grab, release and lift the fuel assemblies so as to put in or take out the fuel assemblies from the reactor. Moreover, in order to shield the strong radioactivity from the fuel assemblies after long-term power operation in the reactor, the refueling machine must have sufficient shielding capability.

17.4 Design of Reactor Refueling

The design of reactor refueling involves reactor pressure vessels, reactor internals, fuel assemblies and associated assemblies, detachable plates, space within the reactor compartment, cranes for fuel loading and unloading, many interfaces between the loading and unloading sites and concrete structures at the storage sites; it shall meet requirements on opening and closing lids of pressure vessels, incoming and outgoing of reactor internals, incoming and outgoing of fuel assemblies and associated assemblies as well as requirements on transportation, storage, cooling, radiation protection, etc.; it must comply with the relevant laws, regulations and standards.

17.4.1 Selection of Materials

(1) General requirements

The materials selected for the fuel handling and storage system shall be compatible with the fuels and environment.

The corrosion resistance shall be fully considered in selection of the materials and parts and components.

(2) Requirements on water quality of the pool

When the temperature is 25 °C, pH value: 4.0–8.0; electrical conductivity: 1 μs/cm; chloride concentration: $\leq$ 0.1 mg/kg; fluoride concentration: 1 mg/kg; calcium ion concentration: $\leq$ 1.0 mg/kg; magnesium ion concentration: $\leq$ 1.0 mg/kg.

(3) Requirements on metal materials

In order to minimize the amount of long half-life radioactive wastes generated in the fuel handling and storage process, metal materials (such as copper, aluminum), which are easy to be activated and have long half-life, shall not been used for facilities and equipment which might be subject to fission neutron exposure.

The objects manipulated by the fuel loading & unloading and storage system are fuel assemblies and associated assemblies, reactor internals and other in-core components, pressure vessel head, control rod drive mechanism and other key components of reactors, all of which have high cleanliness requirements. Therefore, high cleanliness requirements are placed for relevant special equipment and tools and no rust and dirt are allowed. In order to improve the resistance to corrosion, equipment to work in humid condition or water environment shall be made of materials with strong corrosion resistance, such as the stainless steel and nickel base alloy. If materials like carbon steel and iron are used in condition without water, surface electroplating, paint spraying and other surface anti-corrosion measures shall be taken.

(4) Requirements on non-metallic materials

Non-metallic parts or packing materials in contact with stainless steel shall not contain any halogen elements. Non-metallic materials with a pungent smell, fragility, easy to age and crack, or having small tolerance range of temperature shall not be used.

(5) Fire protection requirements for material selection

Noncombustible and thermal insulation materials shall be selected to construct the facilities surrounding the storage racks of fresh fuels.

Cables that may cause and promote a fire shall not be chosen for power supply to the equipment used for handling and storing fresh fuels.

(6) Materials requirements on equipment to store fresh fuels

With regard to the storage rack of fresh fuels, in order to ensure the contact with fuel assemblies is safe, materials compatible with fuel assemble shall be chosen as the surface of the storage unit. The coating easy to crack, flake or vulnerable to damage or detachment during friction and contact shall not be applied on the contact surface.

(7) Material requirements for spent fuel storage pool

The pool cladding shall be made of materials resistant to corrosion and radiation, and be convenient for cleaning and decontamination.

17.4.2 Cooling Design

(1) Cooling during the spent fuel storage

① The cooling, purification and coolant-charging systems of the spent fuel storage pool shall be able to recover the lost forced cooling under any conditions.

② Redundant measures shall be taken for the measurement of the pool water level. The fuel building shall be provided with high-level, low-level and low-low level audible alarms. Any deviation from predetermined limits of the pool water level shall be displayed in the control room.

③ The cooling, purification and coolant-charging systems of the spent fuel storage pool shall be able to make up water for the pool in case of severe accident conditions to maintain the water level above the fuel assemblies stored.
④ The cooling, purification and coolant-charging systems of the spent fuel storage pool shall be able to maintain and recover the water transparency so as to observe and identify the stored fuel assemblies.
⑤ The cooling, purification and coolant-charging systems of the spent fuel storage pool shall make sure the water quality meets the needs for long-term storage of fuels.
⑥ The cooling, purification and coolant-charging systems of the spent fuel storage pool shall be able to exercise the decontamination function under normal operating conditions.
⑦ The design of these systems shall eliminate the siphoning and freezing phenomena of fluid pipes under various conditions.
⑧ The facility to prevent the overfilling of pools shall be provided by design.
⑨ The design of pool cooling system shall comply with the single failure criteria.
⑩ The design of pool cooling, purification and coolant-charging systems shall conform to the requirements of relevant regulations and standards.
⑪ The forced cooling shall be provided to maintain the pool temperature at the stipulated level under normal operating conditions.
⑫ The spent fuel storage rack shall be made of materials with good thermal conductivity and able to form natural circulation and sufficiently cool down the stored fuel assemblies under various conditions. The design shall make sure the pool water will not be discharged by accident. The pool shall be designed in accordance with the zero leakage principle and leak detection measures (such as the partitioned leaked water collection system) shall be provided. The solutions for pool leakage shall be considered in the design.

(2) Core refueling and cooling

① During the refueling of reactor core, the core refueling and cooling system shall be installed to remove the core decay heat so as to ensure the temperature of reactor coolants will not exceed the specified limits during the refueling.
② The core refueling and cooling system must have reliable ultimate heat sink and can provide forced cooling in active ways. It shall provide core cooling during the whole core refueling.

(3) Cooling during fuel transfer

① Sufficient cooling must be provided for fuels during fuel transfer and it is required that residual heat of fuels shall be removed fully by active or passive means.
② The exposure time to air of fuels shall be as short as possible.

17.4.3 Criticality Safety Evaluation

① The criticality safety analysis shall be performed for each of the fuel assembly loading & unloading, transfer and storage systems.
② This criticality safety analysis shall demonstrate that each system will be in subcritical state under the specified limits for various operating conditions.
③ The nuclear criticality safety analysis involves various credible events, for example, toppling and falling of fuel assemblies or dropping of fuels.
④ This analysis shall demonstrate that in each phase of the fuel handling or storage, the subcriticality can be kept even when two independent unusual events occur concurrently.
⑤ The subcriticality characteristics of fuel assemblies shall be verified in a clear way.
⑥ The analysis shall clearly demonstrate that the physical verification carried out after the storage rack is manufactured or constructed conforms to the requirements of subcritical design limits.

17.4.4 Industrial Safety Design

The industrial safety design includes foreign material exclusion, anti-falling, fire control, lighting, communication, ventilation and security.

Special measures shall be designed to ensure materials (including oil) will not fall into the reactor, fuel assembly or spent fuel storage pools during operation or maintenance.

Mechanical and electrical safety devices shall be installed to ensure the operation safety of equipment and industrial safety.

The fire protection design shall ensure the subcriticality of fuels in the event of a fire. The fire extinguishing agents that can moderate neutrons (such as water and foam extinguishing agent) and other fire extinguishing agents shall be marked with obvious signs and warms in the storage area. The design of fuel handling and storage areas shall prevent the fresh fuels from damages exerting an effect on the safety after fire. Combustible materials (such as combustible packaging materials and pipe system transferring combustible materials) shall be limited and controlled in the fresh fuel handling and storage areas. Noncombustible and thermal insulation materials shall be used to construct the facilities around the storage rack of fresh fuels. For cables supplying power to the equipment used for handling and storing fresh fuels, the type selection and the laying of the cables shall not cause or promote a fire.

When fresh fuels are stored in dry conditions, the ventilation system shall be designed to reduce the access of dusts or particles in air.

Measures shall be taken to prevent fuel loss or damage.

17.4.5 Design of Radiation Protection Safety

(1) Spent fuel storage pool

① The design shall guarantee the shielding depth of the pool water.
② Exposure dose to workers staying in this area shall not exceed 0.025 mSv/h under condition I and II.
③ Design measures shall be taken to avoid the adoption of structures that may cause the accumulation of radioactive substances.

(2) Spent fuel storage rack

① The structure and arrangement of spent fuel storage racks shall ensure the subcriticality safety.
② The storage racks shall be designed to enable natural circulation to be established and sufficiently cool down the stored fuel assemblies under various conditions.
③ The materials of storage racks shall be compatible with the fuel and environment and the rack structure shall not cause damage to fuels because of normal handling and storage operations.
④ The racks shall be designed according to the maximum storage capacity required at any time during the reactor lifetime.
⑤ The spent fuel storage racks shall be designed with sufficient safety margins to resist to the single or group loads, such as tilting, swaying, accident overloading, operating basis earthquake (OBE), safe shutdown earthquake (SSE).

Chapter 18
Decommissioning of Marine Nuclear Power Plants

18.1 Overview

The whole lifetime management of a marine nuclear power plant covers the whole process of the research and design, fabrication and manufacture, operation, repair and refueling during its operation, and decommissioning after its shutdown.

A marine nuclear power plant shall be decommissioned when it reaches its design lifetime. The design lifetime of a marine nuclear power plant is generally 30–40 years. The marine nuclear power plants are with high costs, complex manufacturing processes and long cycles. Therefore, in order to make full use of a marine nuclear power plant, its lifetime can be extended even its design lifetime ends, under the premise that the inspection and the safety and reliability analysis and evaluation of key equipment of the nuclear power plant (such as reactor pressure vessels and internals and other non-replaceable components) and the repair or replacement of replaceable equipment prove that its lifetime can be extended. However, even if a nuclear power plant is still within its design lifetime, when the critical equipment is seriously corroded or damaged, or when the repair cost is rather high and the repair cannot ensure the performance, such nuclear power plant shall be decommissioned ahead of its due life-time.

Since the reactors, reactor coolant systems and their auxiliary systems of marine nuclear power plant are radioactive, their decommissioning is a highly difficult, technically complicated and systematic project involving many disciplines, closely related to the radiation exposure safety of workers and the public and the environmental safety, and requiring the comprehensive treatment of radioactive substances. Therefore, the decommissioning of marine nuclear power plant has attracted the worldwide attention. In particular, the countries possessing the marine nuclear power plant have invested many scientific and technical personnel and a large amount of capital to research and develop the decommissioning technologies for marine nuclear power plant. At present, the United States has decommissioned some marine nuclear power plant developed from the 1950s to the 1970s. China has also successfully decommissioned some marine nuclear power plant built in the 20th century.

J. Yu, *Marine Nuclear Power Technology*,
https://doi.org/10.1007/978-981-15-2894-1_18

18.2 Decommissioning Scheme Study

The decommissioning of marine nuclear power plant is generally divided into the following stages: decommissioning feasibility study, decommissioning design and decommissioning implementation. At the feasibility study stage, firstly, the operational history of marine nuclear power plant is investigated, the radionuclides and the residual radioactivity are estimated based on the above investigation results, and the physical radioactivity is measured; then, one to three decommissioning schemes are preliminarily formulated according to national policies, laws, regulations and actual conditions of the marine nuclear power plant, After analysis and evaluation, one of the best schemes is selected; and finally, the safety analysis and environmental pre-evaluation/assessment is performed. At the decommissioning design stage, the decommissioning technology is studied based on the decommissioning schemes, the decommissioning equipment shall be available by R&D or procurement, and the decommissioning processes are established. At the decommissioning implementation stage, the nuclear power plants are dismantled according to the decommissioning technology and processes developed at the design stage, and the radioactive waste disposal is performed.

The decommissioning scheme for marine nuclear power plant shall include the decommissioning timing and decommissioning methods. The decommissioning timing generally includes immediate dismantling, delayed dismantling and permanent storage of nuclear power plant.

The so-called "immediate dismantling" refers to the dismantling of nuclear power plant within 2–5 years after the reactor shutdown. This is mainly related to the last shutdown of the reactor since there is a large amount of residual heat in the reactor for a relatively long time after the shutdown. The unloading of the reactors in the marine nuclear power plant is different from that of NPPs. In a NPP, the reactor and the spent fuel storage pool are closely connected, and the spent fuel discharged from the reactor can be immediately transferred to the storage pool for cooling. Therefore, the reactor unloading can be performed shortly after the reactor shutdown in an NPP. The situation is different for the marine nuclear power plant. The decommissioning of a marine nuclear power plant is carried out in the dock, and the land-based facilities shall be used for the reactor unloading. The reactor spent fuel storage pool is built on the shore, so the spent fuel discharged from the reactor shall be transferred in the air for a certain distance before it reaches the storage pool. In this process, the temperature of spent fuels will increase. As the reactor is at the specified power level and burnup level just before the shutdown of a marine nuclear power plant reactor, it takes 2–5 years to meet this specified requirement for unloading. So the last unloading of the marine nuclear power plant is included in the scope of the decommissioning of the nuclear power plant. Therefore, the decommissioning of marine nuclear power plant begins from the reactor unloading. The immediate dismantling refers to the dismantling immediately performed immediately following the reactor unloading which is conducted within 2–5 years after the shutdown of the marine nuclear power plant.

Delayed dismantling means that the nuclear power plant will be dismantled after 15–20 years since the reactor shutdown The last unloading of the reactor can be carried out within 2–5 years after the reactor shutdown, or before the dismantling of nuclear power plant. In most countries in the world, the reactor unloading is performed within 2–5 years after the reactor shutdown, and the nuclear power plant dismantling is delayed for 15 years after the unloading. This can mitigate the nuclear safety risk of the hull at the storage stage, decrease the radioactivity level during the nuclear power plant dismantling, ensure the radiation protection and safety, reduce the exposure dose of personnel, and reduce the hazards to the public and the environment.

The reactor compartment permanent storage means that, after the reactor unloading, the nuclear power plant will not be dismantled, or only the conventional systems without radioactive contamination will be dismantled, and the reactor, the reactor coolant system and auxiliary systems in the reactor compartment will not be dismantled; instead, the reactor compartment is cut off and separated from the adjacent compartments, sealed and transported to the radioactive waste disposal site for permanent storage, with a storage time longer than hundreds of years.

The decommissioning methods can be generally divided into the following categories.

(1) Bulk shielding packaging of reactor and internals

After the reactor pressure vessels and the internals are dismantled, no further destructive cutting and dismantling will be performed. Instead, the pressure vessels, internals and some in-core components such as control rods and other highly radiative activated and radioactive components will be reassembled according to the shielding packaging requirements to form an large piece of equipment in bulk shielding packaging, and this large piece of equipment will be transported to the radioactive waste disposal site for storage. This method does not involve the cutting and dismantling of pressure vessels, internals, large highly radiative activated and radioactive components (such as the core barrel and holddown barrels) and the highly radiative activated and radioactive components in the core (such as control rods). Therefore, this method can save large amount of funds spent for R&D on the cutting of such items and for the cutting-equipment purchase, eliminate the exposure dose rate for the personnel cutting such highly radioactive components, protect the safety of the public and the environment, and shorten the decommissioning period.

(2) Cutting and dismantling

After the reactor is dismantled, further cutting and dismantling will be performed for the pressure vessels and internals according to the conditioning and packaging requirements of the radioactive waste, so as to divide the pressure vessels and internals into small pieces suitable for packaging. Such small pieces will be sorted, packaged and transported to the radioactive waste disposal site for storage. This will be convenient for packaging, lifting and storage, but it is necessary to develop special cutting machines and tools, and to take the shielding and protection measures during the cutting and dismantling. The cutting of these highly radioactive components requires the development of underwater cutting and dismantling technologies and remote operation technologies, or the use of robots for cutting. Such activities require expensive scientific research funds and equipment purchase funds. Even if the most advanced technologies are used, manual operations will still be needed. Therefore, the radiation exposure of personnel is unavoidable. The cutting will inevitably generate the radioactive particles, so there will be rather high requirements for radiation protection. There is great difficulty in the implementation, and the decommissioning cycle will be rather long. In addition, a lot of shielding packaging boxes shall be processed, which will lead to rather many secondary radioactive wastes.

(3) Bulk treatment of reactor compartment

After the reactor unloading, the reactor compartment will be cut off and separated from the adjacent compartments. After the sealing treatment of both ends, the reactor compartment will be transported to the radioactive waste disposal site for storage, with a storage time longer than 500 years. The scheme does not require the dismantling of radioactive items. Therefore, the decommissioning process will be simplified, the diffusion of radioactive substances will be effectively prevented, since few secondary radioactive wastes will be generated, and both the exposure dose of personnel and the impact on the public and the environment will be very insignificant. However, after the sealing of the both of the compartment ends, the reactor compartment may be even heavier than 1,000 t, which poses higher requirements for lifting, transportation and storage.

18.2.1 Decommissioning Schemes in Foreign Countries

The countries possessing marine nuclear power plant before the 21st century included the United States, Russia, the United Kingdom, France and China. In 1952, the United States constructed the first marine nuclear power plant in the world, for the submarine "Nautilus". In 1958, the former Soviet Union put its first marine nuclear power plant into operation. In 1963, the United Kingdom constructed the marine nuclear power plant for its first attack submarine "HMS Dreadnought". In 1971, France formally put its first marine nuclear power plant of the ballistic missiles submarine "Le Redoutable" into operation.

According to the design lifetime of 30 years for early marine nuclear power plant, all marine nuclear power plant constructed in the early 20th century shall be decommissioned by the end of the same century. Different decommissioning schemes have been developed according to the conditions of different countries.

(1) The United States have decommissioned a large number of marine nuclear power plants. The United States has constructed the Hanford Radioactive Waste Disposal Site with favorable conditions. To be specific, the site is located in a dry desert with little rain and few inhabitants, so it is suitable for the storage of radioactive wastes; in addition, there is a large river nearby, so the water transportation is convenient and suitable for large equipment transportation.

The United States also has the capability to lift kilotons of cargoes. Therefore, during the decommissioning of marine nuclear power plant in the United States, there is no need to dismantle the radioactive items such as reactors and primary loop system, equipment and pipes. Instead, the reactor compartment is cut and separated from the adjacent compartments. After the sealing treatment of both ends of the compartment, the reactor compartment is transported to the Hanford Radioactive Waste Disposal Site by waterway. After the reactor compartment is lifted ashore, it is transported on road for only 20 km to the Hanford Radioactive Waste Disposal Site, where the reactor compartment is buried in the ground.

(2) Russia has the largest number of marine nuclear power plant in the world. After the collapse of the former Soviet Union, since the funds were insufficient and the huge number of marine nuclear power plant to be decommissioned, Russia adopted the delayed disposal and offshore floating storage for the decommissioning of marine nuclear power plant. To be specific, after the reactor unloading, the entire ship or the reactor compartment and its adjacent compartments is cut and separated from the hull to form a two-in-one or three-in-one combined compartment, the non-radioactive items in the adjacent compartments are dismantled, and the combined compartment is sealed and then put into offshore floating storage. This treatment can maintain the stability of the floating state of the combined compartment. The next treatment will be performed after at least 20 years of floating. In recent years, it has been reported that Russia has also adopted the scheme of cutting and separating the reactor compartment from adjacent compartments and dealt with the reactor compartment by long land-storage. This scheme is the same as that for the decommissioning disposal of marine nuclear power plant in the United States.

(3) Since the United Kingdom is an island country and does not have a any radioactive waste disposal site, like Hanford in the United States, so there is no final decommissioning disposal scheme. The United Kingdom was originally planned to sink its first nuclear power submarine "HMS Dreadnought" into the Atlantic. However, UK abandoned the sea disposal plan since it is forbidden to dump radioactive wastes into seas by the *Convention on the Prevention of Marine Pollution by Dumping of Wastes and Other Matter*. The Royal Navy finally performed the reactor unloading and put "HMS Dreadnought" marine power plant into floating storage in the dock of a dockyard of the Royal Navy. Since then, this scheme has been adopted for the decommissioning of subsequent marine nuclear power plant. This is only a temporary transitional scheme, and the final disposal scheme is awaited and studied. It is reported that the United Kingdom is working with the United States to prepare for the implementation of land-storage treatment.

(4) France decommissioned its first marine nuclear power plant in 1991. After the reactor unloading and decontamination, the entire ship, as the national public heritage, was officially opened to the public in May 2002. The packaging and dismantling scheme was adopted for the disposal of subsequent decommissioned marine nuclear power plant. After the reactor unloading, the spent fuel is stored in special storage pools. The dismantled reactors are temporarily stored

in special devices that are resistant to earthquakes and environmental changes. The dismantled equipment, pipes and reactor compartment hull section are cut, disassembled and packaged for temporary dry storage on land, and wait for the final treatment. France is preparing to construct a permanent radioactive waste disposal site.

18.2.2 Decommissioning Scheme in China

China has built the Northwest Radioactive Waste Disposal Site, the natural conditions of which are similar to those of Hanford Radioactive Waste Disposal Site in the United States. This site is located in a dry desert with little rain and few inhabitants, so it is suitable for radioactive waste storage and disposal, but it does not have the conditions of water and land transportation as Hanford in the United States. Therefore, according to the specific conditions of China, the scheme for the decommissioning of marine nuclear power plant is to dismantle the whole ship after the reactor unloading and to perform the bulk shielding packaging of highly activated and radioactive components such as reactor pressure vessel, internals and control rods. The remaining radioactive items is dismantled, sorted, conditioned and packaged, and the spent fuel discharged from the reactor is inspected, canned and sealed for temporary dry storage, and transported on road to the Northwest Radioactive Waste Disposal Site for permanent disposal.

The reactor pressure vessels and the internals in bulk shielding packaging meet the requirements for the highway and railway transportation conditions in China, so this scheme for the decommissioning of marine nuclear power plant conforms to the national conditions of China.

18.3 Study on the Status of Nuclear Power Plant Before Decommissioning

18.3.1 Investigation of Reactor Operation History

The investigation of reactor operation history, as an important task before the decommissioning of marine nuclear power plant, plays an important role in residual radioactivity estimation, decommissioning scheme preparation, and decommissioning process and dismantling method development. Starting from reactor loading to the operation of the nuclear power plant throughout their whole lifetime, all the operation conditions are automatically recorded by computers.

The investigation of reactor operation history involves the reactor operation power level at each stage and in each voyage, the operation duration at each power level, the times of the refueling in the whole lifetime of the reactor, the burnup level of each load of nuclear fuels, the accumulated equivalent full power days for operation of each load of nuclear fuels, times of repair within the whole lifetime of the reactor, times of decontamination, the equipment replaced during each refueling, repair or decontamination, and the radioactivity removed. Therefore, it is a very complicated task.

18.3.2 Calculation and Measurement of Residual Radioactivity

The calculation and measurement of residual radioactivity, which is called the source term investigation in some literatures, is the calculation of the residual radionuclides in the activated reactor components and the contaminated systems and equipment based on the data obtained from the investigation of reactor operation history. The calculations of residual radionuclides and radiation level shall be performed based on the operating power, operation time and burnup level at each stage of the reactor. The physical measurements shall also be carried out to measure the radiation levels of the reactor periphery of the reactor compartment, the primary circuit system equipment, the pipe surfaces and the accessible parts of the reactor compartment space, as well as the contamination levels of the equipment surfaces. The measurement results can be used to verify the accuracy of the radionuclide related calculations. These two tasks are collectively referred to as the source term investigation.

The source term investigation is an important basis for designing the decommissioning processes and formulating the protection measures. The decommissioning of nuclear power plant is a task with radiation exposure, so the key to formulating processes and radiation protection measures is to protect the workers, the public and the environment. Therefore, the source term investigation is particularly important.

The internal diameter and length of the interior of the each compartment of a nuclear power plant are strictly limited, and the reactor coolant system and its auxiliary systems and numerous equipment, pipes and valves for realizing ventilation, air conditioning and negative pressure shall be arranged in the compartment, so the space is very narrow. Therefore, many parts requiring the measurement of radioactivity levels and contamination levels are inaccessible. As a result, it is hard for the radioactivity measurement before decommissioning to correctly reflect the real radiation levels of the reactor compartment. It is necessary to timely perform the tracking measurement of the reactor compartment radiation level along with the progress of the dismantling of radioactive items, so that the decommissioning process and radiation protection measures can be timely adjusted. If conditions permit, the items inaccessible during the repair and the activated items near the reactor shall be sampled and analyzed with the progress of the dismantling of radioactive items, so as to verify the accuracy in the calculation of residual radionuclides.

For the above reasons, the source term investigation shall be made throughout the entire process of decommissioning of nuclear power plant. Before the conduction of each task, the radiation level at the workplace shall be measured for formulating the radiation protection measures to be taken by the workers during the work, especially the discovery of radiation "hot spots" (where the local radiation level is much higher than that in the surrounding areas) by measurement to allow the workers to take special protection measures. The measure generally taken is as follows. Firstly, dismantle the radiation "hot spots"; if such radiation "hot spots" cannot be dismantled because of the obstruction of surrounding equipment and pipes, perform the local shielding of such radiation "hot spots", i.e. wrapping with lead sheath, to reduce the radiation level. Secondly, dismantle the surrounding equipment and pipes to protect the workers from the radiation exposure and reduce the exposure dose of the workers.

18.3.3 Calculation of Reactor Residual Heat Release and Measurement of Related Temperature

After the reactor shutdown, the reactor still releases heat because of the decay of fission products. This phenomenon is called the reactor residual heat release. Therefore, after the reactor shutdown, the reactor coolant shall continue to circulate for a considerable long period of time to remove the reactor residual heat. During the reactor unloading, the reactor coolant cannot continue to circulate any longer, so the heat is removed by the natural circulation of the coolant system. In addition, the spent fuel discharged from the reactor will be transferred to and stored in a storage pool near the berthing wharf. Therefore, in order to ensure the thermal safety of the reactor unloading and the thermal safety of the spent fuel in the air, it is necessary to control the temperature rise of the reactor under natural heat dissipation conditions and the surface temperature of spent fuel elements during the lifting of spent fuel assemblies in the air.

For the selection of reactor unloading time, it is necessary to calculate the reactor residual heat release first. The reactor residual heat release is in direct proportion to the burnup level of the reactor nuclear fuel, the power level before reactor shutdown and operation time at this power level, and in inverse proportion to the time after shutdown. Therefore, the data obtained from the investigation of reactor operation history are the key parameters for calculating the reactor residual heat release.

It can be seen that the accuracy of the calculation of reactor residual heat release is closely related to the accuracy of the data obtained from the investigation of reactor operation history. The deviations of the data obtained from the investigation of reactor operation history will result in certain errors in the calculation of the residual heat release. Therefore, in order to make up for the errors in the calculation of residual heat release, it is necessary to measure the temperature rise of the reactor coolant, so as to determine the best time for reactor unloading.

18.3.4 Inspection of Reactor Control Rod Positions

Compared with land NPP reactors and test reactors, the marine nuclear power plant have compact core layouts, fewer control rods and no specific classification of control rods according to their functions (in the land NPP reactors and test reactors, the control rods are classified into as safety rods, compensation rods, regulating rods and automatic rods). In the land NPP reactors and test reactors, the reactor can be shut down as long as the safety rods are inserted into the core. However, in the marine nuclear power plant, the reactor can be shut down only when all control rods are inserted into the core. In order to ensure the safety of the reactor, a certain margin shall be reserved for the efficiency of control rods. Therefore, the concept of stuck rod criteria is introduced. The stuck rod criteria specify that when a control rod cluster is stuck outside the active area at the normal temperature and the remaining control rods are all at the bottom of the active area with the $K_{\text{eff}} \leq 0.98$, the reactor may become critical if a second control rod cluster is withdrawn from the core at this time.

The inspection of control rod positions shall be performed before the opening of the reactor head. During the inspection of control rod positions, the reactor reactivity shall be monitored. At the normal temperature, each control rod cluster shall be gradually withdrawn to the highest position, i.e. totally outside the active area, thus confirming that the reactor meets the requirement of $K_{\text{eff}} \leq 0.98$. Then this control cluster shall be inserted into the bottom of the active area, and the remaining control clusters shall be inspected in this way one by one.

18.3.5 Tests of Performance of Pumps, Valves and Systems

Before the reactor unloading, reactor core cooling may still be required to remove the residual heat in the core, and thus to reduce the reactor core temperature meeting the unloading requirements. After the reactor unloading, it is necessary to carry out the decontamination of the primary circuit system to decrease the radiation level of the reactor compartment. This requires that all systems in the primary circuit remain intact. In particular, the main reactor coolant pumps, the main valves and the isolation valves of each system shall be operable. Therefore, they shall be inspected, tested and, if necessary, repaired to ensure their operability.

18.4 Reactor Decommissioning

18.4.1 Decommissioning Procedure

According to the structural characteristics of the reactors, the decommissioning shall follow the dismantling principles of "from top to bottom" and "from inside to outside". The reactor decommissioning and dismantling procedure shall be formulated, and the dismantling process shall be developed according to the features of equipment and components to be dismantled. The specific process is as follows (Fig. 18.1).

18.4.2 On-site Condition Preparation

(1) Preparation of site conditions

In order to ensure the nuclear safety of reactor unloading and the radiation safety of dismantling and lifting of internals, two circuits of safe power supplies, the storage pool for storing reactor internals and control rod assemblies, the ventilation facilities, the bridge cranes capable of safely lifting reactor pressure vessels (with a lifting weight of 1.25 times the weight of the reactor pressure vessel) and the passages for the entering and exiting of people and materials shall be available in the decommissioning site; the areas for the decontamination, inspection, contamination measurement, sorting, conditioning and packaging of items dismantled during the decommissioning and the areas for commissioning, repair and storage of decommissioning equipment shall be prepared; and the radioactive liquid waste treatment devices and radioactive solid waste repository shall be ready.

(2) R&D or procurement of special decommissioning equipment

The decommissioning requires the R&D or procurement of the special equipment for reactor head opening, reactor unloading, internals dismantling, spent fuel assembly

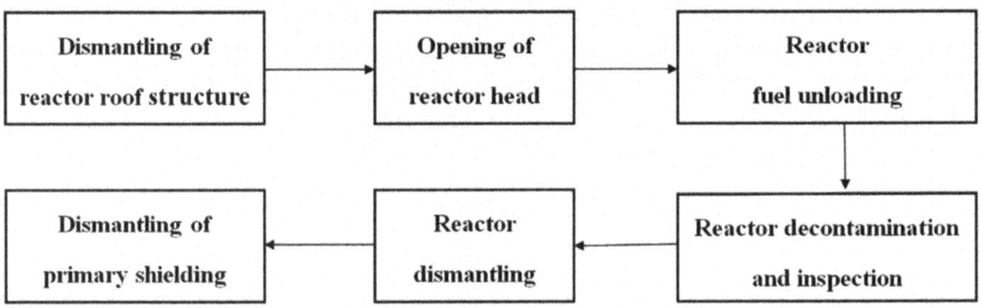

Fig. 18.1 The main procedure of reactor decommissioning

lifting and transportation and spent fuel assembly storage, the special tools for cutting primary pipes, the underwater plasma cutting devices for reactor support skirts, lead shielding dismantling devices and radioactive waste packing containers.

18.4.3 Reactor Unloading

There are two kinds of installation structures for control rods of marine nuclear power plant. In the first kind of structure, the control rods shall be dismantled first and the compression barrels shall then be dismantled before the reactor unloading. Before dismantling the control rods, it is necessary to inject the boron solution into the reactor. According to the relevant national nuclear safety standards of China, the boron concentration in the boron solution injected into the reactor shall ensure that the reactor K_{eff} will not be greater than 0.95 after the control rods are totally removed from the reactor. Therefore, the boron concentration in the boron solution shall be determined by calculation. All boron solution injected into the reactor becomes radioactive waste water. In addition, the treatment process of the boron solution is very complicated. In order to reduce the amount of radioactive waste water, the water in the reactor shall be totally discharged before the boron solution is injected into the reactor. Because of the complexity of the internal structure of the reactor, it is not possible to completely discharge the water in the reactor. In addition, calculation errors may exist for the inaccuracy of input parameters during the calculation. Therefore, the boron concentration in the boron solution injected into the reactor shall be higher than the calculation result, generally 1.5–2 times the calculated value. The reactor criticality safety shall be monitored during the injection of boron solution into the reactor. Based on the change in the neutron counting rate in the reactor during the injection of boron solution, the K_{eff} values at different liquid levels are calculated, until the boron solution floods the active area. The K_{eff} value with the liquid level 300 mm above the active area shall be no more than 0.95.

In the second kind of internal structure of the reactor, it is not necessary to dismantle the control rods during the fuel unloading. Instead, the control rod assemblies will be left in the fuel assemblies. In this case, it is unnecessary to inject boron solution into the reactor.

There are two unloading ways for marine nuclear power plant: one is the container unloading, and the other is to lift the entire barrel of fuel assemblies from the reactor to the storage pool, unload the fuel assemblies in group from the barrel in the storage pool, and store them in the storage pool.

The container unloading has been described in the previous section on reactor repairs, so it will not be described here again. The other unloading method is to lift the entire barrel fuel assemblies from the reactor to the storage pool, remove the fuel assemblies in group from the basket in the storage pool, and store them on the spent fuel storage racks. This method is generally not recommended since it has certain safety risks.

18.4.4 Decontamination

According to the requirements, the decommissioning and decontamination of marine nuclear power plant can be divided into the comprehensive decontamination of the reactor and primary circuit system, and the decontamination of important components.

The reactor coolant of marine nuclear power plant is in long-time operation under high temperature and high pressure. A layer of dense spinel type oxide film is formed on the inner wall of reactor pressure vessel and the internal surfaces of primary circuit system pipes and equipment. A layer of loose corrosion products is deposited on this film. The activated corrosion products are concentrated in these two layers, so these two layers have rather high radiation level. During the decommissioning and dismantling of primary circuit system equipment, the workers have to contact with such equipment at a short distance. In addition, the reactor compartment is narrow and numerous primary circuit system equipment and pipes are of staggered arrangement in three layers in the reactor compartment, making the reactor compartment very crowded. Therefore, the radiation level in the reactor compartment is very high. In order to ensure the radiation safety of the workers and the environmental safety, before the decommissioning of the primary circuit system, thc system decontamination shall be performed to reduce the radiation level of system equipment, pipes and reactor compartment.

At present, the main decontamination method for decommissioned decommissioning of the nuclear power plant used by various countries in the world is the chemical decontamination. To obtain rather high decontamination factor required for the chemical decontamination, three requirements shall be meet: a. reasonable chemical method and reasonable decontaminating solution concentration; b. reasonable temperature; and c. strong scouring force of chemical solution on the decontaminated surfaces to be decontaminated. The layer of dense spinel type oxide film formed on the surfaces of reactor pressure vessels and primary circuit system pipes and equipment is stably and firmly bonded with the metal surfaces, so it is difficult to remove it by a simple chemical dissolution method. The tests and practices have shown that it is helpful to pre-treat with alkaline potassium permanganate solution ($KMnO_4 + NaOH$) before the acid pickling. The reaction between the alkaline potassium permanganate solution and the oxide film on the stainless steel surfaces mainly oxidizes Cr^{3+} to Cr^{6+} soluble in the aqueous alkali, and oxidizes Fe_3O_4 to Fe_2O_3 with a higher solubility. These actions change the valence states of elements forming the oxide film, and make the oxide film loose hence it can be easily dissolved by the acid. Tests have shown that organic acids and their salts, especially the mixtures of ammonium salts, can effectively dissolve the metal oxides and have little corrosiveness. Oxalic acid has a strong oxide dissolving capacity, but the dissolved matters can be easily hydrolyzed into insoluble ferrous oxalate that will re-deposit. Because of the complexity of the structure of primary circuit system, if such deposits exist in the dead angles of system equipment and pipes, they cannot be discharged and result in the radiation level in the reactor compartment still very high, so the decontamination will become meaningless. Ammonium citrate and other complexing agents have the complicating effects on metal ions. For example, ethylene diamine tetraacetic acid (EDTA) can reduce the concentration of iron ions in solution to a level that will not produce oxalate precipitation. Therefore, the so-called two-step method is adopted: Firstly the alkaline potassium permanganate solution is used to pre-treat the reactor pressure vessels and primary circuit system, and then the mixed solution of oxalic acid, ammonium citrate dibasic and EDTA is used for the acid pickling. This is called the two-step method. However, the actual situation is that there is a layer of rather loose radioactive corrosion products on the spinel type oxide film on the pipe surfaces, so the alkaline potassium permanganate solution cannot directly contact with the oxide film, which reduces the effect of the oxide film treatment. However, in practical, the treatment effect will be better if the acid pickling is performed to remove the corrosion products attached to the oxide film before the oxide film treated by alkaline potassium permanganate. The three-step decontamination method of acid pickling—alkaline cleaning—acid pickling (EC–AP–EC) is actually adopted. After each step of decontamination, the system shall be rinsed with clean water before the next decontamination is performed.

Temperature is a crucial factor in improving the decontamination effect. The best decontamination effect can be achieved by the decontamination solution with a temperature of about 90 °C. Similarly, the rinsing water with a temperature of about 90 °C has a rather good effect for dissolving the residual decontamination agent on the metal surfaces.

The reactor, consisting of reactor roof structure, primary shielding at reactor side, reactor pressure vessel, internals and core components, shall be dismantled from top to bottom and from inside to outside. The dismantling sequence is as follows: dismantling the reactor roof structure, opening the reactor head, dismantling the internals and core components, dismantling the pressure vessel shell, and finally dismantling the primary shielding.

18.5 Decommissioning of Circuit Systems and Other Equipment in the Reactor Compartment

The systems inside the reactor compartment generally include reactor coolant system and auxiliary systems, safety systems and reactor compartment negative pressure and air conditioning systems. Dozens of pieces of large equipment include steam generators, pressurizer, main pumps, safety injection pumps, residual heat removal pumps, purification pumps, ion exchangers and heat exchangers in the residual heat removal system and coolant purification system, as well as relevant pipes and valves. The pipes are of criss-cross and the space is narrow, so it is very difficult to dismantle them systematically. The components shall be dismantled in a sequence from top to bottom, from outside to inside, from pipes to equipment and from small equipment to large equipment. However, the "radiation hot spots" with high radiation levels shall be dismantled first. If the "radiation hot spots" cannot be dismantled first, the local shielding shall be performed, and the cold mechanical cutting technology and the enhanced reactor compartment ventilation shall be adopted.

In order to identify the systems and locations of the items to be dismantled, the equipment, pipes and valves in each system shall be marked before the dismantling to ensure the traceability.

18.6 Treatment of Radioactive Wastes

The decommissioning of nuclear facilities is a systematic project for the comprehensive treatment of radioactive substances and the protection of the safety of the human beings and the environment. Therefore, the treatment of decommissioned radioactive items of marine nuclear power plant shall be performed under the framework of law and regulation systems, and follow the relevant principles in the laws, regulations and standards.

(1) Protection of human health

The management and treatment of radioactive wastes shall ensure that their impact on the health of workers and the public is at an acceptable level, and that, after the economic and social factors are considered, the possibility of exposure, the value of personal dose and the number of exposed people are as low as reasonably achievable (ALARA).

(2) Protection of environment

The management and treatment of radioactive wastes shall maintain the environmental protection at an acceptable level, keep the release of radioactive substances to the environment as low as actually achievable, and protect the environment from being jeopardized by such radioactive wastes at present and in the future.

(3) Protection of future generations

The management and treatment of radioactive wastes, especially the disposal of radioactive wastes, the decommissioning and the environmental renovation activities, shall ensure that the expected impact on the health of future generations is not higher than the current acceptable level, and that no undue burdens will be left to the future generations.

(4) Minimization of radioactive wastes

In the decommissioning activities, the amount of radioactive wastes generated shall be controlled so that both the activity and volume are kept as low as actually achievable.

(5) Classification of radioactive wastes

Radioactive wastes are classified as radioactive gaseous wastes, radioactive liquid wastes and radioactive solid wastes, which shall be treated separately. In particular, the radioactive solid wastes shall be strictly sorted according to their physical and chemical behaviors, activity levels and contamination levels, and the mixing of different categories shall be avoided. The precious metals shall be separated from non-activated, non-contaminated and very low level wastes that can be reused under certain limitations to reduce the amount of radioactive wastes.

18.7 Radiation Protection and Safety

During the entire process of the decommissioning of marine nuclear power plant, the radiation protection shall follow three principles for practice justification, protection optimization and personal dose limit: the decommissioning radiation monitoring plan shall be formulated and implemented; during the waste treatment and disposal, the principles of protecting the health of future generations and causing less burdens to future generations shall be followed; and the accidental exposure of personnel and other events caused by improper management of solid wastes shall be prevented.

Before the decommissioning of nuclear power plant, a detailed radiation protection and safety plan shall be formulated. In addition to the calculation and measurement of residual radioactivity, the following contents shall be included in this plan.

18.7.1 Classification and Management of Work Place

The work places shall be divided into controlled areas and supervised areas according to the measurement results for radiation levels and contamination levels of the reactor compartment and the work site. The controlled areas are further divided into high radiation areas and high contamination areas. The operating areas of the workers with an external exposure dose higher than 0.1 mSv/h are defined as the high radiation areas, while those with a surface β-radioactive contamination larger than 40 Bq/cm^2 are defined as high contamination areas. Controlled areas, high radiation areas, high contamination areas and supervised areas shall be equipped with isolation facilities and signs at the entrances to prevent the personnel from entering such areas accidentally and causing unnecessary exposure.

The workers entering the controlled areas shall take special safety protection measures, and the management personnel shall, according to the radiation levels of the controlled areas, arrange the work time and the expected exposure dose for the workers in the controlled area.

18.7.2 Management of Operators

Before the commencement of decommissioning, the comprehensive physical examination shall be performed for the personnel participating in the decommissioning, and relevant examination data shall be kept. The personnel with the physical examination results dissatisfying the requirements should not participate in the decommissioning operation. It is also necessary to perform the background measurement of body radioactive substances of the personnel participating in the decommissioning operation and, after the completion of the operation, the body radioactivity dose of such personnel shall be measured to determine whether there is the internal exposure dose for such personnel.

In the decommissioning practices, the principles of protection and safety optimization shall be implemented to ensure that, after the economic and social factors are considered, the value of personal exposure dose, the number of exposed people and the exposure possibility are kept as low as reasonably achievable. Therefore, during the decommissioning, the effective protective measures shall be taken, and the reasonable exposure dose control value of relevant people shall be determined according to the specification in ICRP60.

18.7.3 Safety Measures for Radiation Protection

In order to ensure the radiation protection safety, effectively reduce the exposure dose of workers and the public and protect the environment, the following protection measures are generally taken during the decommissioning:

① Design a reasonable decommissioning implementation process, as described in Sect. 18.4.1.

② Develop advanced process technologies such as remote control, underwater automatic cutting, water jet cutting and cold mechanical cutting by robots.

③ Provide necessary temporary shielding. For radiation hot spots that cannot be dismantled because of the narrow space, the local shielding can be provided to reduce the radiation to the environment and effectively reduce the exposure dose of the personnel.

④ Isolation and ventilation of work site: the construction site for decommissioning shall be isolated, and the ventilation at the construction site shall be strengthened to reduce the concentration of aerosol. The exhaust air of the construction site shall be filtered by high efficiency filters and discharged from an elevated vent to avoid the impact on the environment.

18.7.4 Strengthening of Radiation Monitoring

(1) Monitoring of sites for decommissioning operation

The monitoring covers the radiation levels of remaining items and construction site, the contamination levels of construction site and the dismantled items, and the aerosol at the construction site. According to the radiation levels and contamination levels of remaining items and construction site in the construction process, the division of controlled areas and supervised areas in the construction site shall be adjusted, the work of the workers shall be reasonably arranged, and the radiation protection measures shall be taken to ensure the safety of the workers and the environment. In addition to the fixed monitoring facilities at the construction site, portable γ monitors and aerosol monitors shall also be used to monitor the γ radiation levels and aerosol concentrations at the positions that are inaccessible for the fixed monitoring facilities, so as to fully understand the distribution of γ radiation field at construction site.

(2) Monitoring of personal exposure dose

For the personal exposure dose monitoring, two kinds of personal radiation dose monitors are generally used. The first kind of monitor is the cumulative dosimeter that reads the accumulated dose of each worker once a month to obtain the monthly accumulated exposure dose of each worker. The other kind of monitor is the direct-reading personal dosimeter with the alarm function that can indicate the exposure dose of each worker and be used to prevent the occurrence of exceeding the personal exposure dose management limit specified in the radiation protection plan.

(3) Environmental monitoring

Generally, environmental monitoring points are arranged within the radius of 5–10 km from the ventilation tower. Before, during and after the decommissioning, the monitoring measurement and sampling analyses shall be performed at such monitoring points. The monitoring includes environmental γ radiation level and gaseous aerosol concentration, total β radioactivity and nuclide analysis of water (seawater, river water and well water), total β radioactivity and nuclide analysis of sediments and soils at the sea bottom and river bottom, and total β radioactivity and nuclide analysis of biological samples. During the decommissioning, if the monitoring results are higher than the background monitoring values before the decommissioning, the causes shall be identified and the measures shall be taken to ensure the environmental safety. The environmental monitoring after the completion of decommissioning is to evaluate the impact of decommissioning on the environment.
Note: The statutory data and regulations in this book are taken from the national standards of China such as GB18871-2002.

Index

J. Yu, *Marine Nuclear Power Technology*,
https://doi.org/10.1007/978-981-15-2894-1

The manufacturer's authorised representative in the EU is Springer Nature Customer Service Centre GmbH, Europaplatz 3, 69115 Heidelberg, Germany. If you have any concerns regarding our products, please contact ProductSafety@springernature.com

Printed and bound by CPI Group (UK) Ltd, Croydon, CR0 4YY
30/09/2026
02225556-0015